Springer Tracts in Modern Physics
Volume 215

Springer Tracts in Modern Physics

Springer Tracts in Modern Physics provides comprehensive and critical reviews of topics of current interest in physics. The following fields are emphasized: elementary particle physics, solid-state physics, complex systems, and fundamental astrophysics.
Suitable reviews of other fields can also be accepted. The editors encourage prospective authors to correspond with them in advance of submitting an article. For reviews of topics belonging to the above mentioned fields, they should address the responsible editor, otherwise the managing editor.
See also springeronline.com

Michael Schulz

Control Theory in Physics and other Fields of Science

Concepts, Tools, and Applications

With 46 Figures

Springer

Michael Schulz
Universität Ulm
Abteilung Theoretische Physik
Albert-Einstein-Allee 11
89081 Ulm, Germany
E-mail: michael.schulz@physik.uni-ulm.de

Library of Congress Control Number: 2005934994

Physics and Astronomy Classification Scheme (PACS):
02.30.Yy, 02.50.Le, 02.70.Rr, 05.20.-y, 05.10.Gg

ISSN print edition: 0081-3869
ISSN electronic edition: 1615-0430
ISBN-10 3-540-29514-3 Springer Berlin Heidelberg New York
ISBN-13 978-3-540-29514-3 Springer Berlin Heidelberg New York

Springer is a part of Springer Science+Business Media
springer.com

Printed in Germany

Typesetting: by the author using a Springer LaTeX macro package
Cover concept: eStudio Calamar Steinen
Cover production: *design &production* GmbH, Heidelberg

Printed on acid-free paper SPIN: 11374343 56/TechBooks 5 4 3 2 1 0

for Beatrix-Mercedes

Preface

Control theory plays an important role in several fields of economics, technological sciences, and mathematics. In principle, empirical concepts of the control of technological devises can already be proved for antiquity. Typical examples are the machines constructed by Archimedes, Philon, or Ktesibios [1, 2]. The highly developed experience in construction and control of machines[1] was never forgotten completely over the following periods of the late antiquity and the early middle age [3], and formed the foundation of the modern engineering. The mathematical scientific interest for the control of mechanical problems starts with the formulation of classical mechanics by Galileo, Galilei, and Isaac Newton. The first mathematically solved nontrivial control problem was the brachistochrone problem formulated by G. Galilei [4] and solved by J. Bernoulli [5].

On the other hand, the control theory is not a common expression in natural sciences, which is all the more surprising, because both scientific theories and scientific experiments actually contain essential features of control theoretical concepts. This appraisal also applies to physics, although especially in this case many subdisciplines, for example mechanics or statistical physics, are strongly related to several ideas of the control theory.

At this point, there is an important warning. The Control Theory for natural sciences is no substitute for the classical application fields of deterministic and stochastic control theories. Initially, an economic control theory differs partially from the control of physical processes. Let us compare for a moment the ideas of control in natural sciences and economics.

Of course, a short definition of economics is not simple, even for seasoned economists. A possible working definition may, however, be: Economics is the study of how people choose to use scarce or limited productive resources to produce various commodities and distribute them to various members of society for their consumption. This definition suggests the large variety of

[1] The name machine comes from the Greek word $\mu\eta\chi\alpha\upsilon\eta$. This name was used originally for lifting devises in Greek theatres.

disciplines combined under the general term "economics": microeconomics, controlling, macroeconomics, finance, environmental economics, and many other scientific branches are usually considered a part of economics. From this short characterization of economics, it is obvious that the aims of control of economic processes and physical or chemical processes are very different. Economic control often means that decisions are made on the basis of all available historical data. In this sense, we may speak about a closed loop control. The current control, given by several economic decisions depends strictly on the early evolution of the system, e.g., a market or a company. Characteristicly, the complete intrinsic dynamics of economic systems is more or less unknown.

Physical processes under control differ essentially from the economic methods discussed above. First of all, the dynamics of a physical or chemical process may usually be described by a sufficiently accurate model independent of the method leading to the model. Therefore, a model may be obtained from first principles or from empirical investigations. But in contrast to the economic models, a physical model can be tested by several experimental methods, which allows us to refine the model by additional terms. Thus, the intrinsic dynamics of a physical system is widely known. From this point of view, the control of a physical system may be computed before the process starts. This is a typical open-loop control.

The approach between both concepts to a common theory is given by the statistical physics of complex systems. The physically mathematical site of the theory of complex systems allows us to derive and formulate evolution laws, limit probability distribution functions, and universal properties. If we have obtained such a suitable theory about the behavior of several complex systems, we may use this knowledge also for the analysis of more complicated systems. However, the description of the evolution of a complex system on the basis of a suitable model implies the neglecting of the dynamics of a set of irrelevant degrees of freedom and therefore of the presence of a more or less pronounced uncertainty. This is the origin of an apparently stochastic behavior observed as a universal property of complex systems. However, the control of such physical complex systems does not really differ from the control of economic systems. We should be aware that the degree of complexity of the economic world is extremely high; however, it is a special part of a global, physical world. Thus, from a mathematical point of view the basic control concepts of complex physical systems are similar to the concepts used for economic systems.

The main goal of this book is to present some of the most useful theoretical concepts and techniques for understanding the ideas of the control theory for physical systems. But it should be noted that the concepts and tools presented are also relevant to a much larger class of problems in the natural sciences, the social sciences, and in engineering.

The central theme of this book is the control of special degrees of freedom as well as of collective and cooperative properties in the behavior of complex systems. This idea allows us to describe control mechanisms on the basis of modern physical concepts, such as Hamiltonian equations, deterministic chaos, self-organization, scaling laws, renormalization group techniques, and complexity, and also traditional ideas of Newtonian mechanics, linear stability, classical field theory, fluctuations, and response theory.

The first chapter covers important notations of the control theory of simple and complex systems. In the subsequent chapter, the basic formulation of the deterministic control theory is presented. On the one hand, the close relationship between the concept of classical mechanics and control theoretical approaches will be demonstrated. On the other hand, several fundamental rules are presented on an immediate rigorous level. This approach requires thorough mathematical language. The main topic in this chapter is the maximum principle of Pontryagin, which allows us to separate the dynamics of deterministic systems under control in an optimization problem and a well-defined set of equations of motion. The third chapter focuses on a frequent class of deterministic control problems: the linear quadratic problems. Such problems occur in a very natural way – if a weak deviation from a given nominal curve should be optimally controlled. Several tools and concepts estimating the stability of controlled systems and several linear regulator problems, which are important especially for the control of technological devises, will also be discussed here.

The control of fields, another mainly physically motivated class of control problems, will be discussed in the next chapter. After a brief discussion of several field theories, the generalized Euler-Lagrange equations for the field control are formulated. Furthermore, the control of physical, and also other fields via controllable sources and boundary conditions, are briefly presented.

Chaos control, controllability, and observability are the key points of the fifth chapter. This part of the book is essentially addressed to dynamic systems with a moderate number of degrees of freedoms and therefore a moderate degree of complexity. In principle, such systems are the link between the deterministic mechanical systems and the complex systems with a pronounced probabilistic character. Systems offering a deterministic chaotic behavior are often observed at mesoscopic spatial scales. In particular, we will present some concepts for stabilization and synchronization of usually unstable deterministic systems.

In the subsequent chapter the basis for the probabilistic description of the control of the complex system is formulated. Whereas all previous chapters focus on the control of deterministic processes, now begins the presentation of control concepts belonging to systems with partial information or several types of intrinsic uncertainties. Obviously, an applicable description of a complex system requires the definition of a set of relevant degrees of freedom. The price one has to pay is that one gets practically no information about the remaining irrelevant degrees of freedom. As a consequence, the theoretical basis used for the analysis of sufficiently complex systems can be described as an essentially probabilistic theory. Chapter 6 gives an introduction to the basics of nonequlibrium physics and the probability theory as far as it is necessary for the subsequent considerations. Some important physical ideas, especially for the derivation of the Nakajima-Zwanzig equation and the Fokker-Planck equation, are used to explain the appearance of stochastic processes on the basis of originally deterministic equations of motion.

In Chapter 7 the basic equations for the open-loop and the feedback control of stochastically driven systems are derived. These equations are very similar to the corresponding relations for deterministic control theories, although the meaning of the involved quantities is more or less generalized. The application of functional integral techniques allows the extension of deterministic control principles to probabilistic concepts, similar to the classical mechanics, to be expanded to the quantum theory on the basis of Feynman's path integrals.

Another important point of the stochastic control theory discussed in Chapter 8 is the meaning of filters and predictors, which may be used to reconstruct at least partially the real dynamics of the system from known historical observations.

From a physical point of view a more exotic topic is the application of game theoretical concepts to control problems. However, these ideas may be helpful for the optimal control of several quantum mechanical experiments. These concepts are discussed in Chapter 9. The difference between deterministic and stochastic games, as well as several problems related to zero-sum games and the Nash equilibrium, will be briefly analyzed in this chapter.

Finally, the last chapter gives a short overview about some general ideas of optimization procedures. This is necessary because most control problems can be splitted into a set of evolution equations and a remaining optimization problem. In this sense, the last chapter of this book may be understood as a tool of stimulations for solving such optimization problems.

This book derives from a course taught at the university at Ulm in the Department of Theoretical Physics, which commenced in 2002. Essentially aimed at students of physics, econophysics, and engineering, the course attracted students, graduate students, and postdoctoral researchers from physics, chemistry, economics, engineering, and financial mathematics. I am indebted to all of them for their interest and their discussions.

First I will thank F. Steiner for his inspiration to prepare the present book. I also wish to thank my colleagues P. Reineker (Ulm), W. Greksch (Halle-Wittenberg), U. Rieder (Ulm), B. M. Schulz (Halle-Wittenberg), R. Wunderlich (Zwickau), and W. Stummer (Erlangen) for valuable discussions. Last, but not least, I wish to express my gratitude to Springer-Verlag, in particular to U. Heuser and J. Lenz for their excellent cooperation.

Ulm
October 2005

Michael Schulz

References

1. S. Sambursky, *The physical world of the Greeks* (London, 1956).
2. O. Neugebauer, *The exact science in Antiquity* (Princeton, 1957).
3. S. Sambursky, *The physical world of the late antiquity* (London, 1962).
4. G. Galilei: *Dialogues concerning two New Sciences*, translated by H. Crew and A. de Salvio (Prometheus Books, Buffalo N.Y., 1998).
5. P Costabel, J. Peiffer: *Die Gesammelten Werke der Mathematiker und Physiker der Familie Bernoulli* (Birkhäuser, Basel, 1988).

Contents

1

Introduction

1.1 The Aim of Control Theory

Control theory plays an important role in several fields of economics, technological sciences, and mathematics. On the other hand, control theory is not a common expression in natural sciences. That is all the more surprising, because both, scientific theories and scientific experiments, actually contain essential features of control theoretical concepts. This appraisal also applies to physics, although especially in this case many subdisciplines, for example, mechanics or statistical physics, are strongly related to several ideas of control theory.

We always speak about control theory in connection with well-defined systems. In order to be more precise, control theory deals with the behavior of dynamic systems over time. The controllability of such systems does not necessarily depend on the degree of complexity. From a general point of view, we should distinguish between external control and intrinsic control mechanisms.

The external control is also denoted as an open loop control. In principle, this control transmits a certain protocol onto the dynamics of the system. In this case, it is unimportant how the practical control is achieved. It may be the result of a personal control by an observer or of a previously fixed program. If the transmission of the program is ended, the system remains in its last state or it follows its own free dynamics further.

As an example, consider the flight of an airplane. In this case, the system is the machine itself. The goal of the control is to bring the airplane from airport A to another airport B. The pilot itself may be interpreted as an external controller of the system. Now let us assume that all the activities of the pilot are recorded in a protocol. A theoretical way to repeat the flight from A to B under the same control is to implement the protocol in the autopilot of the airplane. This is a correct conduct if the airplane is driving perfectly under the same boundary conditions. In reality, the airplane will lose the planned way because direction and strength of the wind, temperature and air pressure vary considerable.

M. Schulz: *Control Theory in Physics and other Fields of Science*
STMP **215**, 1–15 (2006)

Obviously the main disadvantage of an open-loop control type is the lack of sensitivity to the dynamics of the controlled system in its time-dependent environment, because there is no direct connection between the output of the system and its input. Therefore, the external control plays an important role especially for systems with a few degrees of freedom and reproducible boundary conditions.

To avoid the problems of the external control it is necessary to introduce feedback mechanisms. The output of the system is fed back to any change of the current dynamics of the system to a desired reference dynamics. environment. The controller measures the difference between the reference dynamics and the output, i.e., the current error, to change the inputs to the system. This kind of control is also denoted as a closed-loop control or feedback control. In the case of our example, a feedback control may be realized by connecting the autopilot with instruments which measure position, altitude, and flight-direction of the airplane so that each deviation from the course may be immediately corrected. Another possibility of obtaining a closed-loop control is to enlarge formally the current system "airplane" to the more complex system "human pilot and airplane".

A real dividing line between systems which are favored for an exclusive external control or an exclusive feedback control cannot be defined. The choice of an appropriate control mechanism especially for technological systems is the task of control engineering. This discipline focuses on the mathematical modeling of systems of a diverse nature, analyzing their dynamic behavior, and using control theory to make a controller that will cause the systems to behave in a desired manner. The field of control within chemical engineering is often known as process control. It deals primarily with the control of variables in a chemical process in a plant. We expect definitely an increasing number and increasing variety of the intrinsic control mechanisms if the degree of complexity of the system under control increases. The predominant part of the control mechanisms of extremely complex systems is mostly a result of a hierarchical self-organization. Typical examples of such more or less self-controlled systems are biological organisms and social or economic systems with an enormous number of, partially still nonenlightened, control mechanisms.

The control of a system may be realized by different methods. Small physical systems may be sufficiently controlled by a change of boundary conditions and the variation of external fields while more complex systems become controllable via various flexible system parameters or by the injection and extraction, respectively, of energy or matter. But we remark that all possible variable quantities of a system are basically usable for a control. In the framework of control theory all quantities which may be used for a control of the system are defined as input or control function $u(t) = \{u_1(t), \ldots, u_n(t)\}$.

The control mechanisms alone are not the main topic of control theory. This theory connects the system under control and especially their control mechanisms with a certain control aim as an optimalization criterion. In the case of our airplane example, the shortest way, the cheapest way and the safest way between A and B are possible, but not identical control aims. The choice of one of these criteria or of a weighted composition of these possible aims depends on the intentions taken into account by the control designer. The control theory ask for an optimal control in order to find a control law, i.e., an optimum input corresponding to the optimalization criterion. The control aim is often defined by a so-called cost functional which should be minimized to obtain the optimum input $u^*(t)$. It usually takes the form of an integral over the time of a certain function, plus a final contribution that depends on the state in which the system ends up.

The difference between an open-loop control and a closed-loop control can now also be defined within the control function. The optimal input of an open-loop control can be completely determined before the system starts the dynamics. Thus, the control has an a priori character. This concept becomes relevant if the dynamics of the system is deterministic from a theoretical and an experimental point of view. In contrast to this behavior, the control function of a closed-loop control is generated during the evolution of the system. The current state of the system and possibly the history determine the current change of the input with respect to minimize the cost functional. We also denote such behavior an a posteriori control.

1.2 Dynamic State of Classical Mechanical Systems

The determination of an optimum control requires the knowledge of the underlying dynamics of the system under control. In the framework of classical physics, the mechanical state of the system is completely defined by the set of the time-dependent degrees of freedom. The mechanical state of a given system with $2N$ degrees of freedom consists of N generalized coordinates q_i ($i = 1, \ldots, N$) and N generalized momenta p_i conjugate to the q_i. The dynamics can be written in terms of deterministic Hamilton's equations as

$$\frac{\mathrm{d}q_i}{\mathrm{d}t} = \frac{\partial H}{\partial p_i} \quad \text{and} \quad \frac{\mathrm{d}p_i}{\mathrm{d}t} = -\frac{\partial H}{\partial q_i}, \tag{1.1}$$

where $H = H(q, p, u)$ is the Hamiltonian of the system. The Hamiltonian depends on the mechanical state given by the complete set of all $q_i(t)$ and $p_i(t)$, and on the input $u(t)$, defined by the current control law. Formally, the mechanical degrees of freedom can be combined into a $2N$-dimensional vector $\Gamma(t) = \{q_1, \ldots, q_N, p_1, \ldots, p_N\}$. Thus, the whole system can be represented at time t by a point $\Gamma(t)$ in a $2N$-dimensional space, spanned by a reference frame of $2N$ axes, corresponding to the degrees of freedom. This space is called the phase space $\mathbb{P}$. It plays a fundamental role and is the natural framework of the dynamics of classical many-body systems.

The exact determination of all time-dependent degrees of freedom of the system implies the solution of the complete set of the mechanical equations of motion (1.1) of the system. The formally complete predictability of the future evolution of the system is a consequence of the underlying deterministic Newtonian mechanics. In the sense of classical physics, determinism means that the trajectories of all particles can be computed if their momentum and positions are known at an initial time. Unfortunately, this positive result breaks down for real systems with a sufficiently large N. The theory of deterministic chaos [1–3] has shown that even in classical mechanics predictability cannot be guaranteed without absolutely precise knowledge of the initial mechanical configuration of the complete system. This apparent unpredictability of a deterministic, mechanical many-body system arises from the sensitive dependence on the initial conditions and from the fact that the initial conditions can be measured only approximately in practice due to the finite resolution of any measuring instrument.

In order to understand this statement, we state that practically all trajectories of the system through the $2N$-dimensional phase space are unstable against small perturbations. The stability of an arbitrary trajectory to an infinitesimally small perturbation can be studied by the analysis of the so-called Lyapunov exponents. This concept is very geometrical. Imagine an infinitesimally small sphere of radius ε containing the initial position of neighboring trajectories. Under the action of the dynamics, the center of the sphere may move through the phase space $\mathbb{P}$, and the sphere will be distorted. Because the ball is infinitesimal, this distortion is governed by a linearized theory. Thus, the sphere remains an ellipsoid with the $2N$ principal axes $\varepsilon_\alpha(t)$ (Fig. 1.1). Then, the Lyapunov exponents can be defined as

$$\Lambda_\alpha = \lim_{t\to\infty} \lim_{\varepsilon\to 0} \frac{1}{t} \ln \frac{\varepsilon_\alpha(t)}{\varepsilon_\alpha(0)} . \tag{1.2}$$

The limit $\varepsilon \to 0$ is necessary because, for a finite radius ε, as t increases, the sphere can no longer be adequately represented by an ellipsoid due to the increase of nonlinear effects. On the other hand, the long time limit, $t \to \infty$, is important for gathering enough information to represent the entire trajectory. Obviously, the distance between infinitesimal neighboring trajectories diverges if the real part of at least one Lyapunov exponent is positive. If the diameter of the initial sphere has a finite value, then the initial shape is very violently distorted, see Fig. 1.2. The sphere transforms into an amoebalike body that eventually grows out into extremely fine filaments that spread out over the whole accessible phase space. Such a mixing flow is a characteristic property of systems with a sufficiently high degree of complexity [4, 5]. There remains the question of whether Lyapunov exponents with positive real part occur in mechanical systems. We obtain as a direct consequence of the time-reversal symmetry that, for every Lyapunov exponent, another Lyapunov exponent exists with the opposite sign. In other words, we should expect regular behavior only when the real parts of all Lyapunov exponents vanish. This special case

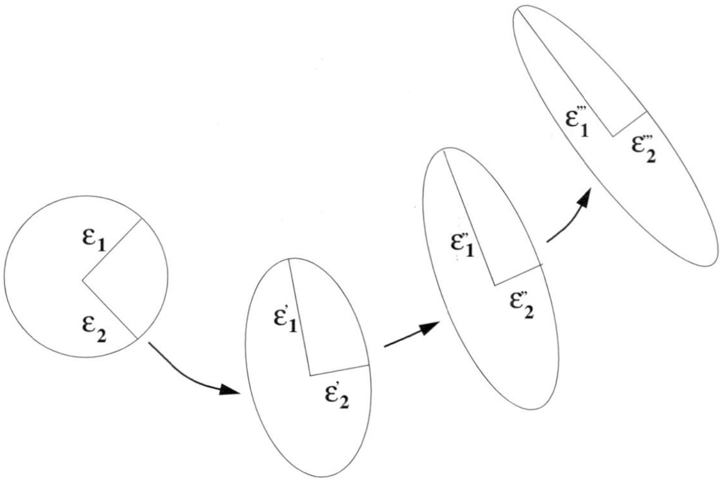

Fig. 1.1. The time evolution of an infinitesimally small ellipsoid of initial principal axis $\varepsilon_1 = \varepsilon_2 = \varepsilon$. With increasing time the initially rotational symmetric region ball is gradually deformed into a pronounced ellipsoid

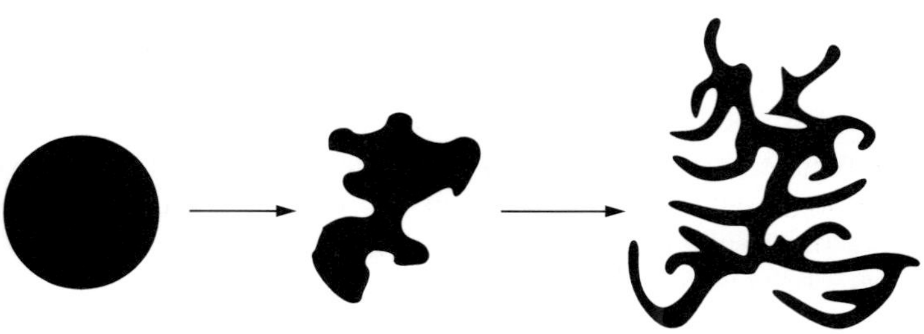

Fig. 1.2. The deformation of a finite sphere of the phase space in the course of its time evolution

is practically excluded for complicated, nonlinear many-body systems. Computer simulations have also demonstrated that relatively simple mechanical systems with a few degrees of freedom already show chaotic behavior[1].

Chaos is not observed in linear systems. In fact, such systems have only Lyapunov exponents with disappearing real part. Mathematically, the signature of a linearity is the superposition principle, which states that the sum of two solutions of the mechanical equations describing the system is again a solution. The theory of linear mechanical systems is fully understood except for some technical problems. The breakdown of the linearity, and therefore the breakdown of the superposition principle, is a necessary condition for

[1] The first rigorous proof of a mixing flow was given by Sinai for a system of N ($N \geq 2$) hard spheres in a finite box [6].

the behavior of a nonlinear mechanical system to appear chaotic. However, nonlinearity alone is not sufficient for the formation of a chaotic regime. For instance, the equation of a simple pendulum is a nonlinear one. The solutions are elliptic functions without any kind of apparent randomness or irregularity.

Standard problems of classical mechanics, such as falling bodies, the pendulum, or the dynamics of planetary systems considering only a system composed of the sun and one planet, require only a few degrees of freedom. These famous examples allowed the quantitative formulation of mechanics by Galileo and Newton. In other words, these famous pioneers of modern physics treated one- or, at most, two-body problems without any kind of chaotic behavior.

The scenario presented in Fig. 1.2 is often also called ergodic behavior. That is true because mixing implies ergodicity. However, ergodicity is not always mixing. Roughly speaking, ergodicity means that the trajectory of a system touches all energetically allowed points of the phase space. But it is not necessary that the distance of initially neighbored trajectories increases rapidly. In other words, the finite initial sphere in Fig. 1.2 is only slightly altered during the motion of an ergodic, but nonmixing, system [7].

If we come back to our control problem, we may conclude that systems with a sufficiently high degree of complexity need other concepts for a successful control as mechanical systems with a few degrees of freedom or with simple linear equations of motion. The previous discussion especially means that the impossibility of a precise determination of the initial conditions of a mechanical system with a sufficiently large number of degrees of freedom prevents purely and simply the open-loop control on the basis of the mechanical equations of motion. Each priori determined control obtained for a well-defined initial condition breaks completely down for an immediately neighbored initial condition because of the instability of the trajectories. That means an effective control of a system with a sufficiently large number of degrees of freedom requires a closed-loop control which is able to adjust weak deviations from the nominal trajectory.

1.3 Dynamic State of Complex Systems

1.3.1 What Is a Complex System?

Control theoretical concepts are not only applied to systems defined on the mechanical level. In that case, the control is usually coupled to such characteristic state variables which seem to influence the dynamics of the system significantly. This empirical concept also allows the control of systems with a strongly pronounced complex structure and dynamics. A system tends to increase its complexity if the number of the degrees of freedom increases. To clarify this statement, we have to discuss what we mean by complex systems. Unfortunately, an exact definition of complex systems is still an open problem. In a heuristic manner, we may describe them as

Complex systems are composed of many particles, or objects, or elements that may be of the same or different kinds. The elements may interact in a more or less complicated fashion by more or less nonlinear couplings.

In order to give this formal definition a physical context, we should qualitatively discuss some typical systems that may be denoted truly complex.

The various branches of science offer us numerous examples, some of which turn out to be rather simple, whereas others may be called truly complex. Let us start with a simple physical example. Granular matter is composed of many similar granules. Shape, position, and orientation of the components determine the stability of granular systems. The complete set of the particle coordinates and of all shape parameters defines the actual structure.

Furthermore, under the influence of external force fields, the granules move around in quite an irregular fashion, whereby they perform numerous more or less elastic collisions with each other. A driven granular system is a standard example of a complex system. The permanent change of the structure due to the influence of external fields and the interaction between the components is a characteristic feature of complex systems.

Another standard complex system is Earth's climate, encompassing all components of the atmosphere, biosphere, cryosphere, and oceans and considering the effects of extraterrestrial processes such as solar radiation and tides. Computers and information networks are interpreted as another class of complex systems. This is especially so with respect to hardware dealing with artificial intelligence, where knowledge and learning processing will be replacing the standard algebra of logic.

In biology, we are again dealing with complex systems. Each higher animal consists of various strongly interacting organs with an enormous number of complex functions and intrinsic control mechanisms. Each organ contains many partially very strong specialized cells that cooperate in a well-regulated fashion.

Probably the most complex organ is the human brain, composed of 10^{11} nerve cells. Their collective interaction allows us to recognize visual and acoustic patterns, to speak, or to perform other mental functions. Each living cell is composed of a complicated nucleus, ribosomes, mitochondria, membranes, and other constituents, each of which contain many further components. At the lowest level, we observe many simultaneously acting biochemical processes, such as the duplication of DNA sequences or the formation of proteins.

This hierarchy can also be continued in the opposite direction. Animals themselves form different kinds of societies. Probably the most complex system in our world is the global human society, especially the economy, with its numerous participants (such as managers, employers, and consumers) its capital goods (such as machines, factories, and research centers), its natural resources, its traffic, and its financial systems, which provides us with another large class of complex systems. Economic systems are embedded in the more comprehensive human societies, with their various human activities and their political, ideological, ethical, cultural, or communicative habits.

All of these systems are characterized by permanent structural changes and a hierarchy of intrinsic, more or less feedback-dominated control mechanisms. A consequent physical concept requires that we have to explain the evolution of a complex system at larger scales starting from the very microscopic level. Definitely, we have to deal with two problems. First, we have to clarify the macroscopic or mesoscopic scales of interest, and then we have to show how the more or less chaotic motion of the microscopic elementary particles of the complex system contributes to pronounced collective phenomena at the emphasized macroscopic scales.

The definition of correct microscopic scales as well as suitable macroscopic scales may sometimes be an ambiguous problem. For instance, in biology we deal with a hierarchy of levels that range from the molecular level through that of animals and humans to that of societies. Formally, we can start from a microscopic, classical many-body system, or alternatively, from the corresponding quantum-mechanical description. But in order to describe a complex system at this ultimately microscopic level, we need an enormous amount of information, which nobody is able to handle.

A macroscopic description allows a strong compression of data so that we are no longer concerned with the microscopic motion but rather with properties at large scales. The appropriate choice of the macroscopic level is by no means a trivial problem. It depends strongly on the question in mind. In order to deal with complex systems, we quite often still have to find adequate variables or relevant quantities to describe the properties of these systems. Each macroscopic system contains a set of usually collective large-scale quantities that may be of interest for the underlying problem. We will denote such degrees of freedom as relevant quantities. The knowledge of these quantities permits the characterization of a special feature of the complex system at the macroscopic level. All other microscopically well-founded degrees of freedom form the huge set of irrelevant variables for the relatively small group of relevant quantities.

The second problem in treating complex systems consists in establishing relations that allow some predictions about the future evolution of the relevant quantities and therefore about the controllability of the system. Unfortunately, the motions of the irrelevant and relevant degrees of freedom of a complex system are normally coupled strongly together. Therefore, an accurate prediction of future values of the relevant degrees of freedom automatically includes the determination of the accurate evolution of the irrelevant degrees of freedom. Here, we need another concept as the above-discussed mechanical approach. The mathematical derivation of this alternative way will be postponed till Chap. 6.

Before we start with a first mathematical treatment of complex systems, let us first try to define them more rigorously. The question of whether a system is complex or simple depends strongly on the level of scientific knowledge.

An arbitrary system of linear coupled oscillators is today an easily solvable problem. In the lifetime of Galileo, without knowledge of the theory of linear differential equations, one surely would have classified this problem as a complex system in the context of our definition specified above.

A modern definition that is independent of the actual mathematical level is based on the concept of algebraic complexity. To this aim, we must introduce a universal computer that can solve any mathematically reasonable problem after a finite time with a program of finite length.

Without going into details, we point out that such a universal computer can be constructed, at least in a thought experiment as was shown by Turing [8]. Of course, there exist different programs that solve the same problem. As a consequence of number theory, the lengths of the programs solving a particular problem have a lower boundary. This minimum length may be used as a universal measure of the algebraic degree of complexity. Unfortunately, this meaningful definition raises another problem. As can be shown by means of a famous theorem by Gödel [9], the problem of finding a minimum program cannot be solved in a general fashion. In other words, we must estimate the complexity of a system in an intuitive way, and we must be led by the level of scientific knowledge.

1.3.2 Relevant and Irrelevant Degrees of Freedom

In a possible, microscopically formulated theory of a complex system all degrees of freedom are equally considered. The mathematical solution of the corresponding system of equations of motion, even if we were able to determine it, would of course be impractical and therefore unusable for the analysis of complex systems. This is because of the large number of contained degrees of freedom and the extreme sensitivity against a change of the initial conditions.

In general, we are interested in the description of complex systems only on the basis of the relatively small number of relevant degrees of freedom. Such an approach may be denoted as a kind of reductionism. Unfortunately, we are not able to give an unambiguous definition of which degree of freedom is relevant for the description of a complex system and which degree of freedom is irrelevant. As we have mentioned in the previous chapter, the relevant quantities are introduced empirically in accordance with the underlying problem.

To proceed, we split the complete phase space $\mathbb{P}$ into a subspace of the relevant degrees of freedom $\mathbb{P}_{\text{rel}}$ and the complementary subspace of the irrelevant degrees of freedom $\mathbb{P}/\mathbb{P}_{\text{rel}}$. Then, every microscopic state Γ may be represented as a combination of the set $X = \{X_1, X_2, \ldots, X_{N_{\text{rel}}}\}$ of N_{rel} relevant degrees of freedom and the set Γ_{irr} of the irrelevant degrees of freedom so that

$$\Gamma = \begin{cases} X \in \mathbb{P}_{\text{rel}} & \text{relevant degrees of freedom} \\ \Gamma_{\text{irr}} \in \mathbb{P}/\mathbb{P}_{\text{rel}} & \text{irrelevant degrees of freedom}\,. \end{cases} \tag{1.3}$$

We may think about this splitting in geometrical terms. The system of relevant degrees of freedom can be represented by a point in the corresponding N_{rel}-dimensional subspace $\mathbb{P}_{\text{rel}}$ of the phase space $\mathbb{P}$. We denote this subspace as the phase space of the relevant degrees of freedom. Obviously, an observer of this reduced phase space $\mathbb{P}_{\text{rel}}$ records apparently unpredictable behavior of the evolution of the relevant quantities. That is because of the fact that the dynamic evolution of the relevant quantities is governed by the hidden irrelevant degrees of freedom on microscopic scales. Thus, different microscopic trajectories in the phase space can lead to the same evolution of the relevant quantities and, vice versa, identical initial configurations in the phase space of the relevant degrees of freedom may develop into different directions.

Unfortunately, there is no theoretical background which allows us to give a particular hint of the preference of a set of relevant dynamic quantities. A possible, but nevertheless heuristic idea is to collect the slow variables in the set of relevant degrees of freedom. We may find some empirical arguments that these quantities substantially determine the macroscopic appearance of the system. However, the choice of which variables are actually slow is largely guided by the problem in mind.

In the subsequent chapters of this book we will demonstrate that the time evolution of the relevant degrees of freedom may be quantitatively expressed by equations of motion of the type

$$\dot{X}_{\alpha} = F_{\alpha}[X, u, t] + \eta_{\alpha}(X, u, t) \quad \alpha = 1, \ldots, N_{\text{rel}} \,. \tag{1.4}$$

Here, $F_{\alpha}[X, u, t]$ is a function or a functional of the relevant degrees of freedom, the above-introduced control function u and possibly the time. The influence of all irrelevant degrees of freedom is collected in $\eta_{\alpha}(X, u, t)$. In contrast to the predictable and usually smooth time-dependence of $F_{\alpha}[X, u, t]$, the unpredictable details of the dynamics of the irrelevant quantities lead to a stochastic or stochastic-like behavior of the time dependence of $\eta_{\alpha}(X, u, t)$. This is the origin that we are not able to predict the evolution of the set of relevant degrees of freedom with an unlimited accuracy even if we know the relevant initial conditions precisely. In other words, the restriction onto the subspace of relevant quantities leads to a permanent loss of information.

We denote the set of relevant degrees of freedom in future as the macroscopic dynamic state $X(t)$ or simply as the dynamic state of the complex system. If the system has no irrelevant degrees of freedom, $X(t)$ is identical to the microscopic state $\Gamma(t)$ and (1.4) degenerates to the canonical system of equations of motion (1.1).

1.3.3 Quasi-Deterministic Versus Quasi-Stochastic Evolution

The control of equations of type (1.4) takes place by the application of several feedback techniques. For this purpose, the further evolution of the complex system is estimated from the history of the dynamic state $\{X(\tau) : t_0 \leq \tau \leq t\}$ and of the history of the control function $\{u(\tau) : t_0 \leq \tau \leq t\}$ and from the available information about the stochastic-like terms $\eta_\alpha(X, u, t)$. This knowledge allows the recalculation of the change of the control function $u(t)$ in such a manner that the control aim will be optimally arrived.

The choice of the control mechanism depends essentially on the mathematical structure of the equations of motion (1.4) and therefore on the degree of complexity of the system under control. We may distinguish between two limiting classes of controlled complex systems, namely quasi-deterministic systems with a dominant deterministic part $F_\alpha[X, u, t]$, i.e., $|F_\alpha[X, u, t]| \gg |\eta_\alpha(X, u, t)|$ and quasi-stochastic systems with a sufficiently strong noise term $\eta_\alpha(X, u, t)$, i.e., $|\eta_\alpha(X, u, t)| \gg |F_\alpha[X, u, t]|$.

The majority of technological systems, for example cars, airplanes, chemical plants, electronic instruments, computers, or information systems, belong to the class of quasi-deterministic systems. This fact is a basic construction principle of engineering in order to obtain a sufficiently high gain of the technological system. Several non-technological systems, for example hydrodynamic experiments, chemical reactions, and diffusion processes are also often predominated by deterministic contributions.

There are different possibilities of suppressing the stochastic-like contributions $\eta_\alpha(X, u, t)$ in the equations of motion (1.4). A popular method used in engineering is the implementation of appropriate filters, for example noise suppressors in electronic instruments, or the utilization of redundant sensors or security systems, for instance in airplanes or in nuclear power stations. These integrated components reduce possible fluctuations and separate random effects from the dynamics of the relevant quantities of the system. As a consequence of these construction principles, the technological system becomes a largely deterministic character. Several technological systems have such a high standard that a temporary open-loop control becomes possible.

Another possibility suppressing the stochastic-like terms $\eta_\alpha(X, u, t)$ of the evolution equations (1.4) is the increase of the number of relevant degrees of freedom. A characteristic ensemble are chemical reactions. Simple kinetic equations with the mean concentration of the reacting components as relevant degrees of freedom have a sufficiently high accuracy and stability for many applications. Although measurable concentration fluctuations exist, we can often neglect these perturbations without serious consequences. In other words, we may assume that chemical reactions at the macroscopic level can be described by complete deterministic equations. But very fast reactions at large spatial scales, for example explosions, and reactions forming spatially and temporally fluctuating structures, e.g., observed for the Belousov-Zhabotinskii reaction [10], show strong fluctuation with an essential influence to the reaction kinetic. However, such fluctuations can be incorporated in deterministic equations if we extend the set of relevant variables by spatial inhomogeneous concentration fields in the evolution equations. Thus, we have to deal now with hydrodynamic reaction-diffusion equations considering local chemical reactions and material transport via diffusion and convection. In other words, inhomogeneous reactions may be described by deterministic evolution equations up to mesoscopic scales, while a description of the same system on the basic of classical space-independent kinetic equations requires the consideration of more or less pronounced fluctuation terms. The fluctuations in the global theory are transformed into deterministic contributions in the refined inhomogeneous theory. On the other hand, simple kinetic equations have only a few relevant degrees of freedom, whereas hydrodynamic reaction–diffusion equations are defined for concentration fields corresponding to a large set of local concentrations. But we remark that the reaction–diffusion equations also contain fluctuation terms originated by the irrelevant degrees of freedom which remain effective at least at microscopic scales. These fluctuations become important under certain conditions, e.g., at low concentrations [11–13] close to dynamic phase transitions [14–16] and for directed percolation problems [17–19].

Quasi-stochastic behavior may be observed for several complex systems with a pronounced self-organized dynamic hierarchy and a variety of intrinsic control mechanisms. For a moment, this statement seems to be surprising. Therefore, let us focus our attention on a biological organism. In a cell, thousands of metabolic processes are going on simultaneously in a well-regulated fashion. In each organ millions of cells cooperate to bring about cooperative blood flow, locomotion, heartsbead, and breathing. Further highly collective processes are well-coordinated motion of animals, the social behavior of animal groups, or the speech and thought in humans. All these well-coordinated processes become possible only through the exchange of information via several control mechanisms organizing the communication between different parts of the organism. However, if we reduce the relevant quantities to a few degrees of freedom, the behavior of the complex biological system becomes unpredictable. For instance, the trajectory of an animal in its local environment or of a shoal of fish [20] is largely a stochastic process. Obviously, the apparently stochastic character is at least partially a result of the choice of the relevant degrees of freedom. Highly hierarchical structured systems also require a large set of relevant variables for a partial elimination of stochastic effects. Because the majority of the interaction of these variables is still open, the precise structure of the deterministic part $F_\alpha[X, u, t]$ of (1.4) remains unknown. The alternative is the restriction on some relevant quantities with the disadvantage of a dominant stochastic-like term in the evolution equations (1.4).

Other examples of quasi-stochastic systems are the system of price fluctuations in financial markets [21–23], earth climate [24] or the seismic activity of the earth crust [25–27]. But relative simple systems may also show quasi-stochastic behavior, e.g., a dice game.

It is quite clear that a control theory of quasi-stochastic complex systems requires other concepts as a control theory of quasi-deterministic or complete deterministic systems.

1.4 The Physical Approach to Control Theory

In the last century, control theory was traditionally connected with engineering and economics. Natural sciences were not primarily interested in control of processes. The classical aim of an experiment was the detection of fundamental laws while a control of the outcome of an experiment was usually not desired. This situation has essentially changed with the development of experimental techniques at mesoscopic scales. The presence of noticeable thermodynamic fluctuations and the partial instability of objects of an order of magnitude of a few nm requires mechanisms to stabilize such sensitive structures. Furthermore, in analogy to the chemically orientated concept of molecular design, physicists would like to design dynamic processes at mesoscopic and microscopic scales. Essentially here, the idea of control theory comes into play.

In the subsequent chapter we start with the basic formulation of deterministic control theory. On one hand, we will demonstrate the close relationship between the concept of classical mechanics and control theoretical approaches. On the other hand, we are interested in the presentation of the fundamental rules on an as soon as possible rigorous level. This approach requires a very mathematical language. The main result of this chapter is the maximum principle of Pontryagin which allows us to separate the dynamics of deterministic systems under control in an optimization problem and a well-defined set of equations of motion.

Chapter 3 focus on a frequently appearing class of deterministic control problems, the linear quadratic problems. Such problems occur in a very natural way, if we wish to control weak deviations from a given nominal curve. But several tools and concepts estimating the stability of controlled systems and several linear regulator problems will also be discussed here.

The control of fields is another, often physically motivated class of control problems which we will study in Chap. 4. After a brief discussion of several field theories, we define generalized Euler–Lagrange equations describing the control of field equations. Furthermore, the control of fields via controllable sources and boundary conditions is discussed.

Chaos control, controllability, and observability are the key points of Chap. 5. This part of the book is essentially addressed to dynamic systems with a moderate number of degrees of freedoms and therefore a moderate degree of complexity. Such systems are often observed at mesoscopic spatial scales. In particular, we will present some concepts for stabilization and synchronization of usually unstable deterministic systems.

Chapter 6 is the basis for the second main part of the book. Whereas all previous chapters focus on the control of deterministic processes, we will start now with the presentation of control concepts belonging systems with partial information or several types of intrinsic uncertainties. The present chapter gives an introduction to the basics of nonequilibrium physics and probability theory necessary for the subsequent considerations. Especially, some physical arguments explaining the appearance of stochastic processes on the basis of originally deterministic equations of motion are presented.

In Chap. 7 we derive the basic equations for the open-loop and the feedback control of stochastic-driven systems. These equations are very similar to the corresponding relations for deterministic control theories, although the meaning of the involved quantities is more or less generalized. However, the deterministic case is always a special limit of the stochastic control equations.

Another important point related to stochastic control problems are the meaning of filters which may be used to reconstruct the real dynamics of the system. Such techniques, as also the estimation of noise processes and the prediction of partially unknown dynamic processes as a robust basis for an effective control, are the content of Chap. 8.

From a physical point of view a more exotic topic is the application of game theoretical concepts to control problems. Several quantum mechanical experiments are eventually suitable candidates for these methods. Chapter 9 explains the difference between deterministic and stochastic games as well as several problems related to zero-sum games and the Nash equilibrium and gives some inspirations how these methods may be applied to the control of physical processes.

Finally, Chap. 10 presents some general concepts of optimization procedures. As mentioned above, most control problems can be split into a set of evolution equations and a remaining optimization problem. In this sense, the last chapter of this book may be understood as a certain tool of stimulations for solving such optimization problems.

References

1. H.G. Schuster: *Deterministic Chaos: An Introduction*, 2nd edn (VCH Verlagsgesellschaft, Weinheim, 1988)
2. K.T. Alligood, T.D. Sauer, J.D. Farmer, R. Shaw: *An Introduction to Dynamical Systems* (Springer, Berlin Heidelberg New York, 1997)
3. R. Balescu: *Equilibrium and Nonequilibrium Statistical Mechanics* (Wiley, New York, 1975)
4. L. Boltzmann: J. f. Math. **100**, 201 (1887)
5. J.W. Gibbs: *Elementary Principles in Statistical Mechanics* (Yale University Press, New Haven, CT, 1902)
6. I. Sinai: Russian Math. Surv. **25**, 137 (1970)
7. V.I. Arnold, A. Avez: *Ergodic Problems of Classical Mechanics, Foundations and Applications* (Benjamin, New York, 1968)
8. A.M. Turing: Proc. London Math. Soc., Ser. 2 **42**, 230 (1936)
9. K. Gödel: Monatshefte für Math. u. Physik **38**, 173 (1931)
10. K.S. Scott: *Oscillations, Waves and Chaos in Chemical Kinetics* (Oxford University Press, New York, 1994)
11. F. Leyvraz, S. Redner: Phys. Rev. Lett. **66**, 2168 (1991)
12. T.J. Cox, D. Griffeath: Ann. Prob. **14**, 347 (1986)

13. C.R. Doering, D. Ben-Avraham: Phys. Rev. A **38**, 3035 (1988)
14. I.M. Lifshitz: Zh. Eksp. Teor. Fiz. **42**, 1354 (1962)
15. I.M. Lifshitz, V.V. Slyozov: J. Phys. Chem. Solids **19**, 35 (1961)
16. C. Wagner: Z. Elektrochem. **65**, 581 (1961)
17. S.R. Broadbent, J.M. Hammersley: Proc. Camb. Phil. Soc. **53**, 629 (1957)
18. R.J. Baxter, A.J. Guttmann: J. Phys. A **21**, 3193 (1988)
19. W. Kinzel: Z. Physik B **58**, 229 (1985)
20. C. Becco: Tracking et modélisation de bancs de poisons. Thesis, University of Liège (2004)
21. M. Schulz: *Statistical Physics and Economics* (Springer, Berlin Heidelberg New York, 2003)
22. W. Paul, J. Baschnagel: *Stochastic Processes: From Physics to Finance* (Springer, Berlin Heidelberg New York, 2000)
23. R.N. Mantegna, H.E. Stanley: Physics investigations Of financial markets. In: *Proceedings of the International School of Physics 'Enrico Fermi', Course CXXXIV* ed by F. Mallamace, H.E. Stanley (IOS Press, Amsterdam, 1997)
24. A. Bunde, Jan F. Eichner, S. Havlin, E. Koscielny-Bunde, H.-J. Schellnhuber, D. Vjushin: Phys. Rev. Lett. **92**, 039801 (2004)
25. B. Berkowitz, H. Scher: Phys. Rev. Lett. **79**, 4038 (1997)
26. D. Sornette, L. Knopoff, Y.Y. Kagan, C. Vanneste: J. Geophys. Res. **101**, 13883 (1996)
27. J.R. Grasso, D. Sornette: J. Geophys. Res. **103**, 29965 (1998)

2

Deterministic Control Theory

2.1 Introduction: The Brachistochrone Problem

In this chapter we focus our attention on the open loop control of deterministic problems. We will see that the language of deterministic control theory is close to the language of classical mechanics. The deterministic control theory requires that the dynamics of the system under control is completely defined by well-defined equations of motion and accurate initial conditions. Although the theoretical description is not influenced by the degree of complexity of the system, the subsequently presented methods are useful if the system has only a few degrees of freedom. The causes of this unpleasant restriction for the practical application of the techniques presented was discussed in Sect. 1.2.

As a very simple introduction, we consider a particle which moves in a two-dimensional space along a fixed curve in a potential $V(x, y)$ without friction. A typical control problem is now the question that what form should the curve have so that for a given initial kinetic energy the particle moves from a given point to another well-defined point in the shortest time? This is the brachistochrone problem formulated originally by Galilei [1] and solved by Bernoulli [2]. In principle, the brachistochrone problem can be formulated on the basis of several concepts. The first way is to interpret the control of the system by the choice of the boundary condition which fixes the particle at the curve. Let $y(x)$ be the form of the curve (Fig. 2.1). The position of the particle is given by the coordinates $\mathbf{x} = (x, y)$ so that the initial position may be defined by $\mathbf{x}_0 = (x_0, y_0)$ with $y_0 = y(x_0)$ while the final position is given by $\mathbf{x}_\mathrm{e} = (x_\mathrm{e}, y_\mathrm{e})$ with $y_\mathrm{e} = y(x_\mathrm{e})$. Furthermore, the conservative force field requires a potential $V(\mathbf{x})$ and we obtain the conservation of the total energy

$$\frac{m}{2}\mathbf{v}^2 + V(\mathbf{x}) = E\,, \tag{2.1}$$

where $\mathbf{v}$ is the velocity of the particle. Thus, the time $\mathrm{d}t$ required for the passage of the curve segment $\mathrm{d}s = \sqrt{\mathrm{d}x^2 + \mathrm{d}y^2}$ is simply

M. Schulz: *Control Theory in Physics and other Fields of Science*
STMP **215**, 17–60 (2006)

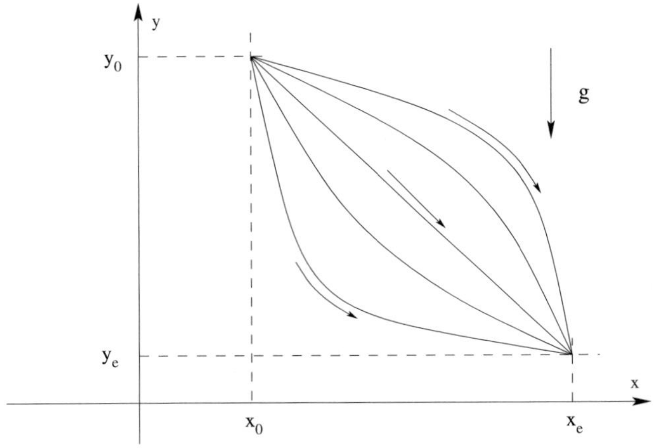

Fig. 2.1. The original brachistochrone problem solved by Bernoulli: which path in a homogeneous gravity field between (x_0, y_0) and (x_e, y_e) is the fastest trajectory?

$$dt = \frac{ds}{|\mathbf{v}|} = \sqrt{m}\frac{\sqrt{1+y'^2(x)}}{\sqrt{2(E - V(\mathbf{x}))}}dx \,, \tag{2.2}$$

and the above-introduced optimum curve minimizing the duration time T between the initial point and the final point follows from the solution of the minimum problem

$$T = \int\limits_{x_0}^{x_e} dx \frac{\sqrt{1+y'^2(x)}}{\sqrt{2(E - V(x, y(x))}} \to \inf \,, \tag{2.3}$$

considering the initial and final conditions $y_0 = y(x_0)$ and $y_e = y(x_e)$, respectively. The solution of (2.3) belongs to a classical variational problem [2,3].

The brachistochrone problem can be also formulated as a optimum control by external forces. To this aim we write the curve in a parametric form $(x(t), y(t))$. The curve $y = y(x)$ may be expressed by the implicit relation $U(x(t), y(t)) = 0$. Thus, the motion of the particle along the curve requires immediately

$$u_x \dot{x} + u_y \dot{y} = 0 \tag{2.4}$$

with

$$u_x = \frac{\partial U(x,y)}{\partial x} \quad \text{and} \quad u_y = \frac{\partial U(x,y)}{\partial y} \,. \tag{2.5}$$

On the other hand, when the particle moves along the curve, two forces act on the particle. The first force, $\mathbf{F} = -\nabla V$, is due to the potential V, the second force is the reaction of support, $\mathbf{u} = (u_x, u_y)$, which is perpendicular to the velocity. Without the knowledge of condition (2.4), the second force cannot be distinguished by physical arguments from an additional external force acting on the free particle in the potential V. Thus, we get the equations of motion

$$\ddot{x} = -\frac{\partial V}{\partial x} + u_x \quad \text{and} \quad \ddot{y} = -\frac{\partial V}{\partial y} + u_y \,, \tag{2.6}$$

and the optimum control problem is now reduced to the minimum problem

$$T \to \inf \quad \text{with} \quad \mathbf{x}(0) = \mathbf{x}_0 \quad \text{and} \quad \mathbf{x}(T) = \mathbf{x}_e \tag{2.7}$$

with the equations of motion (2.6) and condition (2.4) shrinking the external forces $\mathbf{u} = (u_x, u_y)$ on such forces which are equivalent to the reaction of support. As we will see later, representation (2.6) is a characteristic of the optimal control problem via external force fields.

2.2 The Deterministic Control Problem

2.2.1 Functionals, Constraints, and Boundary Conditions

Let us start with a preliminary formulation of an optimal control problem. To this aim, we consider a dynamical system over a certain horizon T, which means we have a problem wherein the time t belongs to an interval $[0, T]$ with $T < \infty$. As discussed in the introduction, each control problem is defined by two groups of variables. The first group are the state variables X with $X = \{X_1, \ldots, X_N\}$. The set of all allowed vectors X spans the phase space $\mathbb{P}$ (or the reduced phase space $\mathbb{P}_{\text{rel}}$) of the underlying system. The physically motivated strict distinction between the phase space $\mathbb{P}$, which contains all degrees of freedom, and the reduced phase space $\mathbb{P}_{\text{rel}}$, which contains only the relevant degrees of freedom, is no longer necessary for the moment. Hence, we use simply the notation 'phase space' for both, $\mathbb{P}$ and $\mathbb{P}_{\text{rel}}$. The second group belongs to the input or control variables $u = \{u_1, \ldots, u_n\}$. The set of all allowed control variables form the control space $\mathbb{U}$.

After this fundamental definitions, we may define the mathematical components of a deterministic control problem. In principle, this problem requires the consideration of constraints, boundary conditions, and functionals.

Boundary conditions are imposed on the end points of the time interval $[0, T]$ considered in the current control problem. These conditions belong only to the trajectory $X(t)$ of the system. Characteristic boundary conditions are

- boundary conditions with fixed end points, i.e., $X(0) = X_0$ and $X(T) = X_e$
- periodic boundary conditions, where the trajectory $X(t)$ has the same values on both end points, i.e., $X(0) = X(T)$, and
- boundary conditions with one ore two free ends.

Functionals define the control aim. These functionals are often denoted as performance or cost functional which should be minimized to obtain the optimum control $u^*(t)$ and a corresponding optimum trajectory $X^*(t)$. There are three standard types of functionals. Integral functionals have the form

$$R[X,u,T] = \int_0^T \mathrm{d}t\phi(t, X(t), \dot{X}(t), u(t)) \,, \tag{2.8}$$

where the integrand $L : \mathbb{R} \times P \times P \times \mathbb{U} \to \mathbb{R}$ is called the performance function or the Lagrangian. We will demonstrate in the subsequent section that this Lagrangian is equivalent under certain conditions to the Lagrangian of classical mechanics. The second type of functionals representing a performance are endpoint functionals. These functionals depend on the terminal values of the trajectory

$$S[X,u,T] = \Phi(X(0), X(T), T) \,. \tag{2.9}$$

Finally, we may consider mixed functionals, defined by a linear combinations of (2.8) and (2.9).

Constraints are either functional equalities, $G_\alpha[t, X(t), u(t)] = 0$, or functional inequalities $G_\alpha[t, X(t), u(t)] \leq 0$, where $\alpha = 1, 2, \ldots$ is the number of constraints. Constraints of the form

$$\dot{X}(t) = F(X, u, t) \tag{2.10}$$

are called differential constraints. These constraints often correspond to the evolution equations, e.g., the deterministic part of (1.4) or the canonical system (1.1). Constraints which do not depend on the derivatives and controls are called geometrical constraints, e.g., $g_\alpha[t, X(t)] = 0$ or $g_\alpha[t, X(t)] \leq 0$. In general, we may conclude that constraints fix at least partially the trajectory of the system through the phase space.

2.2.2 Weak and Strong Minima

The solution of a control problem is equivalent to the determination of the minimum of the corresponding performance functional

$$R[X,u,T] \to \inf \tag{2.11}$$

considering the constraints and the boundary conditions. The solution of this problem is an optimum control $u^*(t)$ and an optimum trajectory $X^*(t)$. We denote (2.11) together with the corresponding constraints and the boundary conditions as a Lagrange problem if $R[X,u,T]$ is an integral. If $R[X,u,T]$ is an endpoint functional, the problem is called the Meier problem and we speak about a Bolza problem in the case of a mixed functional. However, it is simple to demonstrate that this historically motivated distinction is not necessary, because all three apparently different problems are essentially equivalent. For example, the integral representation (2.8) can be transformed into an endpoint functional by introducing a new degree of freedom X_{N+1} with a new equation of motion

$$\dot{X}_{N+1}(t) = \phi(t, X(t), \dot{X}(t), u(t)) \tag{2.12}$$

as an additional constraint and the additional boundary condition $X_{N+1}(0) = 0$. Then, the Lagrange problem (2.8) can be written as a Meier problem

$$R[X, u, T] = \int_0^T dt \dot{X}_{N+1}(t) = X_{N+1}(T) \to \inf \tag{2.13}$$

now with $R[X, u, T]$ as an endpoint functional.

Let us assume that the pair $\{X(t), u(t)\}$ satisfies both the constraints and the boundary conditions. Generally, there exists a noncountable set of such pairs. This may be illustrated by a simple example. We may choose an arbitrary function $u(t)$ and solve the evolution equation (2.10) and the given boundary conditions. Obviously, we will always succeed with this procedure, at least for a sufficiently large class of functions $u(t)$. In the future we define that such a pair $\{X(t), u(t)\}$ is said to be admissible for the control problem.

An admissible pair $\{X^*(t), u^*(t)\}$ yields a local weak minimum (or a weak solution) of the control problem if the inequality

$$R[X, u, T] \geq R[X^*, u^*, T] \tag{2.14}$$

holds for any admissible pairs $\{X(t), u(t)\}$ which satisfy the inequalities

$$\|X - X^*\| \leq \varepsilon \quad \|\dot{X} - \dot{X}^*\| \leq \varepsilon \quad \text{and} \quad \|u - u^*\| \leq \varepsilon \,, \tag{2.15}$$

where we use the maximum norm

$$\|\xi\| = \max_{t \in [0,T]} |\xi(t)| \tag{2.16}$$

for any sufficiently small ε. In what follows we call the small differences $\delta X(t) = X(t) - X^*(t)$ and $\delta u(t) = u(t) - u^*(t)$, respectively, variations around the (weak) minimum $\{X^*(t), u^*(t)\}$. A weak minimum is not necessarily stable against arbitrary velocity variations $\delta\dot{X}(t)$ or strong variations of the control function $\delta u(t)$. We speak about a strong minimum if inequality (2.14) holds for all admissible pairs $\{X(t), u(t)\}$ satisfying the inequality

$$\|X - X^*\| \leq \varepsilon \,. \tag{2.17}$$

In other words, a strong minimum is not affected by arbitrary fluctuations of the velocity $\dot{X}(t)$ and the control function $u(t)$. That means especially that there is no better control function $u(t)$ than $u^*(t)$ for all trajectories close to $X^*(t)$. Each strong minimum is always a weak minimum, but a weak minimum is not necessarily a strong minimum. Finally, if inequality (2.14) holds for all admissible pairs $\{X(t), u(t)\}$, the pair $\{X^*(t), u^*(t)\}$ is called the optimum solution of the control problem.

The general problem of optimum control theory can now be reformulated. In a first step we have to find all extremal solutions of the functional $R[X, u, T]$ considering the constraints and the boundary conditions, and then, we have to check whether these extrema are the optimum solution of the control problem.

2.3 The Simplest Control Problem: Classical Mechanics

2.3.1 Euler–Lagrange Equations

The simplest control problem contains no control function and no constraints. Thus (2.8) reduces to the special functional

$$S[X,T] = \int_0^T \mathrm{d}t L(t, X(t), \dot{X}(t)) \tag{2.18}$$

with fixed points $X(0) = X_0$ and $X(T) = X_\mathrm{e}$ as boundary conditions. From a physical point of view, functional (2.18) can be identified as the well-known mechanical action. Here, the function L is the Lagrangian $L = E_\mathrm{kin} - U$, defined as the difference between the kinetic energy $E_\mathrm{kin} = \sum_i m_i \dot{X}_i^2/2$ and the potential $U = U(X)$ of a conservative mechanical system. Each N-dimensional vector $X \in \mathbb{P}$ denotes the position of system in the phase space. It should be denoted that in the framework of the present problem the phase space $\mathbb{P}$ does not correspond to the standard definition of classical mechanics. Because $\mathbb{P}$ contains only the coordinates of the system and not the momenta, this space is sometimes called the configuration space. The fact that the function L only contains $X(t)$ and $\dot{X}(t)$ but no higher derivatives $\ddot{X}(t), \ldots$ means that a mechanical state is completely defined by the coordinates and the velocities. The Hamilton principle (the principle of least action) requires that the trajectory of the system through the phase space corresponds to the optimum trajectory $X^*(t)$ of the problem $S[X,T] \to \inf$. The solution of this optimum problem leads to the equations of motion of the underlying system which are also denoted as Euler–Lagrange equations.

For the moment we should generalize the physical problem to an arbitrary Lagrangian. The only necessary condition is that the Lagrangian must be continuously differentiable. From this very general point of view, the Euler–Lagrange equations are the necessary conditions that a certain admissible trajectory corresponds to an extremum of the action $S[X,T]$.

Let us now derive the Euler–Lagrange equations in a mathematically rigorous way. That is necessary in order to understand several stability problems which may become important for the subsequent discussion of the general optimal control problem. The solution of this extremum problem consists of three stages.

The initial step is the calculation of the first-order variation. To this aim we assume that $X^*(t)$ is a trajectory corresponding to an extremum of $S[X,T]$ with respect to the boundary conditions. The addition of an arbitrary infinitesimal small variation $\delta X(t)$ with the boundary conditions $\delta X(0) = \delta X(T) = 0$ generates a new trajectory $X(t) = X^*(t) + \delta X(t)$ in the neighborhood of the optimum trajectory. We conclude that all trajectories $X(t)$ are again admissible functions due to the special choice of the boundary conditions for the variation $\delta X(t)$. Thus, we obtain

$$\begin{aligned}\delta S[X^*,T] &= S[X^*+\delta X,T] - S[X^*,T] \\ &= \int_0^T \mathrm{d}t L(t, X^*(t)+\delta X(t), \dot{X}^*(t)+\delta\dot{X}(t)) \\ &\quad - \int_0^T \mathrm{d}t L(t, X^*(t), \dot{X}^*(t)) \\ &= \int_0^T \mathrm{d}t \left[f(t)\delta X(t) + p(t)\delta\dot{X}(t) \right]\end{aligned} \tag{2.19}$$

with the force

$$f(t) = \frac{\partial L}{\partial X^*}(t, X^*, \dot{X}^*) \tag{2.20}$$

and the momentum[1]

$$p(t) = \frac{\partial L}{\partial \dot{X}^*}(t, X^*, \dot{X}^*) \ . \tag{2.21}$$

The second step is the integration by parts. There are two possibilities. Following Lagrange, we have to integrate by parts the second term of (2.19) while following DuBois–Reymond, we integrate by parts the first term. The way of Lagrange assumes a further assumption about the smoothness of the Lagrangian, namely that the generalized momentum (2.21) is continuous differentiable with respect to time. Under this additional condition we obtain

$$\delta S[X^*,T] = \int_0^T \mathrm{d}t \left[f(t) - \dot{p}(t) \right] \delta X(t) \ . \tag{2.22}$$

The integration by parts according to DuBois–Reymond yields

$$\delta S[X^*,T] = \int_0^T \mathrm{d}t \left[p(t) - \int_0^t f(\tau)\mathrm{d}\tau \right] \delta\dot{X}(t) \ . \tag{2.23}$$

Both representations, (2.22) and (2.23), are essentially equivalent.

The assumption that $X^*(t)$ corresponds to an extremum of $S[X,T]$ automatically requires $\delta S[X^*,T] = 0$. Thus, the last stage consists in solving (2.22) and (2.23) considering $\delta S[X^*,T] = 0$. The mathematical proof of this statement is the most difficult part of the derivation. For the sake of simplicity, we restrict our conclusions on some intuitive arguments. It is obvious that in the case of (2.22) the condition $\delta S[X^*,T] = 0$ for all possible variations $\delta X(t)$ automatically requires $\dot{p}(t) = f(t)$ or with (2.20) and (2.21)

[1] Both, force and momentum, are N-dimensional vectors with the components $f_i(t) = \partial L / \partial X_i^*$ and $p_i(t) = \partial L / \partial \dot{X}_i^*$.

$$\frac{\mathrm{d}}{\mathrm{d}t}\left[\frac{\partial L}{\partial \dot{X}^*}(t, X^*, \dot{X}^*)\right] = \frac{\partial L}{\partial X^*}(t, X^*, \dot{X}^*) . \tag{2.24}$$

This is the Euler–Lagrange equation in the Lagrange representation. The second way, originally given by DuBois–Reymond, leads to

$$p(t) = \int_0^t f(\tau)\mathrm{d}\tau + c_0 , \tag{2.25}$$

where c_0 is an arbitrary constant. In other words, if the expression in the brackets of (2.23) has a constant value, the integral vanishes due to the boundary conditions for $\delta X(t)$. Because $\delta S[X^*, T] = 0$ should be valid for all admissible variations, the only solution is (2.25). The explicit form of (2.25) reads

$$\frac{\partial L}{\partial \dot{X}^*}(t, X^*, \dot{X}^*) = \int_0^t \frac{\partial L}{\partial X^*}(\tau, X^*(\tau), \dot{X}^*(\tau))\mathrm{d}\tau + c_0 . \tag{2.26}$$

This equation is also called the DuBois–Reymond representation of the Euler–Lagrange equation. For details of the proof we refer to the literature [5–12].

Physically both (2.24) and (2.26) are equivalent representations of the same problem. Mathematically, (2.26) has the advantage that we need no further assumption about the existence of second-order derivatives of the Lagrangian. But apart from these subtleties, (2.24) and (2.26), respectively, represent the solution of the extremal problem. The solution of the Euler–Lagrange equation, $X^*(t)$ is also called an extremal. In classical mechanics, (2.24) is completely equivalent to the Newtonian equations of motion. The only difference belongs to the boundary conditions. A typical problem of Newtonian mechanics is usually related to differential equations with initial conditions, $X(0) = X_0$ and $\dot{X}(0) = \dot{X}_0$, while the above-derived Euler–Lagrange equations have boundary conditions at both ends of the time interval $[0, T]$.[2]

2.3.2 Optimum Criterion

Weierstrass Criterion

The Euler–Lagrange equations are only a necessary condition for an extremum of the action $S[X, T]$. For the solution of the optimum problem we need an additional criterion which allows us to decide whether an extremal solution $X^*(t)$ corresponds to a local minimum or not. The Weierstrass criterion employs the same smoothness requirement used for the derivation of the Euler Lagrange equations, namely that the Lagrangian is continuously differentiable. The derivation of this criterion is simple and very instructive for our further procedure. To this aim we introduce a special variation (see also Fig. 2.2).

[2] It can be shown that this difference is only an apparent contradiction. Each of the two boundary conditions can be transformed into the other.

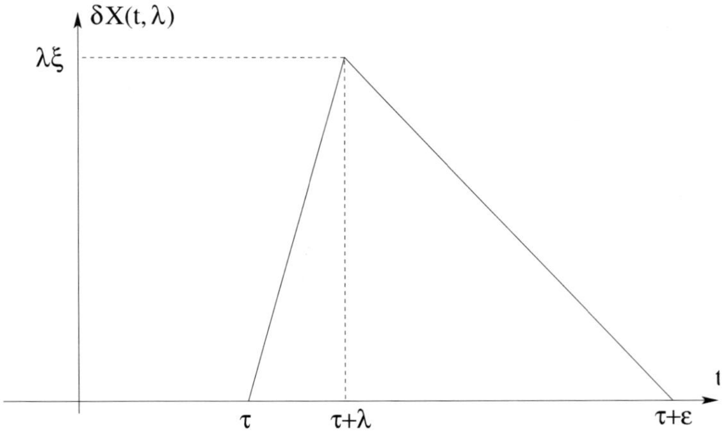

Fig. 2.2. The Weierstrass variation function $\delta X(t,\lambda)$

$$\delta X(t,\lambda)=\begin{cases}(t-\tau)\xi & \tau\leq t<\tau+\lambda\\ \lambda\xi\,(\tau+\varepsilon-t)\,(\varepsilon-\lambda)^{-1} & \tau+\lambda\leq t<\tau+\varepsilon\\ 0 & \text{otherwise}\end{cases} \tag{2.27}$$

with $\varepsilon>\lambda$ and $0<\tau<T-\varepsilon$. This function is not continuously differentiable, but all trajectories $X(t,\lambda)=X^*(t)+\delta X(t,\lambda)$ are admissible in the above-introduced sense. The variation of the velocities are then given by

$$\delta\dot{X}(t,\lambda)=\begin{cases}\xi & \tau\leq t<\tau+\lambda\\ -\lambda\xi\,(\varepsilon-\lambda)^{-1} & \tau+\lambda\leq t<\tau+\varepsilon\\ 0 & \text{otherwise}\,.\end{cases} \tag{2.28}$$

These variations are not necessarily small even for $\varepsilon\to 0$. The Weierstrass velocity variations $\delta\dot{X}(t,\lambda)$ have the form of a needle for small ε and $\lambda\ll\varepsilon$, which gave the reason for calling these variations needlelike variations. The action corresponding to the trajectory $X(t,\lambda)$ is given by

$$\begin{aligned}S[X^*,T,\lambda]&=\int_0^T \mathrm{d}tL(t,X(t,\lambda),\dot{X}(t,\lambda))\\ &=S[X^*,T]-\int_\tau^{\tau+\varepsilon}\mathrm{d}tL(t,X^*(t),\dot{X}^*(t))\\ &+\int_\tau^{\tau+\lambda}\mathrm{d}tL(t,X^*(t)+(t-\tau)\xi,\dot{X}^*(t)+\xi)\\ &+\int_{\tau+\lambda}^{\tau+\varepsilon}\mathrm{d}tL\left(t,X^*(t)+\lambda\xi\frac{\tau+\varepsilon-t}{\varepsilon-\lambda},\dot{X}^*(t)-\frac{\lambda\xi}{\varepsilon-\lambda}\right)\,.\end{aligned} \tag{2.29}$$

Differentiation of $S[X,T,\lambda]$ with respect to the time scale λ and setting $\lambda=0$ leads to

$$\begin{aligned} S'[X^*,T,0] &= L(\tau,X^*(\tau),\dot{X}^*(\tau)+\xi) - L(\tau,X^*(\tau),\dot{X}^*(\tau)) \\ &\quad +\xi\int_{\tau}^{\tau+\varepsilon} \mathrm{d}t\frac{\partial}{\partial X^*}L(t,X^*(t),\dot{X}^*(t))\frac{\tau+\varepsilon-t}{\varepsilon} \\ &\quad -\frac{\xi}{\varepsilon}\int_{\tau}^{\tau+\varepsilon} \mathrm{d}t\frac{\partial}{\partial \dot{X}^*}L(t,X^*(t),\dot{X}^*(t)) + O(\varepsilon)\,. \end{aligned} \tag{2.30}$$

The second term of this relation can be transformed. In the first step we get

$$\begin{aligned} (2) &= \int_{\tau}^{\tau+\varepsilon} \mathrm{d}t\frac{\partial}{\partial X^*}L(t,X^*(t),\dot{X}^*(t))\frac{\tau+\varepsilon-t}{\varepsilon} \\ &= \frac{1}{2}\int_{\tau}^{\tau+\varepsilon} \mathrm{d}t\frac{\partial}{\partial X^*}L(t,X^*(t),\dot{X}^*(t)) + O(\varepsilon) \end{aligned} \tag{2.31}$$

and in the second step we take into account that $X^*(t)$ is an extremum solution which satisfies the Euler–Lagrange equation. We obtain especially from the DuBois–Reymond representation (2.26)

$$\begin{aligned} \int_{\tau}^{\tau+\varepsilon} \mathrm{d}t\frac{\partial}{\partial X^*}L(t,X^*(t),\dot{X}^*(t)) &= \frac{\partial L}{\partial \dot{X}^*}(\tau+\varepsilon,X^*(\tau+\varepsilon),\dot{X}^*(\tau+\varepsilon)) \\ &\quad - \frac{\partial L}{\partial \dot{X}^*}(t,X^*,\dot{X}^*)\,. \end{aligned} \tag{2.32}$$

The substitution of (2.30) and (2.31) in (2.30) and passing to the limit $\varepsilon\to 0$ yields

$$\begin{aligned} S'[X^*,T,0] &= L(\tau,X^*(\tau),\dot{X}^*(\tau)+\xi) - L(\tau,X^*(\tau),\dot{X}^*(\tau)) \\ &\quad -\xi\frac{\partial}{\partial \dot{X}^*}L(\tau,X^*(\tau),\dot{X}^*(\tau))\,. \end{aligned} \tag{2.33}$$

A minimum of the action $S[X,T]$ for $X(t)=X^*(t)$ requires $S[X^*,T,\lambda]\geq S[X^*,T]$ for all $\lambda\neq 0$ and therefore $S'[X^*,T,0]\geq 0$. The latter implies the inequality

$$\begin{aligned} &L(\tau,X^*(\tau),\dot{X}^*(\tau)+\xi) - L(\tau,X^*(\tau),\dot{X}^*(\tau)) \\ &\qquad\qquad -\xi\frac{\partial}{\partial \dot{X}^*}L(\tau,X^*(\tau),\dot{X}^*(\tau)) \geq 0\,, \end{aligned} \tag{2.34}$$

which must hold for all ξ and $\tau\in[0,T]$. Inequality (2.34) is the so called Weierstrass criterion, see Fig. 2.3, which is a necessary condition for the action $S[X,T]$ to have a strong local minimum for $X(t)=X^*(t)$. If (2.34) is valid only for $|\xi|<\xi_{\max}<\infty$, the Weierstrass criterion indicates a weak local

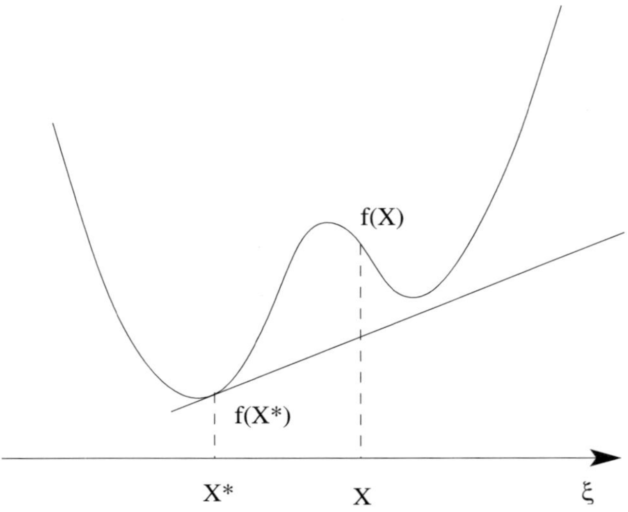

Fig. 2.3. One dimensional geometrical interpretation of the Weierstrass criterion

minimum. For the classical mechanics, we have $L = T - U$, where the kinetic energy is a simple quadratic form of the velocities, $T = \sum_i m_i \dot{X}_i^2/2$ and the potential depends only on the coordinates X and the time. Thus we obtain

$$\frac{1}{2}\sum_{i=1}^{N} m_i \xi_i^2 \geq 0 ,$$

where we have explicitly used the component representation of the N-dimensional vector quantity ξ. We conclude that the Weierstrass criterion for a strong local minimum is always satisfied for the standard Newtonian mechanics. Moreover, the criterion holds if the Lagrangian is a convex function in the velocities.

Legendre Criterion

The Legendre criterion is a second-order condition which follows from the second variation of functional (2.18). But in contrast to the derivation of the Euler–Lagrange equations we now consider finite variations $\|\delta X(t)\| < \varepsilon$ and $\|\delta \dot{X}(t)\| < \varepsilon$. Thus, the expansion of the difference $S[X^* + \delta X, T] - S[X^*, T]$ in terms of the fluctuations leads to

$$S[X^* + \delta X, T] - S[X^*, T] = \delta^1 S[X^*, T] + \delta^2 S[X^*, T] + o(\varepsilon^3) . \tag{2.35}$$

The first-order variation $\delta^1 S[X^*, T]$ vanishes identically because $X^*(t)$ is an extremal. Thus, the first nonvanishing contribution is the second-order variation $\delta^2 S[X^*, T]$. For sufficiently small variations we expect a local minimum of the functional $S[X, T]$ for $X(t) = X^*(t)$ if $\delta^2 S[X^*, T] > 0$. Before we evaluate $\delta^2 S[X^*, T]$ as a functional of the variations, we have to assume that the

Lagrangian is at least twice continuously differentiable. Under these assumption, the second variation of (2.18) is given by[3]

$$\delta^2 S[X^*,T] = \int_0^T \mathrm{d}t \sum_{i,j=1}^N \left[\delta\dot{X}_i(t)A_{ij}(t)\delta\dot{X}_j(t) + 2\delta\dot{X}_i(t)B_{ij}(t)\delta X_j(t)\right] + \int_0^T \mathrm{d}t \sum_{i,j=1}^N \left[\delta X_i(t)C_{ij}(t)\delta X_j(t)\right] \tag{2.36}$$

with the components

$$A_{ij}(t) = \frac{\partial^2 L}{\partial\dot{X}_i^*\partial\dot{X}_j^*} \quad B_{ij}(t) = \frac{\partial^2 L}{\partial\dot{X}_i^*\partial X_j^*} \quad C_{ij}(t) = \frac{\partial^2 L}{\partial X_i^*\partial X_j^*} \tag{2.37}$$

of the matrix functions A, B, and C. A necessary condition for a weak local minimum of $S[X,T]$ at $X(t) = X^*(t)$ is that the matrix

$$\begin{pmatrix} A & B \\ B^T & C \end{pmatrix} \tag{2.38}$$

is positive definite for all $t \in [0,T]$. As a consequence, the matrix A must be also positive definite. This is the necessary Legendre criterion for a minimum of the action at $X^*(t)$. It is a criterion for a weak local minimum because the quadratic form of the second-order variation (2.36) requires small fluctuations $\|\delta X(t)\| < \varepsilon$ and $\|\delta\dot{X}(t)\| < \varepsilon$, otherwise higher contributions of the expansion (2.35) remain effective and destroy the positive definite character of the second-order variation.

The Jacobi Criterion

Both criteria discussed above have a local character in the sense that the criteria must be verified for each point of time of the extremal $X^*(t)$. But local conditions are not enough to decide whether a trajectory is an optimum or not. This may be illustrated by a simple example. The shortest distance between two points of a sphere is an arc of a great circle. Because of the topology, there exists two different complementary parts of the great circle connecting these points, but only the shorter one is the optimum solution. On the other hand, each point of both lines satisfies the local criteria.

The Jacobi criterion is a global criterion for a weak local minimum. We focus on the one-dimensional case, i.e., the trajectories $X(t)$ are simple scalar functions of time. We start from the second variation (2.36) and require that the Legendre criterion holds strictly, i.e., $A(t) > 0$. Integrating by parts, we obtain

[3] In order to avoid confusion we use here the component representation.

$$\delta^2 S[X^*,T] = \int_0^T \mathrm{d}t \left[A(t)\delta\dot{X}(t)^2 + \left[C(t) - \frac{\mathrm{d}}{\mathrm{d}t}B(t) \right] \delta X(t)^2 \right]$$
$$= \int_0^T \mathrm{d}t \left[\left[C(t) - \frac{\mathrm{d}}{\mathrm{d}t}B(t) \right] \delta X(t) - \frac{\mathrm{d}}{\mathrm{d}t}\left(A(t)\delta\dot{X}(t) \right) \right] \delta X(t) \,. \quad (2.39)$$

Now we consider a special variation which solves $\delta^2 S[X^*,T] = 0$. One possible trajectory may be obtained from the Jacobi equation

$$\left[C(t) - \frac{\mathrm{d}}{\mathrm{d}t}B(t) \right] \delta X(t) - \frac{\mathrm{d}}{\mathrm{d}t}\left(A(t)\delta\dot{X}(t) \right) = 0 \quad (2.40)$$

or

$$\delta\ddot{X}(t) + \frac{\dot{A}(t)}{A(t)}\delta\dot{X}(t) - \left[\frac{C(t)}{A(t)} - \frac{\dot{B}(t)}{A(t)} \right] \delta X(t) = 0 \quad (2.41)$$

with the boundary conditions $\delta X(0) = \delta X(T) = 0$. This equation is an ordinary second-order differential equation, which has an existing and up to a constant prefactor unique solution $\delta X^{(J)}(t)$, the Jacobi trajectory. We assume that this solution is not trivial. The zeros of this solution distinct from the initial point $t = 0$ and the end point T are called the points conjugate to the initial point. These points play an important role for the Jacobi criterion. It is simple to show that each polygonal extremal which follows from the Jacobi trajectory by a substitution of the curve segment between two neighboring Jacobi zeros by $\delta X(t) \equiv 0$ is also a nontrivial variation with $\delta^2 S[X^*,T] = 0$. Let us assume there exists an additional zero for the time τ $(0 < \tau < T)$ such that $\delta X^{(J)}(\tau) = 0$. Now we may construct a new variation with $\delta X^{(J,0)}(t) = \delta X^{(J)}(t)$ for $t < \tau$ and $\delta X^{(J,0)}(t) = 0$ for $t > \tau$. Furthermore, we introduce a slightly different variation (Fig. 2.4)

$$\delta X^{(J,\lambda)} = \begin{cases} \delta X^{(J)}(t) & 0 \le t < \tau - \lambda \\ \rho\,(\tau + \mu\lambda - t)\,\delta X^{(J)}(\tau - \lambda) & \tau - \lambda \le t < \tau + \mu\lambda \\ 0 & \tau + \mu\lambda \le t < T \end{cases} \quad (2.42)$$

(with $\rho^{-1} = \lambda + \mu\lambda$) which differs from $\delta X^{(J,0)}(t)$ only in the small interval $\tau - \lambda \le t < \tau + \mu\lambda$. Now we are able to calculate the difference between the corresponding second variations (2.39), namely

$$\Delta\delta^2 S[X^*,T] = \int_{\tau-\lambda}^{\tau+\mu\lambda} \mathrm{d}t A(t) \left(\delta\dot{X}^{(J,\lambda)}(t)^2 - \delta\dot{X}^{(J,0)}(t)^2 \right)$$
$$+ \int_{\tau-\lambda}^{\tau+\mu\lambda} \mathrm{d}t \left[C(t) - \dot{B}(t) \right] \left(\delta X^{(J,\lambda)}(t)^2 - \delta X^{(J,0)}(t)^2 \right) . \quad (2.43)$$

The integrals can be estimated for a sufficiently small λ by the mean value theorem of integral calculus and the expansion of $\delta X^{(J)}(\tau - \lambda)$ and $\delta X^{(J,0)}(t)$

around the Jacobi zero τ for $t < \tau$, i.e., $\delta X^{(J)}(\tau - \lambda) = -\delta \dot{X}^{(J)}(\tau)\lambda + O(\lambda^2)$ and $\delta X^{(J,0)}(t) = \delta \dot{X}^{(J)}(\tau)(t - \tau) + O(\lambda^2)$. Hence, we arrive at

$$\Delta \delta^2 S[X^*, T] = -\frac{\lambda}{2} A(\tau) \delta \dot{X}^{(J)}(\tau)^2 + o(\lambda^2) \tag{2.44}$$

because the contribution of the second term in (2.43) is of an order of magnitude of λ^3. From here we conclude that the second variation $\delta^2 S[X^*, T]$ with respect to $\delta X^{(J,\lambda)}(t)$ becomes negative since the second variation with respect to the $\delta X^{(J,0)}(t)$ vanishes. In other words, if a nontrivial solution of the Jacobi equation has points conjugate to the initial point, there always exists a certain variation $\delta X^{(J,\lambda)}(t)$ with a negative $\delta^2 S[X^*, T]$ if, as mentioned above, the Legendre criterion holds, $A(\tau) > 0$. This is the Jacobi criterion. It is also a necessary condition for a weak minimum.

2.3.3 One-Dimensional Systems

Several well-known one-dimensional models of classical mechanics are very instructive for understanding the problems related to the classical calculus of variations. We will not discuss several physical applications which the reader may find in standard textbooks [13–15]. Here, we focus our attention on the characterization of the optimum trajectory and not on the solution of the Euler–Lagrange equations.

We start with a free particle. The corresponding action is given by

$$S[X, T] = \frac{m}{2} \int_0^T dt \dot{X}^2(t) \quad \text{with} \quad X(0) = x_0 \quad \text{and} \quad X(T) = x_e\,. \tag{2.45}$$

The Euler–Lagrange equation is $\ddot{X}(t) = 0$ and we obtain a solution that satisfies the boundary conditions of a motion with the constant velocity $X^*(t) = x_0 + (x_e - x_0)(t/T)$. Obviously, this solution is unique and corresponds to the optimum of the problem. Especially, a simple check shows that

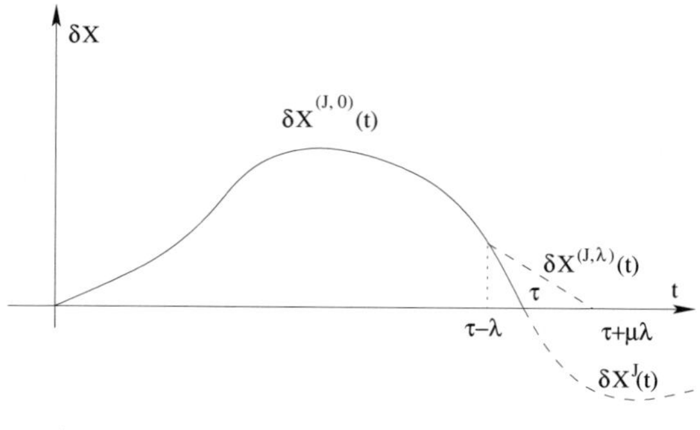

Fig. 2.4. Schematic representation of the Jacobi variation functions $\delta X^J(t)$, $\delta X^{(J,0)}(t)$, and $\delta X^{(J,\lambda)}(t)$ close to a Jacobi zero

both the Weierstrass and the Legendre criteria are fulfilled. The Jacobi equation reads $\delta \ddot{X}(t) = 0$ and has only a trivial solution.

Another situation occurs in the case of a linearly velocity-dependent mass, $m = m_0 + \alpha \dot{X}(t)$. The action

$$S[X,T] = \int_0^T dt \frac{m_0 + 2\alpha \dot{X}(t)}{2} \dot{X}^2(t) \qquad X(0) = x_0 \quad X(T) = x_e \tag{2.46}$$

leads now to the Euler–Lagrange equation $(m_0 + 3\alpha\dot{X}(t))\ddot{X}(t) = 0$ with the same solution $X^*(t) = x_0 + (x_e - x_0)(t/T)$ as in the case of a free particle. A real physical situation corresponds to a positive mass. Thus, we have to consider only such time intervals T and distances $\Delta x = x_e - x_0$ which satisfy the inequality $m_0 T > \max(-2\alpha\Delta x, 0)$. The Legendre criterion, $m_0 + 3\alpha\dot{X}(t) = m_0 + 3\alpha\Delta x/T > 0$, requires a stronger condition for the parameters leading to a weak minimum, namely $m_0 T > \max(-3\alpha\Delta x, 0)$. In this case, the Jacobi equation, $[m_0 + 3\alpha\Delta x/T]\,\delta\ddot{X}(t) = 0$, has always a trivial solution so that no conjugated point exists. On the other hand, the Weierstrass criterion leads to the necessary inequality $[3\alpha\Delta x/T + \alpha\xi + m_0/2]\,\xi^2 \geq 0$ which is violated for sufficiently negative values of ξ. That means the extremal solution is not strong minimum. In fact, a small change of the extremal trajectory, $X^*(t) \to X^*(t) + \delta X(t)$ with $\delta X(t) = -\Omega\Delta x(t/T)$ for $0 \leq t \leq T\Omega^{-2}$ and $\delta X(t) = \left[(t/T - 1)/(\Omega^2 - 1)\right]\Omega\Delta x$ for $T\Omega^{-2} \leq t \leq T$ leads to the following asymptotic behavior of the action

$$\begin{aligned} S[X^* + \delta X, T] &= S[X^*, T] - \frac{1}{T^2}\alpha\Delta x^3\Omega \\ &\quad + \frac{1}{2T^2}\Delta x^2(m_0 T + 6\alpha\Delta x) + O\left(\Omega^{-1}\right) . \end{aligned} \tag{2.47}$$

Whereas the simple case $\alpha = 0$ always yields $S[X^* + \delta X, T] \geq S[X^*, T]$, we find for $\alpha \neq 0$ and sufficiently large Ω trajectories in the neighborhood of the extremal with $S[X^* + \delta X, T] < S[X^*, T]$. Although the maximum norm of the trajectory variations, $\|\delta X\| = |\Delta x|/\Omega$, may be chosen sufficiently small, the corresponding variation of the velocity diverges $\|\delta\dot{X}\| = \Omega|\Delta x|/T$. Thus, the extremal of (2.46) is not a strong minimum of the action.

Let us proceed for our examples with the action

$$S[X,T] = \frac{m}{2}\int_0^T dt e^{g(t)}\dot{X}^2(t) \quad \text{with} \quad X(0) = x_0 \quad X(T) = x_e . \tag{2.48}$$

Such an action breaks the time translation symmetry, but mechanical actions of type (2.48) are sometimes used to incorporate friction into mechanical equations of motion. The Euler–Lagrange equation reads $\ddot{X}(t) + \dot{g}(t)\dot{X}(t) = 0$ and $\dot{g}(t)$ may be interpreted as a (time-dependent) friction coefficient. The special choice $g(t) = g_0 + \gamma t$ leads to the classical Newtonian friction law.

Here, we will study another friction type given by $g(t) = \beta \ln t$ with $\beta < 1$. The solution of the Euler–Lagrange equation is $X^*(t) = (x_e - x_0)(t/T)^{1-\beta} + x_0$. It is easy to verify that the extremal $X^*(t)$ yields the optimum of the problem. Unfortunately, the solution is not continuously differentiable for $t \to 0$. That means, the condition for a strong minimum, $\|\dot{X} - \dot{X}^*\| \leq \varepsilon$, remains indefinable.

The situation becomes more complicated for $\beta \geq 1$. The Euler–Lagrange equation now yields the general solution $X^*(t) = c_1 t^{1-\beta} + c_0$ but no curve of this family satisfies the boundary conditions. On the other hand, the lowest value of $S[X,T]$ is zero. This can be checked by the following approach. If we take a minimizing sequence $X_n(t) = (x_e - x_0)(t/T)^{1/n} + x_0$ or $X_n(t) = (x_0 - x_e)(1 - nt/T)_+ + x_e$ (with $(\xi)_+ = \xi$ for $\xi > 0$ and $(\xi)_+ = 0$ for $\xi < 0$), then we find that $S[X,T] \to 0$ for $n \to \infty$. However, the above-introduced sequences do not converge continuously to the limit function $X_\infty(t)$ so that the extremal solution is not continuously differentiable.

Finally, we discuss the action of a harmonic oscillator of frequency ω. Especially, we ask for periodic solutions $X(0) = X(T) = 0$. Thus, we have the problem

$$S[X,T] = \frac{m}{2} \int_0^T \mathrm{d}t \left(\dot{X}^2(t) - \omega^2 X^2(t) \right) \to \inf \quad X(0) = X(T) = 0\,. \quad (2.49)$$

The Euler–Lagrange equation now reads $\ddot{X}(t) + \omega^2 X(t) = 0$. The extremal solution is $X^*(t) = 0$ for $\omega T < \pi$ and $X^*(t) = X_0 \sin(\omega t)$ for $\omega T = \pi$. Since the Lagrangian is of the standard form $L = T - U$, the Weierstrass criterion suggests a strong minimum for these extremal solutions. We obtain for both types of extremals $S[X^*,T] = 0$. The following algebraic transformations

$$\begin{aligned}
S[X,T] &= \frac{m}{2} \int_0^T \mathrm{d}t \left(\dot{X}^2(t) - \omega^2 X^2(t) \right) \\
&= \frac{m}{2} \int_0^T \mathrm{d}t \left[\dot{X}^2(t) + \omega^2 \left(\tan^{-2} \omega t - \sin^{-2} \omega t \right) X^2(t) \right] \\
&= \frac{m}{2} \int_0^T \mathrm{d}t \left[\dot{X}^2(t) + X^2(t) \omega^2 \tan^{-2} \omega t - 2\dot{X}(t) X(t) \omega \tan^{-1} \omega t \right] \\
&= \frac{m}{2} \int_0^T \mathrm{d}t \left[\dot{X}(t) - X(t) \omega \tan^{-1} \omega t \right]^2 \qquad (2.50)
\end{aligned}$$

show that $S[X^*,T] = 0$ is in fact the lower limit of the action. But it should be remarked that (2.50) holds only for $\omega T \leq \pi$, because the expression $X(t) \tan^{-1} \omega t$ has no relevant singularities as long as $0 \leq \omega T \leq \pi$. Note that the singularities for $t = 0$ and $T = \pi\omega^{-1}$ are cancelled due to the boundary conditions. In other words, there are a unique solution, $X^*(t) = 0$ for $\omega T < \pi$, and an infinite number of extremal solutions, $X^*(t) = X_0 \sin(\omega t)$ for $\omega T = \pi$, and all of them yield the optimum of problem (2.49).

For $\omega T = n\pi$, $n > 1$, the Euler–Lagrange equation again yields the extremal $X^*(t) = X_0 \sin(\omega t)$ with an arbitrary amplitude X_0. The corresponding Jacobi equation reads $\delta\ddot{X}(t) + \omega^2 \delta X(t) = 0$, i.e., we get the solution $\delta X(t) \sim \sin(\omega t)$. The zeros of this solution are the conjugates of the initial point. Since the first conjugate point, $t = \pi\omega^{-1}$, now belongs to the interval $[0, T]$, the Jacobian criterion suggests that all extremals obtained from the Euler–Lagrange yield neither a strong, nor a weak minimum.

It remains the extremal $X^*(t) = 0$ which is the unique solution of the Euler–Lagrange equations for $\omega T > \pi$, $\omega T \neq n\pi$. The corresponding action of this extremal is $S[X^*, T] = 0$. However, (2.50) fails for $\omega T > \pi$, and trajectories with a negative action, $S[X, T] < 0$, become possible. For an illustration, we compute explicitly the action for the trajectory $\overline{X}(t) = \varepsilon \sin(\pi t/T)$. We obtain

$$S[\overline{X}, T] = -\frac{m}{4}\varepsilon^2 \frac{\pi^2}{T}\left(\frac{\omega^2 T^2}{\pi^2} - 1\right), \tag{2.51}$$

which has always negative values for $\omega T > \pi$. The distance between $\overline{X}(t)$ and $X^*(t)$, namely $\|\overline{X} - X^*\| = \varepsilon$, can be chosen arbitrarily close to zero. This means that the extremal $X^*(t) = 0$ for $\omega T > \pi$ no longer yields even a weak minimum.

These examples show that the strong formulation of the principle of least action, $S \to \inf$, originally defined by Hamilton is not suitable as a fundamental physical principle. Therefore, the modern physical literature prefers a Hamilton principle which is weakened to the more appropriate claim $S \to$ extr. under the simultaneous assumption of a sufficiently smoothness of the trajectories.

2.4 General Optimum Control Problem

2.4.1 Lagrange Approach

Basic Equations

We now consider a generalized functional of the integral form

$$R[X, u, T] = \int_0^T \mathrm{d}t \phi(t, X(t), u(t)) \,. \tag{2.52}$$

As mentioned in Sect. 2.2.1, the minimization of this performance functional defines the control aim. Furthermore, we have demonstrated in the same chapter that all other types of control problems, e.g., endpoint functionals or mixed types may be rewritten into (2.52). The time t belongs to the interval $[0,T]$ with $T < \infty$. The state variable $X = X(t)$ with $X = \{X_1, \dots, X_N\}$ represents a trajectory through the N-dimensional phase space $\mathbb{P}$ of the underlying system. The second group is the set of the control variables $u = u(t)$ with $u = \{u_1, \dots u_n\}$. The set of all allowed control variables form the control space $\mathbb{U}$.

Furthermore, we consider some constraints, which may be written as a system of differential equations

$$\dot{X}(t) = F(X, u, t) . \tag{2.53}$$

In principle, these equations can be interpreted as the evolution equations of the system under control. We remark that functional (2.8) can be easily transformed into (2.52) by introducing N additional control variables and setting

$$\dot{X}_\alpha(t) = u_{n+\alpha}(t) \quad \text{for} \quad \alpha = 1, \dots, N . \tag{2.54}$$

In this sense, the mechanical equations of motion discussed above mathematically in details can be reformulated. The Lagrangian $L = T(\dot{X}) - U(X)$ now becomes the form $L = T(u) - U(X)$ and we have to consider N constraints $\dot{X}(t) = u$. But the application of the concept defined by functional (2.52) and the evolution equations (2.53) is much larger as the framework of classical mechanics. Equations (2.53) may also represent the kinetics of chemical or other thermodynamic nonequilibrium processes, the time-dependent changes of electrical current and voltage in electronic systems or the flow of matter, energy, or information in a transport network. But many other applications are also possible.

Another remark belong to the control functions. These quantities should be free in the sense that the control variables have no dynamic constraints. This means that a reasonable control problem contains no derivatives of the control functions $u(t)$. In other words, if a certain problem contains derivatives of n' control functions, we have to declare these functions as additional degrees of freedom of the phase space. Thus, the reformulated problem has only $n - n'$ independent control variables, but the dimension of the phase space is extended to $N + n'$. On the other hand, state variables the dynamics of which is not defined by an explicit evolution equation of type (2.53) are not real dynamical variables. These free variables should be declared as control variables. Finally, constraints of the form of simple equalities, $g(t, X(t), u(t)) = 0$, should be used for the elimination of some free state variables or control functions before the optimization procedure is carried out. That means, m independent constraints of the simple equality type reduce the dimension of the common space $\mathbb{P} \times U$ from $N + n$ to $N + n - m$.

In summary, the control problems considered now are defined by functional (2.52), by N evolution equations of type (2.53) for the N components of the state vector X, and by n free control functions collected in the n-dimensional vector u. Such problems occur in natural sciences as well as in technology, economics, and other scientific fields.

In order to complete the control problem, we have to consider the boundary conditions for state variables X. We introduce conditions for the initial point and the end point by equations of the type

$$b_a[X(0)] = 0 \quad \text{and} \quad \bar{b}_a[X(T)] = 0\,, \tag{2.55}$$

where a runs over all conditions we have taken into account. For the following discussion we assume N independent initial and N independent final conditions. These conditions fix the start and end points of the trajectory $X(t)$ completely. However, the number of boundary conditions may be less than $2N$. In this case, we have at least partially free boundary conditions. The subsequently derived concept also works in this case.

Our aim must be the derivation of necessary conditions for an optimal solution of the system under control. In future, we will not stress the mathematical accuracy as strongly as in the previous chapters. We refer to the extensive specialized literature [16–19] for specific and rigorous proofs.

Lagrange Multipliers and Generalized Action

The basic idea of combining constraints (2.53) with functional (2.52) to a common optimizable functional is the application of Lagrange multipliers. Let us start with an illustration of Lagrange's idea. To this aim we consider a function $f(x)$ mapping the d-dimensional space on a one-dimensional manifold, f: $\mathbb{R}^d \to R$. Furthermore, we have p constraints, $C_k(x) = 0$, $k = 1, \ldots, p$. Now we ask for an extremal solution of $f(x)$. Without constraints, we have to solve the extremal conditions

$$\frac{\partial f(x)}{\partial x_\alpha} = 0 \quad \text{for} \quad \alpha. = 1, \ldots, d\,. \tag{2.56}$$

With constraints, we construct the Lagrange function

$$l(x, \lambda) = f(x) + \sum_{i=1}^{p} \lambda_p C_p(x)\,. \tag{2.57}$$

The new variables λ_p are called the Lagrange multipliers. The Lagrange principle now consists in the determination of the extremum of $l(x, \lambda)$ with respect to the set of variables $(x, \lambda) \in \mathbb{R}^d \times R^p$. In other words, we have to solve the extended extremal conditions

$$\frac{\partial l(x, \lambda)}{\partial x_\alpha} = 0 \quad \text{and} \quad \frac{\partial l(x, \lambda)}{\partial \lambda_k} = 0 \tag{2.58}$$

for $\alpha = 1, \ldots, d$ and $k = 1, \ldots, p$. The first group of these equations explicitly reads

$$\frac{\partial f(x)}{\partial x_\alpha} = -\sum_{i=1}^{p} \lambda_p \frac{\partial C_p(x)}{\partial x_\alpha} \quad \text{for} \quad \alpha = 1, \ldots, d \tag{2.59}$$

while the second group reproduces the constraints, $C_k(x) = 0$, $k = 1, \ldots, p$.

It is easy to extend this principle on functionals and constraints of the type (2.53). The only difference is that each point of the d-dimensional vector x must be replaced by an infinite-dimensional vector with components labeled by infinitely many points of time and the number of the corresponding degrees of freedom. Thus, functional (2.52) and constraints (2.53) can be combined to a generalized 'Lagrange function'

$$R'[X,u,T,p] = R[X,u,T] + \int_0^T \mathrm{d}t \left(\dot{X}(t) - F\left(X,u,t\right)\right) P(t) \tag{2.60}$$

with the N-dimensional vector function $P(t) = \{P_1(t), P_2(t), \ldots, P_N(t)\}$ as generalized Lagrange multipliers. The vector $P(t)$ is sometimes called the adjoint state vector or the generalized momentum. The set of all admissible vectors $P(t)$ forms the N-dimensional adjoint phase space $\overline{\mathbb{P}}$. Finally, we can also introduce Lagrange multipliers for the boundary conditions. Because these conditions are declared only at the end points of the interval $[0,T]$, we need only a finite number of additional Lagrange multipliers which are collected in the two vectors Λ and $\overline{\Lambda}$. Thus, the complete 'Lagrange function' now reads

$$\begin{aligned} S[X,P,u,T,\Lambda,\overline{\Lambda}] = {} & R[X,u,T] + \int_0^T \mathrm{d}t \left(\dot{X}(t) - F\left(X,u,t\right)\right) P(t) \\ & + b(X(0))\Lambda + \overline{b}(X(T))\overline{\Lambda}\,. \end{aligned} \tag{2.61}$$

In future, we call functional (2.61) generalized action. To proceed, we write this action in the standard form

$$\begin{aligned} S[X,P,u,T,\Lambda,\overline{\Lambda}] = {} & \int_0^T \mathrm{d}t L\left(t, X(t), \dot{X}(t), P(t), u(t)\right) \\ & + b(X(0))\Lambda + \overline{b}(X(T))\overline{\Lambda} \end{aligned} \tag{2.62}$$

with the generalized Lagrangian

$$L\left(t, X, \dot{X}, P, u\right) = \phi(t,X,u) + P\left(\dot{X} - F\left(X,u,t\right)\right). \tag{2.63}$$

It is important to notice that for each trajectory satisfying the constraints and the boundary condition, the generalized action $S[X,P,u,T,\Lambda,\overline{\Lambda}]$ approaches the performance functional $R[X,u,T]$. The formulation of (2.62) is the generalized first step of Lagrange's concept corresponding to the formulation of the Lagrange function (2.57). The second step, namely the derivation of the necessary conditions for an extremal solution corresponding to (2.58), leads to generalized Euler–Lagrange equations.

Euler–Lagrange Equations

The general control aim is to minimize functional (2.52) considering constraints (2.53) and the boundary conditions (2.55). The above-discussed extension of Lagrange's idea of functionals means that the minimization of the generalized action (2.61) with respect to the state $X(t)$, the control $u(t)$, the adjoint state $P(t)$ (corresponding to an infinitely large set of Lagrange multipliers fixing the constraints) and the Lagrange multipliers Λ and $\overline{\Lambda}$ (which considers the boundary conditions) is completely equivalent to the original problem. In other words, we have to find the optimum trajectory $(X^*(t), P^*(t), u^*(t))$ through the space $\mathbb{P} \times \overline{P} \times U$.

To solve this problem, we consider for the moment the action

$$S[X, P, u, T, \Lambda, \overline{\Lambda}] = \int_0^T \mathrm{d}t \widetilde{L}\left(t, X(t), \dot{X}(t), P(t), u(t), \Lambda, \overline{\Lambda}\right), \tag{2.64}$$

which we wish to minimize. The solution of the optimum problem, $S \to \inf$, can be obtained by the above-discussed calculus of variations, but now for the generalized 'state' (X, P, u) instead of the classical state X. Formally, we obtain three groups of Euler–Lagrange equations

$$\frac{\mathrm{d}}{\mathrm{d}t} \frac{\partial \widetilde{L}}{\partial \dot{X}^*} = \frac{\partial \widetilde{L}}{\partial X^*} \quad \frac{\partial \widetilde{L}}{\partial P^*} = 0 \quad \frac{\partial \widetilde{L}}{\partial u^*} = 0 \,. \tag{2.65}$$

Additionally, we have to consider the minimization with respect to the two vectors Λ and $\overline{\Lambda}$. Here, we simply obtain the necessary conditions

$$\frac{\mathrm{d}S[X, P, u, T, \Lambda, \overline{\Lambda}]}{\mathrm{d}\Lambda} = 0 \quad \frac{\mathrm{d}S[X, P, u, T, \Lambda, \overline{\Lambda}]}{\mathrm{d}\overline{\Lambda}} = 0 \,. \tag{2.66}$$

Now, we identify action (2.64) with action (2.63) which belongs to the control problem. This requires that the Lagrangian $\widetilde{L}$ has the special structure

$$\widetilde{L} = L + \delta(t) b(X) \Lambda + \delta(t - T) \bar{b}(X) \overline{\Lambda} \,, \tag{2.67}$$

where L is the Lagrangian (2.63). Here, $\delta(t)$ is Dirac's δ-function. Let us write the Euler–Lagrange equations (2.65) in a more explicit form considering (2.67) and (2.63). The first group of (2.65) leads to

$$\frac{\mathrm{d}}{\mathrm{d}t} \frac{\partial L}{\partial \dot{X}^*} = \frac{\partial L}{\partial X^*} + \delta(t) \frac{d\left(b(X^*) \mid \Lambda\right)}{\mathrm{d}x^*} + \delta(t - T) \frac{d(\bar{b}(X^*) \mid \overline{\Lambda})}{\mathrm{d}x^*}. \tag{2.68}$$

The expression $(A \mid B)$ indicates the scalar product between the two vectors A and B, which we have simply written up to now as AB. We have introduced this agreement to avoid confusions with respect to the presence of more then two vectors, and we will use this notation only if it seems to be necessary.

We conclude from (2.68) that with the exception of the initial and the end points of the time interval $[0, T]$, the equations obtained are identical to the classical Euler–Lagrange equations for an extremal evolution of the state vector, namely

$$\frac{\mathrm{d}}{\mathrm{d}t} \frac{\partial L}{\partial \dot{X}^*} = \frac{\partial L}{\partial X^*} \,. \tag{2.69}$$

However, considering (2.63), these equations are the evolution equations for the adjoint state vector

$$\dot{P}^* = \frac{\partial}{\partial X^*}\phi(t, X^*, u^*) - \frac{\partial}{\partial X^*}\left(F\left(X^*, u^*, t\right) \mid P^*\right) \,. \tag{2.70}$$

These equations are also called adjoint evolution equations. The additional contributions due to the boundary conditions of the state $X(t)$ can be transformed into the boundary conditions for the adjoint state. To this aim we integrate the complete equation (2.68) over a small time interval $[-\varepsilon, \varepsilon]$ and $[T-\varepsilon, T+\varepsilon]$, respectively. Carrying out the limit $\varepsilon \to 0$, we arrive at

$$\left.\frac{\partial L}{\partial \dot{X}^*}\right|_{t=0} = b'(X^*(0))\Lambda \quad \text{and} \quad \left.\frac{\partial L}{\partial \dot{X}^*}\right|_{t=T} = -\overline{b}'(X^*(T))\overline{\Lambda} \tag{2.71}$$

with the matrices b' and $\overline{b}'$ having the components

$$b'_{\alpha a} = \frac{\partial b_a(X)}{\partial X_\alpha} \qquad \text{and} \qquad \overline{b}'_{\alpha a} = \frac{\partial \overline{b}_a(X)}{\partial X_\alpha} \,. \tag{2.72}$$

The index a runs over the N initial and final, respectively, boundary conditions and α runs over the N components of the state vector. With (2.68), the boundary conditions (2.71) can be explicitly written as

$$P^*(0) = b'(X^*(0))\Lambda \quad \text{and} \quad P^*(T) = -\overline{b}'(X^*(T))\overline{\Lambda} \,. \tag{2.73}$$

These relations are usually called transversality conditions. The second group of (2.65) leads together with (2.67) and (2.63) to

$$\frac{\partial L}{\partial P^*} = 0 \quad \text{or} \quad \dot{X}^* = F\left(X^*, u^*, t\right) \tag{2.74}$$

i.e., this group reproduces constraints (2.53) describing the evolution of the state variables. The last group of (2.65) yields with (2.67) and (2.63)

$$\frac{\partial L}{\partial u^*} = 0 \quad \text{or} \quad \frac{\partial}{\partial u^*}\left(F\left(X^*, u^*, t\right) \mid P^*\right) - \frac{\partial}{\partial u^*}\phi(t, X^*, u^*) = 0 \,. \tag{2.75}$$

Finally, we have to consider the extremal conditions (2.66). These equations reproduce the boundary conditions (2.55) for the state vector X. The complete set of equations defining the extremals of the general control problem consists of N first-order differential equations for the N components of the state vector and N first-order differential equations for the N components of the adjoint state. Furthermore, we have n algebraic equations for the n control functions. The $2N$ differential equations require $2N$ boundary conditions. On the other hand, (2.55) and (2.73) yield $4N$ boundary conditions. This overestimation is only an apparent effect, because (2.55) and (2.73) also contain $2N$ free components of the vectors Λ and $\overline{\Lambda}$ which can be fixed by the $2N$ surplus boundary conditions[4].

[4] In the case of partially free boundary conditions, we may have only $2N - \alpha$ boundary conditions for X. That automatically requires that there are also only

Isoperimetric Problems

A special case of control problems occurs if one or more constraints are integrals. However, these problems can also be reduced to the above introduced general case. As an example, let us minimize the action

$$S[X,T] = \int_0^T \mathrm{d}t L_0 \left(t, X(t), \dot{X}(t)\right) \to \inf \tag{2.76}$$

under the constraints

$$\int_0^T \mathrm{d}t g_\alpha \left(t, X(t), \dot{X}(t)\right) = \rho_\alpha \quad \text{for} \quad \alpha = 1, \ldots, m \tag{2.77}$$

and $2N$ boundary conditions of type (2.55). Such a problem is called an isoperimetric problem. In order to transform this problem into the standard form, we introduce N control functions via the additional constraints

$$\dot{X}_\alpha(t) = u_\alpha(t) \quad \text{for} \quad \alpha = 1, \ldots, N \tag{2.78}$$

and we extend the state vector by m new components via the constraints

$$\dot{X}_{N+\alpha}(t) = g_\alpha \left(t, X(t), u(t)\right) \quad \text{for} \quad \alpha = 1, \ldots, m \tag{2.79}$$

and the additional boundary conditions

$$X_{N+\alpha}(0) = 0 \quad \text{and} \quad X_{N+\alpha}(T) = \rho_\alpha \quad \text{for} \quad \alpha = 1, \ldots, m\,. \tag{2.80}$$

Thus, we obtain the generalized Lagrangian

$$\begin{aligned} L = {} & L_0 \left(t, X(t), u(t)\right) + P(t) \left(\dot{X}(t) - u(t)\right) \\ & + P'(t) \left(\dot{X}'(t) - g\left(t, X(t), u(t)\right)\right) \end{aligned} \tag{2.81}$$

for N control functions and $N + m$ state variables and $N + m$ adjoint state variables. For the seek of simplicity, we have split the state vector as well as the adjoint state vector, in two subvectors, $X = \{X_1, \ldots, X_N\}$ and $X' = \{X_{N+1}, \ldots, X_{N+m}\}$ as well as $P = \{P_1, \ldots, P_N\}$ and $P' = \{P_{N+1}, \ldots, P_{N+m}\}$. Thus we obtain the following set of evolution equations for the extremal solution

$$\dot{P}^*_\alpha = \frac{\partial}{\partial X^*_\alpha} L_0 \left(t, X^*, u^*\right) - \sum_{\beta=N+1}^{N+m} P^*_\beta \frac{\partial}{\partial X^*_\alpha} g_\beta \left(t, X^*, u^*\right) \tag{2.82}$$

$2N - \alpha$ Lagrange multipliers. This situation is similar to the statement that α multipliers in (2.71) are simply set to zero. On the other hand, we still have $2N$ boundary conditions even due to (2.71). That means there are $4N-\,\alpha$ necessary boundary conditions for X and P which contain $2N - \alpha$ free parameters (the multipliers). Thus, $2N$ boundary conditions remain effective, which are necessary to get a unique and complete solution of the system of differential equations (2.69) and (2.74).

for $\alpha = 1, \ldots, N$ and

$$\dot{P}^*_\alpha = 0 \tag{2.83}$$

for $\alpha = N+1, \ldots, N+m$ and the boundary conditions

$$P^*_\alpha(0) = \sum_{\beta=1}^{N} \left. \frac{\partial b_\beta(X)}{\partial X_\alpha} \right|_{X=X^*(0)} \Lambda_\beta \tag{2.84}$$

and

$$P^*_\alpha(T) = -\sum_{\beta=1}^{N} \left. \frac{\partial \bar{b}_\beta(X)}{\partial X_\alpha} \right|_{X=X^*(T)} \overline{\Lambda}_\beta \tag{2.85}$$

for $\alpha = 1, \ldots, N$ and

$$P^*_\alpha(0) = \Lambda_\alpha \quad \text{and} \quad P^*_\alpha(T) = -\overline{\Lambda}_\alpha \tag{2.86}$$

for $\alpha = N+1, \ldots, N+m$. The second group of evolution equations are given by (2.78) with the boundary conditions (2.55), (2.79), and (2.80). The third group of the generalized Euler–Lagrange equations of the isoperimetric problem are the N algebraic relations

$$\frac{\partial}{\partial u^*_\alpha} L_0\left(t, X^*, u^*\right) = P^*_\alpha + \sum_{\beta=N+1}^{N+m} P^*_\beta \frac{\partial}{\partial u^*_\alpha} g_\beta\left(t, X^*, u^*\right) . \tag{2.87}$$

Isoperimetric control problems become relevant for systems with global conservation laws, for instance, processes consuming a fixed amount of energy or matter.

2.4.2 Hamilton Approach

In course of the formulation of the classical mechanics on the basis of the Lagrangian and the corresponding Euler–Lagrange equations, the mechanical state is described by coordinates and velocities. However, such a description is not the only possible one. The application of momenta instead of velocities presents several advantages, in particular, for the investigation of general problems of classical mechanics. This alternative concept is founded on the canonical equations of classical mechanics (1.1) which follow directly from the Euler–Lagrange equations. Therefore, it is also desirable to transform the Euler–Lagrange equations of the generalized control problem into a canonical system. The first relation we need follows directly from the Lagrangian (2.63)

$$P = \frac{\partial L}{\partial \dot{X}} \,. \tag{2.88}$$

This relation is exactly the same as the definition of the momentum, well-known from classical mechanics. That is the reason that we also call the adjoint state the generalized momentum. The total derivative of the Lagrangian $L\left(t, X, \dot{X}, P, u\right)$ is

$$dL\left(t,X,\dot{X},P,u\right)=\frac{\partial L}{\partial X}dX+Pd\dot{X}+\frac{\partial L}{\partial P}dP+\frac{\partial L}{\partial u}du+\frac{\partial L}{\partial t}dt \tag{2.89}$$

where we have used (2.88). The total derivative $\mathrm{d}L^*$ for an extremal trajectory reduces to

$$\mathrm{d}L^*=\mathrm{d}L\left(t,X^*,\dot{X}^*,P^*,u^*\right)=\dot{P}^*\mathrm{d}X^*+P^*\mathrm{d}\dot{X}^*+\frac{\partial L^*}{\partial t}\mathrm{d}t\,, \tag{2.90}$$

where we have considered (2.69), (2.75), (2.74), and (2.88). This equation can now be transformed into

$$\mathrm{d}H^*=\dot{X}^*\mathrm{d}P^*-\dot{P}^*\mathrm{d}X^*-\frac{\partial L^*}{\partial t}\mathrm{d}t \tag{2.91}$$

with the Hamiltonian $H=P\dot{X}-L$. Because of the structure of the total derivative (2.91), the Hamiltonian satisfies the canonical equations

$$\dot{X}^*=\frac{\partial H^*}{\partial P^*}\quad\text{and}\quad\dot{P}^*=-\frac{\partial H^*}{\partial X^*} \tag{2.92}$$

for the extremal solution, and furthermore we have the relations

$$\frac{\partial H}{\partial t}=-\frac{\partial L}{\partial t}\quad\text{and}\quad\frac{\partial H}{\partial u}=-\frac{\partial L}{\partial u}\,. \tag{2.93}$$

The explicit form of the Hamiltonian corresponding to the Lagrangian (2.63) is

$$H\left(t,X,P,u\right)=PF\left(X,u,t\right)-\phi(t,X,u)\,. \tag{2.94}$$

With this representation it is easily to check relations (2.93). Especially for the extremal trajectory we obtain the necessary condition

$$\frac{\partial H^*}{\partial u^*}=0\,, \tag{2.95}$$

which is due to (2.93) equivalent to (2.75). Condition (2.95) completes the set of canonical equations (2.92) with respect to the solution of the underlying minimization problem. In fact, we may verify that the first group of the canonical equations reproduces constraints (2.53) while the second group corresponds to the evolution equations for the adjoint state vector, i.e., the momenta, (2.70) of the extremal solution. The boundary conditions (2.55) and (2.73), respectively, for the state X and the momenta P, respectively, remain unchanged. This also implies, that the Lagrange multipliers in (2.73) are free quantities in order to compensate the apparent overestimation of the set of boundary conditions.

Autonomous systems are characterized by an explicit time-independent Lagrangian and due to (2.93) also a time-independent Hamiltonian. In this case, the Hamiltonian of the extremal trajectory is constant. This statement follows directly from

$$\frac{\mathrm{d}H\left(X^*,P^*,u^*\right)}{\mathrm{d}t}=\frac{\partial H^*}{\partial X^*}\dot{X}^*+\frac{\partial H^*}{\partial P^*}\dot{P}^*+\frac{\partial H^*}{\partial u^*}\dot{u}^*=0\,, \tag{2.96}$$

where we have used the canonical equations (2.92) and the extremum condition for the control functions (2.95). We remark that the invariance of an autonomous Hamiltonian along the extremal solution, $H(X^*, P^*, u^*) = \text{const.}$ corresponds to the conservation of energy in the classical mechanics.

2.4.3 Pontryagin's Maximum Principle

The Hamilton and the Lagrange approach lead to equivalent conditions necessary for the optimum control of a given system. Furthermore, the Euler–Lagrange equations and the corresponding canonical Hamilton equations are very close to the related equations of classical mechanics. The main difference is that both, the Lagrangian and the Hamiltonian, contain a set of control functions $u(t)$ besides the variables describing the motion of the system through the phase space ($X, \dot{X}$ or X, P). On the other hand, the extremal trajectory is defined by a set of differential equations for X and P, while the solution of the optimum control follows from a set of algebraic equations

$$\frac{\partial H^*}{\partial u^*} = 0 \quad \text{or} \quad \frac{\partial L^*}{\partial u^*} = 0 \,. \tag{2.97}$$

From a physical point of view, the control functions are not dynamical variables. These properties suggest that the initial elimination of the control functions before applying the Euler–Lagrange or Hamilton equations is desirable. To this aim, we return to action (2.62). The optimum control problem $S \to \inf$ requires that the action attains the minimum over all admissible controls $u(t)$ for the optimal control $u^*(t)$. Because the action contains no derivatives of $u(t)$ and furthermore there is no multiplicative coupling between the derivatives of the state vector X and the control functions,[5] the minimum of S with respect to $u(t)$ is reached for the minimum of $L\left(t, X, \dot{X}, P, u\right)$ with respect to u. This statement follows from the obvious formula

$$\min_{u(t)} \int \eta(t, u(t))dt = \int \min_{u} \eta(t, u)dt \tag{2.98}$$

and corresponds directly to the second equation of (2.97). In other words, the optimum control function $u^*(t)$ can be obtained as a function of X, $\dot{X}$, P, and t by minimizing the Lagrangian $L\left(t, X, \dot{X}, P, u\right)$ with respect to u. This condition is much stronger than (2.97) because the latter condition indicates only an extremal solution. Furthermore, the admissible control can be easily extended to control variables u restricted to an arbitrary, possibly time-dependent region $\mathbb{U}'(t) \subset \mathbb{U}$.

This basic concept is called the Pontryagin maximum principle [20]. The maximum principle[6] allows the determination of each global minimum solution u^* of the Lagrangian for each time t and each configuration $X, \dot{X}, P$ of

[5] In other words, the variation calculus yields no differential equations for $u(t)$.
[6] The name maximum principle belongs to the maximization of the Hamiltonian, see below.

the dynamical state of the system. In other words, the application of Pontryagin's maximum principle leads to an optimum solution with respect to the control functions. We denote the solution of $L\left(t, X, \dot{X}, P, u\right) \to \inf$ as the preoptimal control $u^{(*)}(X, \dot{X}, P, t)$, i.e., $u^{(*)}(X, \dot{X}, P, t)$ fulfils for a given time t and a given state the inequality

$$L\left(t, X, \dot{X}, P, u^{(*)}(X, \dot{X}, P, t)\right) \leq L\left(t, X, \dot{X}, P, u\right) \tag{2.99}$$

for all $u \in \mathbb{U}'(t) \subset U$. Furthermore, we call the Lagrangian

$$\begin{aligned} L^{(*)}\left(t, X, \dot{X}, P\right) &= L\left(t, X, \dot{X}, P, u^{(*)}(X, \dot{X}, P, t)\right) \\ &= \min_{u \in \mathbb{U}'(t) \subset U} L\left(t, X, \dot{X}, P, u\right) \end{aligned} \tag{2.100}$$

the preoptimized Lagrangian. The Lagrangian $L\left(t, X, \dot{X}, P, u\right)$ is said to be regular if for each admissible value of X, $\dot{X}$, P, and t a unique and absolute minimum exists. We consider as an example the free particle problem with the mechanical action $S = 1/2 \int \mathrm{d}t\dot{X}^2 \to \inf$. This problem may be rewritten into the generalized control problem $R = 1/2 \int \mathrm{d}tu^2$ with the constraint $\dot{X} = u$. Thus the generalized Lagrangian of this simple problem reads $L = u^2/2 + (\dot{X} - u)P$. This Lagrangian has a unique and absolute minimum with respect to u for the preoptimal control $u^{(*)} = P$. Thus, the preoptimized Lagrangian is $L^{(*)} = \dot{X}P$-$P^2/2$. On the other hand, the obvious non-physical action $S = \int \mathrm{d}t\dot{X}^3$ leads to a generalized Lagrangian $L = u^3 + (\dot{X} - u)P$ so that $L \to -\infty$ for $u \to -\infty$, i.e., this Lagrangian is not regular. But it can be regularized by a suitable restriction of u, for instance $u > 0$.

We have two possible ways to solve the generalized optimum problem on the basis of Pontryagin maximum principle:

- We may start from the Lagrangian and determine the solution of the Euler–Lagrange equation for arbitrary, but admissible control functions. As a result, we obtain preextremal trajectories $X^{(*)} = X^{(*)}[t, u(t)]$ and $P^{(*)} = P^{(*)}[t, u(t)]$ for each control function $u(t)$. Afterwards, we substitute the solutions $X^{(*)}$ and $P^{(*)}$ in the Lagrangian and determine the optimum control $u^*(t)$ by the minimization of $L(t, X^{(*)}, \dot{X}^{(*)}, P^{(*)}, u)$ with respect to u for all time points $t \in [0, T]$. The disadvantages of this way are that the computation of $\dot{X}^{(*)}$ eventually requires some assumptions about the smoothness of the control functions $u(t)$ and that the preextremal trajectories $X^{(*)}$ and $P^{(*)}$ are usually complicated functionals of $u(t)$.
- The alternative approach starts from a minimization of the Lagrangian $L(t, X, \dot{X}, P, u)$ with respect to the control functions $u(t)$. The result is the preoptimal control $u^{(*)}(X, \dot{X}, P, t)$. In contrast to the first way, $u^{(*)}(X, \dot{X}, P, t)$ is a simple function of $X, \dot{X}, P$. In a subsequent step we substitute $u^{(*)}$ in the Lagrangian and determine the optimal trajectory X^*

and the other dynamic quantities $\dot{X}^*$ and P^* from the preoptimized Lagrangian $L^{(*)}(t, X, \dot{X}, P)$. The optimal control follows by inserting these solution in $u^{(*)}$, i.e., we have $u^*(t) = u^{(*)}(X^*, \dot{X}^*, P^*, t)$. The disadvantage of this way is that the explicitly formulated Euler–Lagrange equations may become a complicated structure.

The Pontryagin maximum principle is also applicable in the case of the Hamilton approach. Due to the Legendre transformation, $H = P\dot{X} - L$, we now have to search for the maximum of the Hamiltonian with respect to the control function. This maximum problem can be interpreted as a strong extension of (2.95), which indicates only an extremum of the Hamiltonian with respect to the optimal control. We call the solution of $H(t, X, P, u) \to \sup$ again the preoptimal control $u^{(*)}(X, P, t)$, which is defined by the inequality

$$H\left(t, X, P, u^{(*)}(X, P, t)\right) \geq H(t, X, P, u) \quad \text{for all } u \in \mathbb{U}'(t) \subset U\,. \tag{2.101}$$

The Hamiltonian

$$\begin{aligned} H^{(*)}(t, X, P) &= H\left(t, X, P, u^{(*)}(X, P, t)\right) \\ &= \max_{u \in \mathbb{U}'(t) \subset U} H(t, X, P, u) \end{aligned} \tag{2.102}$$

is said to be the preoptimized Hamiltonian. It is a regular function if for each admissible value of X, $\dot{X}$, P and t a unique and absolute maximum exists. For the above-discussed free particle problem, the Hamiltonian is given by $H = Pu - u^2/2$. The Hamiltonian is regular and yields the preoptimal control $u^{(*)} = P$ and therefore the preoptimized Hamiltonian $H^{(*)} = P^2/2$.

The maximum principle often allows a very simple approach to general statements of the control theoretical calculus. A typical example is the derivation of the Weierstrass criterion (2.34). We start from the Lagrangian $L(t, X, \dot{X})$ and transform this expression in the standard form $L(t, X, u)$ by introducing the constraints $\dot{X} = u$. The corresponding Hamiltonian is then $H = Pu - L(t, X, u)$. The maximum principle requires that the optimal control u^* satisfies the special version of inequality (2.101)

$$H(t, X^*, P^*, u^*) \geq H(t, X^*, P^*, u) \quad \text{for all } u \in \mathbb{U}'(t) \subset U\,, \tag{2.103}$$

and therefore

$$P^* u^* - L(t, X^*, u^*) \geq P^* u - L(t, X^*, u)\,. \tag{2.104}$$

On the other hand, the maximum of H is defined by $\partial H(t, X^*, P^*, u^*)/\partial u^* = 0$ which leads to $P^* = \partial L(t, X^*, u^*)/\partial u^*$. Thus we obtain from (2.104) considering the constraint for the optimum solution $\dot{X}^* = u^*$

$$L(t, X^*, u) - L(t, X^*, \dot{X}^*) \geq \frac{\partial L(t, X^*, \dot{X}^*)}{\partial \dot{X}^*}\left(u - \dot{X}^*\right)\,, \tag{2.105}$$

which is the above-discusses Weierstrass criterion (2.34).

As in the calculus of variations, we can encounter most diverse situations which occur during the solution of control problems with the aid of Pontryagin's maximum principle. Such problems are the lack of solutions, the necessary smoothness of solutions, or the existence of a set of admissible trajectories which satisfy the maximum principle and are not optimal. Because a large class of optimal control problems concerns bounded sets of admissible controls one often get the impression that such problems are always soluble. But this is not correct. A typical counter example are sliding processes which cannot solved straightforwardly by the application of Pontryagin's maximum principle; see below.

2.4.4 Applications of the Maximum Principle

In the following chapter we present some simple, but instructive examples of the application of Pontryagin's maximum principle. Of course, these examples have more or less an academic character, but they should show the large variety of optimum control problems, which can be solved by using the Hamilton approach together with the maximum principle. More applications and also realistic examples can be found in the comprehensive literature [21–26].

Linear Control Problems

All terms of a linear control problem contain the control functions up to the first-order. Such problems are very popular in several problems of natural sciences. Important standard problems are additive and multiplicative controlled processes. Let us illustrate the typical problems related to these types of optimal control by some simple examples.

Sliding Regimes

We first study two simple processes with additive control. The first example is a so-called $1d$-sliding process,

$$\int_0^T dt X^2 \to \inf \quad \dot{X} = u \quad |u| = \alpha \quad X(0) = 0 \quad X(T) = \theta .$$

The corresponding Hamiltonian of this problem is $H = Pu - X^2$ and we obtain the preoptimal control $u^{(*)} = \alpha$ for $P > 0$ and $u^{(*)} = -\alpha$ for $P < 0$. The preoptimized Hamiltonian is simply $H^{(*)} = \alpha |P| - X^2$. Thus, we obtain the canonical equations $\dot{P}^* = 2X^*$ and $\dot{X}^* = \alpha$ for $P^* > 0$ and $\dot{X}^* = -\alpha$ for $P^* < 0$. Furthermore, we introduce the initial condition $P^*(0) = P_0$. This relation may be justified by the transversality conditions (2.73). Considering all initial conditions, we find $X^*(t) = \alpha t$ and $P^*(t) = P_0 + \alpha t^2$ if $P_0 > 0$ and $X^*(t) = -\alpha t$ and $P^*(t) = P_0 - \alpha t^2$ for $P_0 < 0$, i.e., an initially positive (negative) momentum remains positive (negative) over the whole time interval. The behavior for $P_0 = 0$ is undefined within the framework of the maximum principle. The final condition requires $\alpha T = |\theta|$, i.e., we find a unique and optimal solution only for $\alpha = |\theta|/T$. In this case, we have the optimal control $u^*(t) = |\theta|/T \operatorname{sign}\theta$ and the optimal trajectory $X^*(t) = \theta t/T$. No admissible control exists for $\alpha T \neq |\theta|$.

On the other hand, it is easy to see that a positive value of the functional results from any admissible trajectory. The set of admissible trajectories is empty only for $\alpha T < |\theta|$. For example, Fig. 2.5 shows a set of admissible trajectories $X_k(t)$ for $\theta = 0$. The corresponding value of the functional tends to zero on the sequence $X_1(t)$, $X_2(t)$, ... Furthermore, it is easy to show that the trajectories $X_k(t)$ converge uniformly to $X_\infty(t) = 0$. In contrast, the sequence of controls converges to anything.

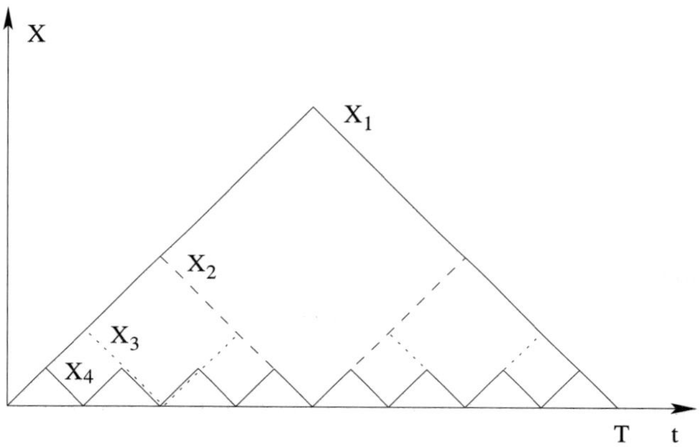

Fig. 2.5. The first four trajectories $X_1, \ldots, X_4$ of a set of admissible trajectories for $\theta = 0$ converging to the limit trajectory $X(t) = 0$ for all $t \in [0, T]$

As a second example we consider a particle of mass $m = 1$ moving on a straight line under the effect of a unique force u and a Newtonian friction $-\mu\dot{x}$ from the position $x(0) = 0$ to $x(T) = x_e$. The initial and final velocities should vanish, $\dot{x}(0) = \dot{x}(T) = 0$. Then we have a two-dimensional state $X = (x, p)$, where $p = \dot{x}$ is the momentum. The equations of motion are given by $\dot{x} = p$ and $\dot{p} = -\mu p + u$. The total amount of work injected into to system is simply given by

$$R = \int_0^T dtpu \,. \tag{2.106}$$

A possible control problem is now to minimize the total work, $R \to \inf$, where u is restricted to the interval $-u_0 \leq u \leq u_0$. To this aim we introduce the generalized momentum $P = (q, r)$ and construct the Hamiltonian

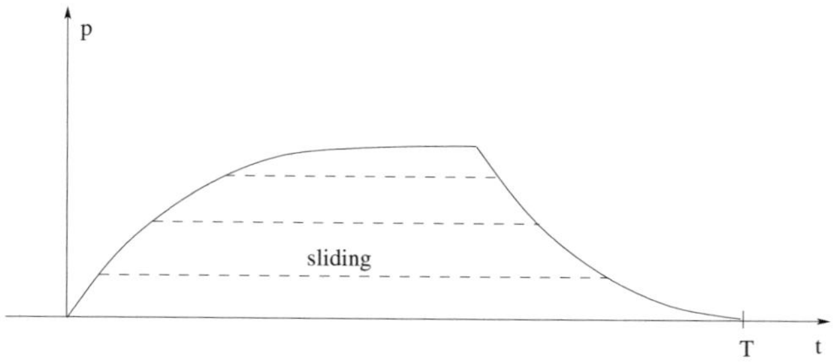

Fig. 2.6. Optimum momentum as function of time for different distances $x_e \leq x_{crit}$. An initial acceleration regime followed by the braking regime exists for $x_e = x_{crit}$. Shorter distances show an intermediate sliding regime with a constant velocity. The sliding velocity decreases with decreasing x_e

$$H = qp + r(u - \mu p) - up \ . \tag{2.107}$$

The preoptimal control function is $u^{(*)} = u_0 \text{sign}\,(r - p)$, and the preoptimized Hamiltonian is now $H^{(*)} = qp + u_0 |r - p| - \mu pr$. Thus we get the canonical equations $\dot{x} = p$, $\dot{q} = 0$, $\dot{r} = -q + \mu r + u_0 \text{sign}\,(r - p)$, and $\dot{p} = -\mu p + u_0 \text{sign}\,(r - p)$. There exists a unique solution only for $x_e = x_{crit}$ with

$$(2 \ln(1 + e^{\mu T}) - 2 \ln 2 - \mu T) u_0 = \mu^2 x_{crit} \ , \tag{2.108}$$

which corresponds to an acceleration regime with $u^* = u_0$ for $0 \leq t < \mu^{-1} \ln \left[(1 + e^{\mu T})/2 \right]$ and a subsequent braking regime with $u^* = -u_0$ for $\mu^{-1} \ln \left[(1 + e^{\mu T})/2 \right] < t \leq T$. The largest velocity, $p_{max} = \mu^{-1} u_0 \tanh \mu T/2$, is reached for the crossover between both regimes. No solution exists for $x_{crit} < x_e$, while a sliding regime occurs for $x_{crit} > x_e$. Here, we also have an initial acceleration up to the velocity $p_0 < p_{max}$ with $u^* = u_0$, followed by the sliding regime, defined by a constant velocity $p(t) = p_0 = \mu^{-1} u^*$ and a final braking from p_0 to $p(T) = 0$ with $u^* = -u_0$.

The crossover between the three regimes and the value of p_0 are determined by unique solutions of algebraic equations which concern the total time T as the sum of the duration of the three regimes and the total length x_e of the path from the initial point to the end point as sum of the three subpaths. We remark that the velocity is a continuous function also for both crossover points.

Multiplicative Coupled Control

Typical multiplicative controlled processes are chemical reactions of the type $A + C \to 2C$ and $C + B \to 0$ where the concentrations of A and C are external changeable quantities which may be used to control the chemical creation or annihilation of molecules of type B. The kinetics of the mentioned reactions can be described by the balance equation for the concentration X of the component C, $\dot{X}(t) = u(t) X(t)$, where the control function $u(t) =$

$k_1 c_A(t) - k_2 c_B(t)$ depends on the external changeable concentrations $c_A(t)$ and $c_B(t)$ of the components A and B. The kinetic coefficients k_1 and k_2 of both reactions are assumed to be constant. The control function is constrained by the maximum concentrations of the A and B components. For the sake of simplicity we assume $-1 \leq u \leq 1$. A possible control aim is the minimization of the final concentration $X(T)$. Hence, we have to solve the problem

$$R = \int_0^T \mathrm{d}t\dot{X}(t) = X(T) - X_0 = \int_0^T \mathrm{d}tu(t)X(t) \to \inf \ . \tag{2.109}$$

The problem has the Hamiltonian $H = (P(t) - 1)u(t)X(t)$ and the preoptimized control is simply $u^*(t) = \operatorname{sign}((P(t) - 1)X(t))$. Thus, we get $H = |(P(t) - 1)X(t)|$ and therefore

$$\dot{X}(t) = X(t)u^*(t) \quad \text{and} \quad \dot{P}(t) = (1 - P(t))u^*(t) \ . \tag{2.110}$$

The free boundary condition for $t = T$ requires $P(T) = 0$ due to the transversality condition (2.73). The evolution equations (2.110) prevent that neither X nor $1 - P$ can be 0. Thus, the trajectory is defined by the solution of the equation

$$\frac{dX}{dP} = \frac{X}{1 - P} \ . \tag{2.111}$$

We get the solution $(1 - P)X = \text{const.}$ and therefore $u^*(t) = u_0 = \text{const.}$ From here, it immediately follows from (2.110) that $X(t) = X_0 \exp(u_0 t)$ and $P(t) = 1 - \exp(u_0(T - t))$. Finally, we obtain the optimum control law $u^*(t) = -\operatorname{sign} X_0$.

Although realistic applications of the maximum principle on additive or multiplicative controlled problems are much complicated as the simple examples suggest, the typical feature is a linear dependence of the Hamiltonian on the control function. Thus, an unlimited range of the control, $-\infty < u < \infty$, leads usually to an undefined preoptimized Hamiltonian and therefore to a lack of solutions.

Time Optimal Control

A large class of problems are minimum time problems. Basically, an optimum time problem consists in steering the system in the shortest time from a suitable initial point of the phase space to an allowed final state. The functional to be minimized is in this case simply

$$R = T = \int_0^T \mathrm{d}t \tag{2.112}$$

and the Hamiltonian (2.94) reduces to

$$H(t, X, P, u) = PF(X, u, t) - 1 \ . \tag{2.113}$$

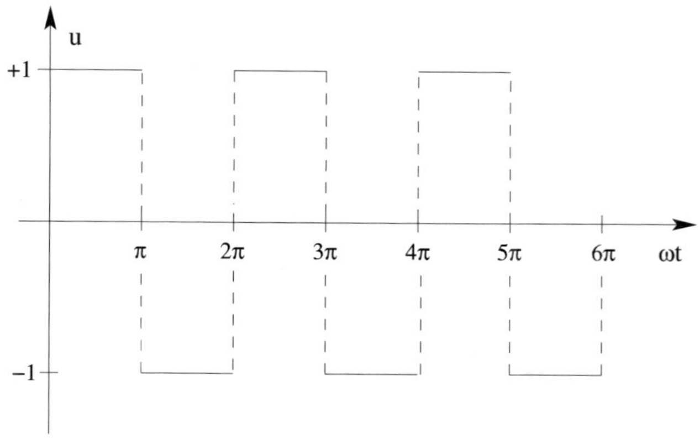

Fig. 2.7. Optimal control function for $u_0 = 1$

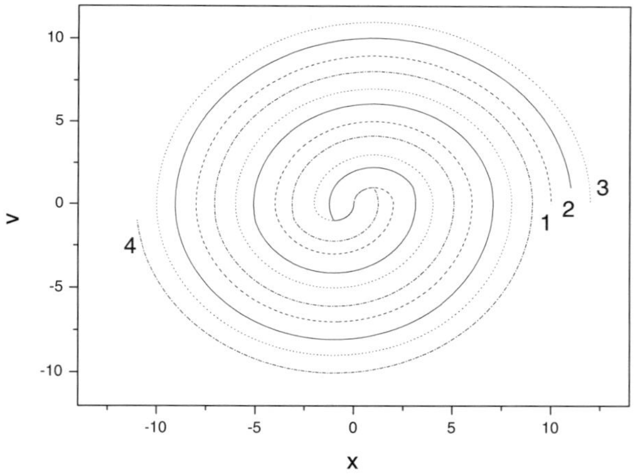

Fig. 2.8. Several trajectories of the optimum oscillator problem in the position–velocity diagram with the initial condition $x_0 = v_0 = 0$. The phase φ_0 of the control function is π (for curve 1), $\pi/2$ (2), 0 (3) and $-\pi/2$ (4)

As an example we consider a harmonic oscillator, $\dot{x} = p$, $\dot{p} = -\omega^2 x + u$, with the state vector $X = (x, p)$. The external force u is restricted by $-u_0 < u < u_0$ and may be used for the control of the system. The corresponding Hamiltonian reads $H = qp + r(u - \omega^2 x) - 1$ and the preoptimized control is $u^{(*)} = u_0 \operatorname{sign} r$. The canonical equations of the control problem are simply given by the above-introduced mechanical equations of motion, $\dot{x} = p$, $\dot{p} = -\omega^2 x + u_0 \operatorname{sign} r$, and the adjoint set of differential equations, $\dot{q} = \omega^2 r$, $\dot{r} = -q$. We obtain $\ddot{r} + \omega^2 r = 0$ with the unique solution $r = r_0 \cos(\omega t + \varphi_0)$. Thus, the optimal control function u^* is a periodic step function with the step length $\tau = \pi/\omega$ and amplitude $\pm u_0$ (Fig. 2.7). Finally, the solution of $\ddot{x} + \omega^2 x = u^*(t)$ yields the optimum trajectory. In principle, the optimum solution $x^*(t)$ has four free parameters, namely the initial position, $x_0 = x(0)$, the initial velocity, $v_0 = v(0)$, the phase φ_0 of the control function and finally the time T. This allows us to determine the minimum time for a transition from any initial state (x_0, p_0) to any final state (x_e, p_e). The trajectories starting from a given initial point can be parametrized by the phase φ_0 of the control function. Obviously, the set of all trajectories covers the whole phase space, see Fig. 2.8.

Complex Boundary Conditions

Problems with the initial state and final state, respectively, constrained to belong to a set $\mathbb{X}_0$ and $\mathbb{X}_e$, respectively, become important, if the preparation or the output of processes or experiments allows some fluctuations. We refer here to a class of problems with partially free final states. A very simple example [27] is the control of a free Newtonian particle under a control force u, $-u_0 < u < u_0$. The initial state is given, while the final state should be in the target region $-\xi_e \leq x_e \leq \xi_e$ and $-\eta_e \leq p_e \leq -\eta_e$. We ask for the shortest time to bring the particle from its initial state to one of the allowed final states. The equations of motion, $\dot{x} = p$, $\dot{p} = u$, require the Hamiltonian, $H = qp + ru - 1$. Thus, the preoptimized control is $u^{(*)} = u_0 \operatorname{sign} r$. The canonical equations of the control problem are given by the equations of motion, $\dot{x} = p$, $\dot{p} = u_0 \operatorname{sign} r$, and the adjoint equations, $\dot{q} = 0$, $\dot{r} = -q$. Thus, we obtain $\ddot{r} = 0$ with the general solution $r = r_0 + Rt$ and $q = -R$. The linearity of $r(t)$ with respect to the time suggests that $u^{(*)}$ switches at most once during the flight of the particle from the initial point to the target region. First, we consider all trajectories which reach an allowed final state without switch. These trajectories are given by $x(t) = x_e + p_e t \pm u_0 t^2/2$ and $p(t) = p_e \pm u_0 t$, and therefore, $x \mp p^2/2u_0 = x_e \mp p_e^2/2u_0$. Hence, the primary basin of attraction with respect to the target is the gray-marked region in Fig. 2.9. All particles with initial conditions inside this region move under the correct control but without any switch of the control directly to the target. All other particles are initially in the secondary basin of attraction. They move along parabolic trajectories through the phase space into the primary basin of attraction. If the border of this basin was reached, the control switches as the particle moves now along the border into the target region.

Complex Constraints

In the most cases discussed above, the constraints were evolution equations of type (2.53). But there are several other possible constraints. One of these possibilities is isoperimetric constraints where some functions of the state and the control variables are subject to integral constraints; see Sect. 2.4.1. Other

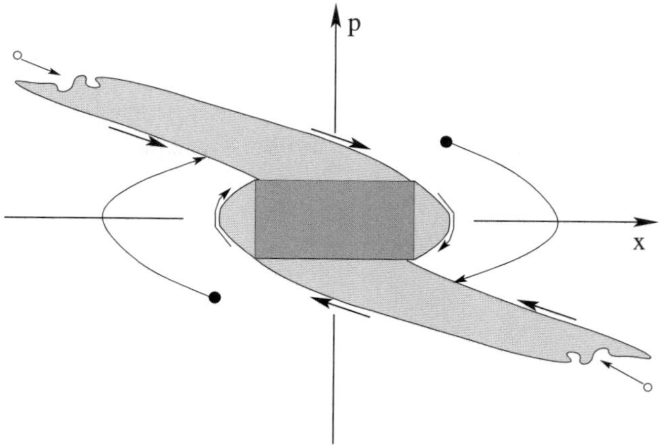

Fig. 2.9. The structure of the primary and secondary basins of attraction. The particles move in the direction of the arrows

cases are constraints where some functions of the state and the control functions must satisfy instantaneous constraints over the whole control interval $0 \leq t \leq T$

$$g(t, X(t), u(t)) = 0 \quad \text{or} \quad G[t, X(t), u(t)] \leq 0 . \tag{2.114}$$

The first class of these constraints can be used to eliminate some state variables or control functions from the optimum control problem before the optimization procedure is carried out.

The inequality constraints can be transformed into an equality constraint by addition of a new control variable $\widetilde{u}(t)$

$$G[t, X(t), u(t)] + \widetilde{u}(t) = 0 \quad \text{with} \quad \widetilde{u}(t) \geq 0 . \tag{2.115}$$

Then we may proceed as in the case of an equality constraint. Thus, the new control variable enters the original optimum control problem and we can apply Pontryagin's maximum principle as discussed above. In the same way, we may consider evolution inequalities

$$\dot{X}(t) \leq F(X, u, t) . \tag{2.116}$$

Relations of this type are very popular in nonequilibrium thermodynamics [28–30]. Another important class of constraints, as in several branches of natural sciences, are problems where the state variables must satisfy equality (or inequality) constraints at M isolated subsequent time points, i.e., constraints of the form

$$g_i(t_i, X(t_i), u(t_i)) = 0 \quad \text{for} \quad 0 < t_1 < t_2 < \cdots < t_M < T . \tag{2.117}$$

Typical examples are rendezvous problems where particles collide at a certain time, or where space shuttles and space stations meet at a certain time. We remark that when these constraints are present, the control functions, the

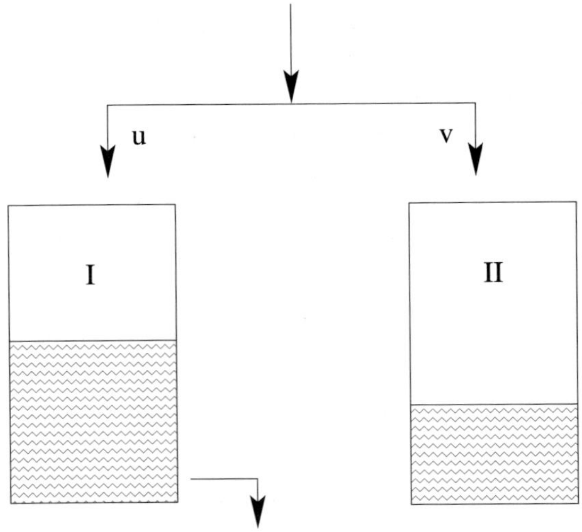

Fig. 2.10. Two tanks with common input

generalized momenta as well as the Hamiltonian may be discontinuous at the isolated time points t_i.

We finish this chapter with a simple example [27] related to a control under global instantaneous equality constraints. To this aim we consider a system of two tanks; see Fig. 2.10. The outgoing flow of tank I is proportional to the volume of the liquid, while tank II is closed. The two tanks are fed through a constant input flow, which can be divided in any way, $u+v = \text{const.}$ where u and v are the both subflows. Obviously, the evolution equations of this system are given by $\dot{x} = -x + u$ and $\dot{y} = v$, where x and y are the heights of the liquid in tank I and tank II, respectively. The problem is to drive the system in the shortest time from the initial state (x_0, y_0) to the final state $(x_{\mathrm{e}}, y_{\mathrm{e}})$. The Hamiltonian of this problem is $H = q(u-x) + pv - 1$. Considering the equality constraint, we obtain the reduced Hamiltonian $H = q(u-x) + p(1-u) - 1$ with the control $u \in [0,1]$ and the adjoint states (q,p). Thus, we find the preoptimal control $u^{(*)} = (\operatorname{sign}(q-p)+1)/2$ and therefore $H = |q-p|/2 + q(1/2-x) + p/2 - 1$. The corresponding canonical equations are $\dot{x} = -x + (\operatorname{sign}(q-p)+1)/2$, $\dot{y} = (1 - \operatorname{sign}(q-p))/2$, and $\dot{q} = q$, $\dot{p} = 0$. Thus we obtain the solution $p = p_0$ and $q = q_0 \exp t$. Hence, $q - p$ changes the sign at most one. Therefore, we have four scenarios:

1. $u^{(*)} = 0$ for $t \in [0,T]$: This regime requires $y = y_0 + t$ and $x = x_0 \exp\{-t\}$. In other words, the final conditions require $y_{\mathrm{e}} - y_0 = \ln x_0/x_{\mathrm{e}}$ and the final time is simply $T = y_{\mathrm{e}} - y_0$.
2. $u^{(*)} = 1$ for $t \in [0,T]$: Here, we get $y = y_0 = y_{\mathrm{e}}$ and $x = 1 + (x_0 - 1)\exp\{-t\}$. A unique solution exists only for $1 < x_{\mathrm{e}} < x_0$, or $1 > x_{\mathrm{e}} > x_0$, and the minimum time is $T = \ln(x_0 - 1)/(x_{\mathrm{e}} - 1)$. Obviously, this

scenario is included in the first and the both subsequent cases as the special realization for $y_0 = y_e$.

3. $u^{(*)} = 1$ for $0 < t < \tau$ and $u^{(*)} = 0$ for $\tau < t < T$: In this case, we get the final conditions $x_e = \exp\{-T + \tau\} + (x_0 - 1)\exp\{-T\}$ and $y_e = y_0 + T - \tau$. Thus, $\tau = \ln(x_0 - 1) - \ln(x_e \exp\{y_e - y_0\} - 1)$ and $T = \tau + y_e - y_0$. A positive τ exists for (i) $y_e - y_0 < \ln x_0/x_e$, $x_e \exp\{y_e - y_0\} > 1$ and $x_0 > 1$ and for (ii) $y_e - y_0 > \ln x_0/x_e$, $x_e \exp\{y_e - y_0\} < 1$ and $x_0 < 1$, but the final time T of case (ii) is larger than T of the subsequent control regime. Thus, there remains only case (i).
4. $u^{(*)} = 0$ for $0 < t < \tau$ and $u^{(*)} = 1$ for $\tau < t < T$: In this case we obtain the final conditions $y_e = y_0 + \tau$ and $x_e = 1 + (x_0 - \exp\tau)\exp\{-T\}$. That means we have $\tau = y_e - y_0$ and $T = \ln(x_0 - \exp\tau) - \ln(x_e - 1)$. The relation $\tau < T$ requires (i) $y_e - y_0 > \ln x_0/x_e$, $x_e < 1$ and $x_0 < \exp\{y_e - y_0\}$ or (ii) $y_e - y_0 < \ln x_0/x_e$, $x_e > 1$ and $x_0 > \exp\{y_e - y_0\}$, but the elapsed time T of case (ii) is larger than T of the previous control regime.

Figure 2.11 shows an illustration of the obtained regimes.

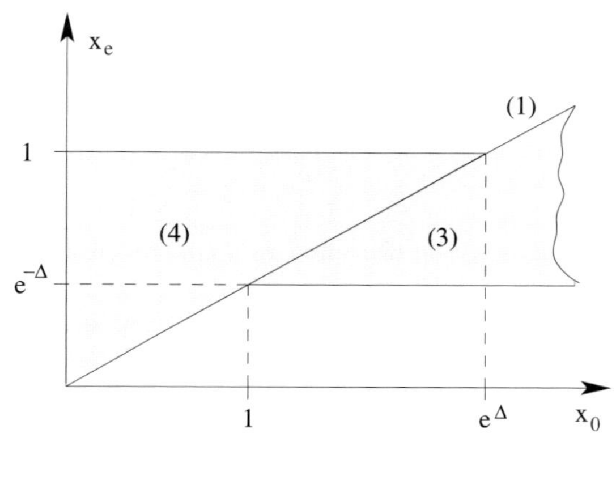

Fig. 2.11. The regions of existing optimal solutions with $\Delta = y_e - y_0$. The first regime corresponds to the straight line separating regime 3 from regime 4

2.4.5 Controlled Molecular Dynamic Simulations

A large class of numerical studies of molecular systems are so-called molecular dynamic simulations. In principle, these techniques numerically solve the set of Newtonian equations of motion corresponding to the system in mind. Such a solution leads to a microcanonical description of the system which is characterized by the conservation of the total energy of the system. In general, molecular dynamic methods are always related to an appropriate set of deterministic evolution equations. The introduction of the temperature requires the consideration of a thermodynamic bath which can be interpreted as the source of stochastic forces driving the system. The corresponding evolution equations now become a stochastic character and they are no longer an object of molecular dynamics methods.

However, sometimes it is reasonable to simulate the bath by deterministic equations. This can be done by two standard methods, namely a combination of molecular dynamics equations with additional constraints, or the formal extension of the original system. The first case [31, 32] considers additional constraints, for example, the conservation of the kinetic energy

$$E_{\text{kin}} = \sum_{i=1}^{M} \frac{m\dot{x}_i^2}{2} = \frac{3}{2} MT \tag{2.118}$$

with M the number of particles and T the desired temperature. The corresponding equations of motion follow from the above-discussed variational principle

$$\ddot{x}_i = F_i - \lambda \dot{x}_i \quad \text{with} \quad i = 1, \ldots, M \tag{2.119}$$

with F_i the current force[7] acting on particle i and the Lagrange multiplier

$$\lambda = \frac{\sum\limits_{i=1}^{M} \dot{x}_i F_i}{m \sum\limits_{i=1}^{M} \dot{x}_i^2} . \tag{2.120}$$

In principle, this result may be classified as a control problem with one additional algebraic constraint.

The second type of generalized molecular dynamic simulations belongs to an extension of the equations of motion [33–35]. These equations may be interpreted as a typical result of the control theory. Here, we present a very simple version. Let us assume that we have the canonical equations of motion

$$\dot{x}_i = \frac{p_i}{m} \quad \text{and} \quad \dot{p}_i = F_i - u p_i , \tag{2.121}$$

where we have introduced an additional 'friction' term up_i with the scalar control function u. The implementation of u is a violation of the originally conservative structure of the equations of motion. This contribution should simulate the existence of the thermodynamical bath. Furthermore, we introduce the performance

$$R = \int\limits_0^T \mathrm{d}t \left[\frac{1}{2} \left(\sum_{i=1}^{M} \frac{p_i^2}{2m} - \frac{3}{2} MT \right)^2 + \frac{C}{2} u^2 \right] , \tag{2.122}$$

[7] This force is, of course, generated by the interaction of particle i with all other particles.

where $C > 0$ is a free parameter. That means we are interested in small fluctuations of the kinetic energy around their thermodynamically expected average $3/2MT$ and simultaneously in small friction coefficients. From here, we obtain the generalized Hamiltonian (2.94)

$$H = \sum_{i=1}^{M} \left[P_i(F_i - up_i) + Q_i \frac{p_i}{m} \right] - \frac{1}{2} \left(\sum_{i=1}^{M} \frac{p_i^2}{2m} - \frac{3}{2} MT \right)^2 - \frac{C}{2} u^2 \quad (2.123)$$

with the generalized momenta Q_i (corresponding to $\dot{x}_i$) and P_i (corresponding to $\dot{p}_i$). From here, we obtain the preoptimized control

$$u^{(*)} = -\frac{1}{C} \sum_{i=1}^{M} P_i p_i \,. \quad (2.124)$$

Hence, we get the evolution equations

$$\dot{x}_i = \frac{p_i}{m} \quad \text{and} \quad \dot{p}_i = F_i + \frac{p_i}{C} \sum_{j=1}^{M} P_j p_j \quad (2.125)$$

and the corresponding set of adjoint evolution equations

$$\dot{Q}_i = -\sum_{j=1}^{M} P_j \frac{\partial F_j}{\partial x_i} \quad (2.126)$$

and

$$\dot{P}_i = -\frac{P_i}{C} \sum_{j=1}^{M} P_j p_j - \frac{Q_i}{m} + \left(\sum_{j=1}^{M} \frac{p_j^2}{2m} - \frac{3}{2} MT \right) \frac{p_i}{m} \,. \quad (2.127)$$

The numerical solution of the extended system of evolution equations[8] (2.125), (2.126), and (2.127) now yields a deterministic substitute process for the evolution of a many-particle system in a thermodynamic bath.

2.5 The Hamilton–Jacobi Equation

Up to now, we have considered the generalized action (2.62) or the cost functional (2.52) as the starting point for the application of the variational calculus. The central aim of our previous analysis was the determination of the optimum trajectory and the optimum control. But sometimes it is necessary to know the value of the functional along the optimum trajectory. Of course, one can compute functional (2.52) directly from the optimal curve $X^*(t)$ and the optimum control $u^*(t)$. We will derive an alternative way which allows us to determine the performance functional without the knowledge of X^* and

[8] Note that we now have $4dM$ instead of $2dM$ differential equations for the evolution of the model system in a d-dimensional space.

u^*. First, the performance functional (2.52) and the generalized action (2.62) are identical at the optimum curve,

$$S[X^*, P^*, u^*, T, \Lambda^*, \overline{\Lambda}^*] = R[X^*, u^*, T], \tag{2.128}$$

because the optimal solution satisfies both the boundary conditions (2.55) and constraints (2.53). Especially the Lagrangian (2.63) simply becomes

$$L\left(t, X^*(t), \dot{X}^*(t), P^*(t), u^*(t)\right) = \phi(t, X^*(t), u^*(t)) . \tag{2.129}$$

On the other hand, $S[X^*, P^*, u^*, T, \Lambda^*, \overline{\Lambda}^*]$ is simply a function of the boundary conditions and the time T, i.e., we may write $S[X^*, P^*, u^*, T, \Lambda^*, \overline{\Lambda}^*] = S(X_0, X_e, T)$. We emphasize again that $S(X_0, X_e, T)$ means here the optimum action. Let us now determine the change in $S(X_0, X_e, T)$ for a small change in the final boundary conditions, $X_e \to X_e + \delta X_e$. The change in the boundary conditions also changes the optimal trajectory. Formally, we obtain

$$\begin{aligned}\delta S(X_0, X_e, T) &= S(X_0, X_e + \delta X_e, T) - S(X_0, X_e, T)\\ &= \int_0^T \mathrm{d}t \frac{\partial L^*}{\partial X^*} \delta X^* + \int_0^T \mathrm{d}t \frac{\partial L^*}{\partial \dot{X}^*} \delta \dot{X}^*\\ &\quad + \int_0^T \mathrm{d}t \frac{\partial L^*}{\partial P^*} \delta P^* + \int_0^T \mathrm{d}t \frac{\partial L^*}{\partial u^*} \delta u^* \end{aligned} \tag{2.130}$$

with δX^*, $\delta \dot{X}^*$, δP^*, and δu^* being the changes of the optimal trajectories of the state, the momenta, and the control due to the change in the boundary condition. The boundary terms, $b(X_0)\Lambda$ and $\overline{b}(X_e)\overline{\Lambda}$, does not contribute to the change in the action because the initial conditions satisfy optimal curves $b(X_0) = 0$ and the change in the final boundary condition (which satisfies $\overline{b}(X_e) = 0$) implies a change in the functional structure $\overline{b} \to \overline{b}'$ with $\overline{b}'(X_e + \delta X_e) = 0$. In other words, all boundary terms are separately canceled in $S(X_0, X_e, T)$ as well as in $S(X_0, X_e + \delta X_e, T)$. The second term in (2.130) is now integrated by parts. Considering (2.75), (2.74), and (2.69), we arrive at

$$\begin{aligned}\delta S(X_0, X_e, T) &= \left.\frac{\partial L^*}{\partial \dot{X}^*} \delta X^*\right|_0^T + \int_0^T dt \left[\frac{\partial L^*}{\partial X^*} - \frac{d}{dt}\frac{\partial L^*}{\partial \dot{X}^*}\right] \delta X^*\\ &= \left.\frac{\partial L^*}{\partial \dot{X}^*}\right|_{t=0} \delta X_0 + \left.\frac{\partial L^*}{\partial \dot{X}^*}\right|_{t=T} \delta X_e \end{aligned} \tag{2.131}$$

and therefore with (2.88) and $\delta X_0 = 0$

$$\delta S(X_0, X_e, T) = P_e^* \delta X_e . \tag{2.132}$$

We conclude from the relation that

$$P_e^* = \frac{\partial S(X_0, X_e, T)}{\partial X_e} . \tag{2.133}$$

On the other hand, the functional structure (2.62) implies the relation

$$\frac{\mathrm{d}S(X_0, X_\mathrm{e}, T)}{\mathrm{d}T} = L\left(T, X_\mathrm{e}, \dot{X}_\mathrm{e}^*, P_\mathrm{e}^*, u_\mathrm{e}^*\right) . \tag{2.134}$$

The total derivative may also be written as

$$\begin{aligned}\frac{\mathrm{d}S(X_0, X_\mathrm{e}, T)}{\mathrm{d}T} &= \frac{\partial S(X_0, X_\mathrm{e}, T)}{\partial T} + \frac{\partial S(X_0, X_\mathrm{e}, T)}{\partial X_\mathrm{e}} \dot{X}_\mathrm{e}^* \\ &= \frac{\partial S(X_0, X_\mathrm{e}, T)}{\partial T} + P_\mathrm{e}^* \dot{X}_\mathrm{e}^* .\end{aligned}$$

Thus, we obtain

$$\begin{aligned}\frac{\partial S(X_0, X_\mathrm{e}, T)}{\partial T} &= L\left(T, X_\mathrm{e}, \dot{X}_\mathrm{e}^*, P_\mathrm{e}^*, u_\mathrm{e}^*\right) - P_\mathrm{e}^* \dot{X}_\mathrm{e}^* \\ &= -H\left(T, X_\mathrm{e}, P_\mathrm{e}^*, u_\mathrm{e}^*\right) .\end{aligned} \tag{2.135}$$

The optimum control u_e^* can be substituted by the preoptimized control, $u_\mathrm{e}^* = u_\mathrm{e}^{(*)}(T, X_\mathrm{e}, P_\mathrm{e}^*)$. Finally, we replace the momentum P_e^* by (2.133) and

$$\frac{\partial S}{\partial T} + H\left(T, X_\mathrm{e}, \frac{\partial S}{\partial X_\mathrm{e}}, u_\mathrm{e}^{(*)}\left(T, X_\mathrm{e}, \frac{\partial S}{\partial X_\mathrm{e}}\right)\right) = 0 . \tag{2.136}$$

This nonlinear first-order partial differential equation defines the action $S = S(X_0, X_\mathrm{e}, T)$. Equation (2.136) is called the Hamilton–Jacobi equation. In principle, (2.136) solves the above-introduced problem. Unfortunately, the general solution of a partial differential equation of first-order depends on arbitrary functions. The specific structure of these functions is usually fixed by suitable boundary conditions.

For many applications in optimal control theory, the knowledge of these functions is secondary. The leading role is played by so-called complete integrals. In our case, this is a solution of (2.136) with $N + 1$ arbitrary, but independent constants[9]. Since (2.136) contains only derivatives of S, one of these constants is additive. The general structure of a complete integral is given by

$$S_\mathrm{comp} = f(T, X_\mathrm{e}, \widetilde{P}) + C_0 \tag{2.137}$$

with the constants $\widetilde{P} = (\widetilde{P}_1, \ldots, \widetilde{P}_N)$ and C_0. The condition that S_comp contains independent constants is $\det \partial^2 f / \partial X_\mathrm{e} \partial \widetilde{P} \neq 0$. We remark that the general solution can be obtained from the complete integrals by the construction of the corresponding envelope. We now use $f(T, X_\mathrm{e}, \widetilde{P})$ as the generating function for the canonical transformation

$$P = \frac{\partial f}{\partial X_\mathrm{e}} \quad \widetilde{X} = \frac{\partial f}{\partial \widetilde{P}} \quad \widetilde{H} = H + \frac{\partial f}{\partial T} = H + \frac{\partial S_\mathrm{comp}}{\partial T} , \tag{2.138}$$

[9] The number of independent constants in a complete integral is equivalent to the number of independent variables. In the present case, we have N state variables $X_1, X_2, \ldots, X_N$ and the time T.

where the new $\widetilde{H}$ satisfy the canonical equations (2.92) with the new coordinates $\widetilde{X}$ and the new momenta $\widetilde{P}$. On the other hand we get $\widetilde{H} = 0$ because S_{comp} is a solution of (2.136). Thus we obtain $\mathrm{d}\widetilde{X}/\mathrm{d}t = \mathrm{d}\widetilde{P}/\mathrm{d}t = 0$ and therefore $\widetilde{X} = \text{const}$. Hence, the solution of $\widetilde{X} = \partial f/\partial \widetilde{P}$ with respect to the final state is a function $X_{\mathrm{e}}(T, \widetilde{X}, \widetilde{P})$ of the time T and $2N$ independent constants. On the other hand, the trajectory of X_{e} is identical to the optimum path[10] of the system X^*. Therefore, $X_{\mathrm{e}}(t, \widetilde{X}, \widetilde{P})$ is a general solution of the optimum problem. The constants $\widetilde{X}$ and $\widetilde{P}$ may be used to fix the initial and final boundaries of this solution.

In this sense we may reformulate the concept of the Hamilton–Jacobi theory: a complete integral $S(t, X, \widetilde{P}) + C_0$, which satisfies the Hamilton–Jacobi equation

$$\frac{\partial S}{\partial T} + H\left(t, X, \frac{\partial S}{\partial X}, u^{(*)}\left(t, X, \frac{\partial S}{\partial X}\right)\right) = 0\,, \tag{2.139}$$

allows us to construct the general solution $X^*(t, \widetilde{P}, \widetilde{X})$ of the system of canonical equations (2.92) by solving the algebraic equations $\widetilde{X} = \partial S(t, X, \widetilde{P})/\partial \widetilde{P}$ for X and to determine the open parameters $\widetilde{X}$ and $\widetilde{P}$ in such a way that the boundary conditions are fulfilled.

Let us finally demonstrate this concept for a very simple example which concerns the performance functional of a free particle

$$R = \int_0^T \frac{u^2}{2}\mathrm{d}t \to \inf \tag{2.140}$$

for the simple constraint $\dot{x} = u$, $u \in (-\infty, +\infty)$ and the boundary conditions $x(0) = x_0$ and $x(T) = x_{\mathrm{e}}$. We obtain the Hamiltonian $H = qu - u^2/2$ with the generalized momentum q. The preoptimal control is simply $u^{(*)} = q$ and the preoptimized Hamiltonian is $H = q^2/2$. This leads to the Hamilton–Jacobi equation

$$\frac{\partial S}{\partial t} + \frac{1}{2}\left(\frac{\partial S}{\partial x}\right)^2 = 0\,, \tag{2.141}$$

and the separation of variables leads to the complete integral $S = -c^2 t/2 + cx$ and therefore to $\partial S/\partial c = x - ct = b$ with the free constants b and c. The boundary conditions require $b = x_0$ and $c = (x_{\mathrm{e}} - x_0)/T$ and the optimal solution is $x^* = x_0 + (x_{\mathrm{e}} - x_0)(t/T)$.

[10] This statement is a direct consequence of the time-local structure of the Hamiltonian and the Lagrangian, which means the sequence of the optimum trajectory from X_0 to X_1 and the subsequent optimum trajectory from X_1 to X_{e} yields the optimum trajectory from X_0 to X_{e}.

References

1. G. Galilei: *Dialogues concerning two New Sciences*, translated by H. Crew, A. de Salvio (Prometheus Books, Buffalo, NY, 1998)
2. P. Costabel, J. Peiffer: *Die Gesammelten Werke der Mathematiker und Physiker der Familie Bernoulli* (Birkhäuser, Basel, 1988)
3. B. Singh, R. Kumar: Indian J. Pure Appl. Math. **19**, 575 (1988)
4. T. Koetsier: The story of the creation of the calculus of variations : the contributions of Jakob Bernoulli, Johann Bernoulli and Leonhard Euler (Dutch), in *1985 Holiday Course : Calculus of Variations*, Eds. A. M. H. Gerards, J. W. Klop (CWI, Amsterdam, 1985), 1–25.
5. A.D. Ioffe, V.M. Tihomirov: *Theory Extremal Problems* (North-Holland, Amsterdam, 1979)
6. R. Bulirsch, A. Miele, J. Stoer, K. Well: *Optimal Control* (Birkhäuser, Basel, 1998)
7. D.A. Carlson, A.B. Haurie, A. Leizarowitz: *Infinite Horizon Optimal Control* (Springer, Berlin Heidelberg New York, 1991)
8. P. Whittle: *Optimal Control: Basics and Beyond* (Wiley, Chichester, 1996)
9. J.H. Davis: *Foundations of Deterministic and stochastic Control* (Birkhäuser, Boston, 2002)
10. V.I. Arnold: *Mathematical Methods of Classical Mechanics* (Springer, Berlin Heidelberg New York, 1989)
11. T.W. Kibble: *Classical Mechanics* (Imperial College Press, London, 2004)
12. M.G. Calkin: *Lagrangian and Hamiltonian Mechanics* (World Scientific Publishing, Singapore, 1997)
13. G.R. Fowles: *Analytical Mechanics* (Brooks/Cole Publishing Co., Pacific Grove, 1998)
14. D. Kleppner: *An Introduction to Mechanics* (McGraw-Hill, New York, 1973)
15. T.L. Chow: *Classical Mechanics* (Wiley, Chichester, 1995)
16. K.H. Hoffmann, I. Lasiecka, G. Leugering, J. Sprekels, F. Tröltzsch: *Optimal Control of Complex Structures.* International Series of Numerical Mathematics, vol. 139 (Birkhäuser, Basel, 2001)
17. A. Strauss: *An Introduction to Optimal Control Theory* (Springer, Berlin Heidelberg New York, 1968)
18. D.A. Carlson, A.B. Haurie, A. Leizarowitz: *Infinite Horizon Optimal Control* (Springer, Berlin Heidelberg New York, 1991)
19. J. Borggaard, J. Burkhardt, M. Gunzburger: *Optimal Design and Control* (Birkhäuser, Basel, 1995)
20. L.S. Pontryagin, V.G. Boltyanskii, R.V. Gamkrelidze, E.F. Mishchenko: *The Mathematical Theory of Optimal Processes* (Interscience Publishers, New York, 1962)
21. A.E. Bryson, Y.C. Ho: *Applied Optimal Control* (Hemisphere Publishing Co., Washington, 1975)
22. M. Athans, P.L. Falb: *Optimal Control* (McGraw-Hill, New York, 1966)
23. G. Knowles: *An Introduction to Applied Optimal Control* (Academic, New York, 1981)
24. D.J. Bell, D.H. Jacobson: *Singular Optimal Control Problems* (Academic, New York, 1975)
25. R. Burlisch, D. Kraft: *Computational Optimal Control* (Birkhäuser, Basel, 1994)

26. J. Gregory: *Constraint Optimization in the Calculus of Variations and Optimal Control Theory* (Van Nostrand Reinhold, New York, 1992)
27. A. Locatelli: *Optimal Control* (Birkhäuser, Basel, 2001)
28. D. Zubarev, D. Zubarev, G. Röpke: *Statistical Mechanics of Nonequilibrium Processes: Basic Concepts, Kinetic Theory* (Akademie-Verlag, Berlin, 1996)
29. R. Zwanzig: *Nonequilibrium Statistical Mechanics* (Oxford University Press, Oxford, 2001)
30. G.F. Mazenko: *Nonequilibrium Statistical Mechanics* (Wiley, Chichester, 2005)
31. D.J. Evans, G.P. Morriss: Comp. Phys. Rep. **1**, 297 (1984)
32. D.J. Evans, W.G. Hoover, B.H. Failor, B. Moran: Phys. Rev. A **28**, 1016 (1983)
33. S. Nosé: J. Chem. Phys. **81**, 511 (1984)
34. W.G. Hoover: Phys. Rev. A **31**, 1695 (1985)
35. S. Toxvaerd: Mol. Phys. **72**, 159 (1991)

3

Linear Quadratic Problems

3.1 Introduction to Linear Quadratic Problems

3.1.1 Motivation

Suppose we have a deterministic system under control, described by dynamical equations of motion for an N-dimensional state vector $X(t)$. Usually, these equations can be written as a set nonlinear first-order differential equations (2.53) which are essentially influenced by the control function $u(t)$. Furthermore, let us assume that we have obtained the optimum trajectory $X^*(t)$ and the optimum control $u^*(t)$ by the methods described in the previous chapter. We also denote $X^*(t)$ as a nominal state of the system and $u^*(t)$ as a nominal input. Unfortunately, we must expect that unavoidable uncertainties in the system description and disturbances are acting on the system, so that the real trajectory $X(t)$ shows some deviations from the optimal trajectory $X^*(t)$.

The determination $X^*(t)$ and $u^*(t)$ and the application of these solutions on a real experiment or a real system may be interpreted as an open loop control scheme as discussed in Sect. 1.1. This concept is sufficient as far as the optimal trajectories and controls are stable against disturbances. But it may be possible that a small deviation $Y(t) = X(t) - X^*(t)$ of the system state from the optimum trajectory decreases rapidly and the system becomes unstable in comparison to the desired nominal state. In this case, it seems rather reasonable to steer against a small deviation $Y(t)$ by a small correction $w(t) = u(t) - u^*(t)$. This can be done by a controller which measures the deviation $Y(t)$ of the current state from the nominal state and supplies the evolution of the system by the correction $w(t)$ in order to make $Y(t)$ small (Fig. 3.1). If the deviations are sufficiently small, the effect $Y(t)$ and $w(t)$ can be evaluated through the linearized equations of motion, following from a series expansion of the usually nonlinear equations of motion (2.53)

$$\dot{Y}(t) = \frac{\partial F\left(X^*, u^*, t\right)}{\partial X^*} Y(t) + \frac{\partial F\left(X^*, u^*, t\right)}{\partial u^*} w(t) \,. \tag{3.1}$$

M. Schulz: *Control Theory in Physics and other Fields of Science*
STMP **215**, 61–92 (2006)

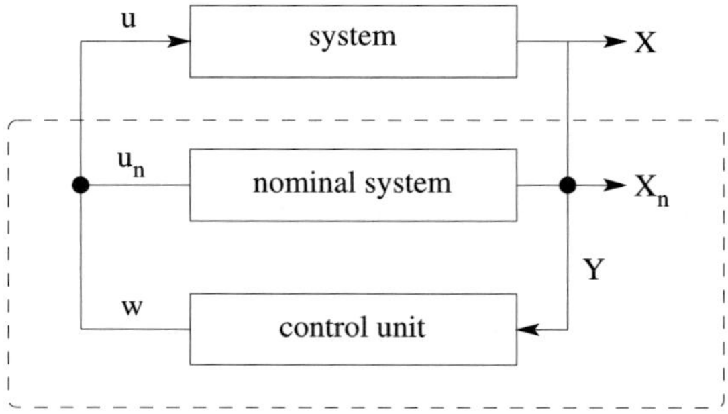

Fig. 3.1. The formal relations between a system under control, the nominal system and the controller

This system of linear differential equations with possibly time-dependent coefficients defines the linear response of the system on an arbitrary but small change of the control function $u(t)$ against the optimum control $u^*(t)$. Although the evolution of the linearized system can be completely determined from (3.1), the corresponding initial conditions and a certain control aim which is still to be defined, the general situation described here is rather a closed loop control than an open loop scheme.

3.1.2 The Performance Functional

The control problem for the above-introduced linear system of equations of motion must be completed by the performances functional in order to declare the control aim. Here we will give a heuristic motivation for the typical structure of these important quantity. We start from the original performance functional (2.52). The minimization of this functional together with the corresponding constraints and boundary conditions supplied the optimal trajectory $X^*(t)$ and the optimum control $u^*(t)$. Thus, we obtain

$$\int_0^T \mathrm{d}t\phi(t, X^*(t), u^*(t)) \leq \int_0^T \mathrm{d}t\phi(t, X'(t), u'(t)) , \tag{3.2}$$

where $X'(t)$ and $u'(t)$ are trajectories and controls which strictly satisfy the constraints, i.e., the evolution equations of the underlying system and the boundary conditions. We substitute $X'(t) = X^*(t) + Y(t)$ and $u'(t) = u^*(t) + w(t)$ and consider that $Y(t)$ and $w(t)$ are small deviations which can be described by the linearized equations of motion (3.1). An expansion of the performance functional in terms of Y and u up to the second-order leads to

$$0 \leq \int_0^T dt \frac{\partial \phi^*}{\partial X^*} Y(t) + \int_0^T dt \frac{\partial \phi^*}{\partial u^*} w(t) + \frac{1}{2} \int_0^T dt Y(t) \frac{\partial^2 \phi^*}{\partial X^{*2}} Y(t)$$

$$+\int_0^T dtY(t)\frac{\partial^2\phi^*}{\partial X^*\partial u^*}w(t)+\frac{1}{2}\int_0^T dtw(t)\frac{\partial^2\phi^*}{\partial u^{*2}}w(t) \qquad (3.3)$$

with $\phi^* = \phi(t, X^*, u^*)$. The linear terms disappear[1] and it remains only an inequality for a squared form. This squared form is often estimated by the stronger inequality

$$\int_0^T \mathrm{d}tY(t)\frac{\partial^2\phi^*}{\partial X^{*2}}Y(t)+\int_0^T \mathrm{d}tw(t)\frac{\partial^2\phi^*}{\partial u^{*2}}w(t)\geq 0 \qquad (3.4)$$

so that we obtain a performance functional of a quadratic type

$$J[Y,w]=\frac{1}{2}\int_0^T \mathrm{d}t\left[Y(t)Q(t)Y(t)+\int_0^T \mathrm{d}tw(t)R(t)w(t)\right]. \qquad (3.5)$$

The matrix functions $Q(t)$ and $R(t)$ can be identified with the second-order derivatives in (3.4). We remark that for the most applications and especially for sufficiently complicated systems the rigorous derivation of $Q(t)$ and $R(t)$ is often replaced by a more or less empirically chosen quantities. The only condition one must take into account is that $Q(t)$ and $R(t)$ are symmetric and positive definite matrices. In this case, functional (3.5) has only one global minimum which is reached for $Y = w = 0$ when the constraints are neglected. However, the minimization of (3.5) under consideration of the equations of motion (3.1) still guarantees the smallest possible corrections $Y(t)$ and $w(t)$ to a nominal state $X^*(t)$ and a nominate control $u^*(t)$.

The minimization of the quadratic performance functional (3.5) under the constraint of linear evolution equations (3.1) is called a linear quadratic problem. As we have illustrated, such problems arise in a fairly spontaneous and natural way. In principle, the linear quadratic problem is again a deterministic optimum control problem not for the optimal trajectory and the optimal control, but for the deviations from the optimum solution. Linear quadratic problems have a broad spreading in natural, technological, and economic sciences.

3.1.3 Stability Analysis

Linear Stability Analysis

First of all, we will check whether a further control of a system close to the nominal state is actually necessary. To this aim, we consider a possible

[1] It should be remarked that the linear terms vanish only for such $Y(t)$ and $w(t)$ which satisfy the constraints. Therefore, we have $\int \mathrm{d}t\,(\partial\phi^*/\partial X^*)\,Y(t) = 0$ and $\int \mathrm{d}t\,(\partial\phi^*/\partial u^*)\,w(t) = 0$ but not necessarily $\partial\phi^*/\partial X^* = 0$ and $\partial\phi^*/\partial u^* = 0$. The latter would correspond to the variational derivative of the performance functional, where all admissible functions, $Y(t)$ and $w(t)$, are taken into account independently from their constraints.

deviation $Y(t)$ of the current state $X(t)$ against the optimum state $X^*(t)$, but we allow no additional control as the optimum control $u^*(t)$, i.e., we have $w(t) = 0$. For the further investigations we may write the original nonlinear system of evolution equations (2.53) in the form

$$\begin{aligned}\dot{Y}(t) &= F\left(X^*(t) + Y(t), u^*(t), t\right) - F\left(X^*(t), u^*(t), t\right) \\ &= H\left(Y(t), t\right) .\end{aligned} \tag{3.6}$$

If this equation approaches the optimum state, i.e., $Y(t) \to 0$ for $t \to \infty$, a further control is not necessary for stabilizing the system. Such a system is called self-stabilized. Otherwise, if an initially small deviation becomes sufficiently large or it diverges in the long time limit, the system is unstable and needs definitely an additional control.

An apparently simple case occurs for autonomous differential equations which depend not explicitly on time. In this case, we have the explicitly time-independent version of the evolution equations (3.6):

$$\dot{Y} = H\left(Y\right) \tag{3.7}$$

with a (stable or instable) fixed point $Y = 0$ in the N-dimensional phase space. The linear stability analysis requires the linearized evolution equation

$$\dot{Y} = AY , \tag{3.8}$$

where the matrix A has the components

$$A_{\alpha\beta} = \left.\frac{\partial H_\alpha\left(Y\right)}{\partial Y_\beta}\right|_{Y=0} \quad \text{with} \quad \alpha, \beta = 1, \ldots, N . \tag{3.9}$$

A standard method for the characterization of the stability of (3.7) is analysis of the linear equation (3.8). In particular, the equation is called stable with respect to a fixed point $Y = 0$ if the real part of all eigen-values of A is negative.

Resonances

The linear stability analysis does not always yield sufficient indications for the stability or instability of a given system. In order to avoid misleading conclusions with respect to the results of the linear stability analysis, we expand $H\left(Y\right)$ in terms of a Taylor series with respect to Y. Then we can write (3.7) in the form

$$\frac{\partial Y}{\partial t} = AY + \psi^{(r)}(Y) . \tag{3.10}$$

The rest function is given by $\psi^{(r)}\left(Y\right) = H\left(Y\right) - AY$. The leading term of the function $\psi^{(r)}\left(Y\right)$ is of an order of magnitude $\left|Y\right|^r$ with $r \geq 2$. Let us now introduce a transformation $z = Y + h(Y)$ where h is a vector polynomial with the leading order 2 so that $h(0) = \partial h/\partial Y|_{Y=0} = 0$. Thus we obtain

$$\frac{\mathrm{d}z}{\mathrm{d}t} = \frac{\partial z}{\partial Y}\frac{\mathrm{d}Y}{\mathrm{d}t} = \left(1 + \frac{\partial h}{\partial Y}\right)\frac{\mathrm{d}Y}{\mathrm{d}t} = \left(1 + \frac{\partial h}{\partial Y}\right)\left[AY + \psi^{(r)}\right]$$
$$= AY + \frac{\partial h}{\partial Y}AY + \psi^{(r)} + \frac{\partial h}{\partial Y}\psi^{(r)}$$
$$= Az - \left[Ah - \frac{\partial h}{\partial Y}AY - \psi^{(r)}\right] + \frac{\partial h}{\partial Y}\psi^{(r)} . \tag{3.11}$$

We determine the open function h by setting

$$\hat{L}_A h = Ah - \frac{\partial h}{\partial Y}AY = \psi^{(r)} . \tag{3.12}$$

This equation has a unique solution if the eigenvalues of the introduced operator $\hat{L}_A$ are nonresonant. To understand this statement, we consider that the matrix A has the set of eigenvalues $\lambda = \{\lambda_1, \ldots, \lambda_N\}$ and the normalized eigenvectors $\{e_1, \ldots, e_N\}$. The vector Y can be expressed in terms of these bases, $Y = \eta_1 e_1 + \cdots + \eta_N e_N$. The eigenvectors of $\hat{L}_A$ are the following vector monomials

$$\varphi_{m,\gamma} = \eta_1^{m_1} \ldots \eta_N^{m_N} e_\gamma \tag{3.13}$$

with $m = \{m_1, \ldots, m_N\}$. The m_α are nonnegative integers satisfying $m_1 + \cdots + m_N \geq 2$. Note that $\hat{L}_A$ acts in the space of functions which have an asymptotic behavior $h \sim |Y|^r$ with $r \geq 2$ for $|Y| \to 0$. We remark that $A\varphi_{m,\gamma} = \lambda_\gamma \varphi_{m,\gamma}$ and

$$\sum_{\alpha,\beta} \frac{\partial \varphi_{m,\gamma}}{\partial Y_\alpha} A_{\alpha\beta} Y_\beta = \sum_\beta \frac{\partial \varphi_{m,\gamma}}{\partial \eta_\beta} \lambda_\beta \eta_\beta = (m, \lambda)\, \varphi_{m,\gamma} , \tag{3.14}$$

where (m, λ) is the euclidean scalar product between the vectors m and λ. Thus we find

$$\hat{L}_A \varphi_{m,\gamma} = [\lambda_\gamma - (m, \lambda)]\, \varphi_{m,\gamma} , \tag{3.15}$$

i.e. the operator $\hat{L}_A$ has the eigenvalues $\lambda_\gamma - (m, \lambda)$. If all eigenvalues of $\hat{L}_A$ have nonzero values, (3.12) has a unique solution. That requires $(m, \lambda) \neq \lambda_\gamma$. Otherwise, we have a so-called resonance $\lambda_\gamma = (m, \lambda)$, and $\hat{L}_A$ is not reversible.

Suppose that no resonances exist. Then the solution of (3.12) defines the transformation function $h(y)$ such that

$$\dot{z} = Az + \frac{\partial h}{\partial Y}\psi^{(r)}(Y) \tag{3.16}$$

Comparing the order of the leading terms of h and $\psi^{(r)}$ we find that the product $\psi^{(r)}(Y)\,\partial h/\partial Y$ is of an order $r+1$ in $|Y|$. Considering the transformation between z and Y, we arrive at

$$\dot{z} = Az + \psi^{(r+1)}(z) , \tag{3.17}$$

where $\psi^{(r+1)}(z)$ is a nonlinear contribution with a leading term proportional to $|z|^{r+1}$. The repeated application of this formalism generates an increasing order of the leading term.

In other words, the nonlinear differential equation approaches step by step a linear differential equation. This is the content of the famous theorem of Poincaré [2]. In the case of resonant eigenvalues the Poincaré theorem must be extended to the theorem of Poincaré and Dulaque [2]. Here, we get the following differential equation instead of (3.17):

$$\dot{z} = Az + w(z) + \psi^{(r+1)}(z) \ , \tag{3.18}$$

where $w(z)$ contains the resonant monomials. The convergence of this procedure depends on the structure of the eigen value spectra of the matrix A. If the convex cover of all eigenvalues $\lambda_1, \ldots, \lambda_N$ in the complex plane does not contain the origin, the vector $\lambda = \{\lambda_1, \ldots, \lambda_N\}$ is an element of the so-called Poincaré region of the corresponding $2N$-dimensional complex space. Otherwise, the vector is an element of the Siegel region [3].

If λ is an element of the Poincaré region, the above-discussed procedure is convergent and the differential equation (3.10) or (3.7) can be mapped formally onto a linear differential equation for nonresonant eigenvalues or onto the canonical form (3.18). In the first case, the stability of the original differential equation (3.7) is equivalent to the stability of the linear differential equation $\dot{z} = Az$. That means especially that, because of (3.7), the linearized version of the original differential equation system is sufficient for the determination of the stability of the fixed point $Y = 0$. In the second case, we have to analyze the nonlinear normal form (3.18) for a study of the dynamics of the original system in the neighborhood of the fixed point $Y = 0$. If λ is an element of the Siegel region, the convergence cannot be guaranteed.

The Poincaré theorem allows a powerful analysis of the stability of systems of differential equations which goes beyond the standard method of linear approximation. In particular, this theorem can be a helpful tool classifying the above-discussed self-stabilization of a system and many other related problems.

In the case of a one-dimensional system only one eigen value $\lambda = A$ exists. Then the fixed point $Y = 0$ corresponds to a stable state for $\lambda < 0$ and to an unstable state for $\lambda > 0$. Special investigations considering the leading term of the nonlinear part of (3.10) are necessary for $\lambda = 0$.

Another situation occurs for a two-dimensional system. Here we have two eigenvalues, λ_1 and λ_2. If resonances are excluded, the largest real part of the eigenvalues determines the stability or instability of the system. A resonance exists if $\lambda_1 = m_1\lambda_1 + m_2\lambda_2$ or $\lambda_2 = m_1\lambda_1 + m_2\lambda_2$ where m_1 and m_2 are nonnegative integers. In this case we expect a nonlinear normal form (3.18) containing the resonant monomials.

Let us illustrate the formalism by using a very simple example. The eigenvalues $\lambda_1 = -\lambda_2 = \mathrm{i}\Omega$, obtained from the linear stability analysis, are usually identified with a periodic motion of the frequency Ω. But this case contains two resonances, namely, $\lambda_1 = 2\lambda_1 + \lambda_2$ and $\lambda_2 = \lambda_1 + 2\lambda_2$. Thus the stationarity of the evolution of the corresponding nonlinear system of differential equations

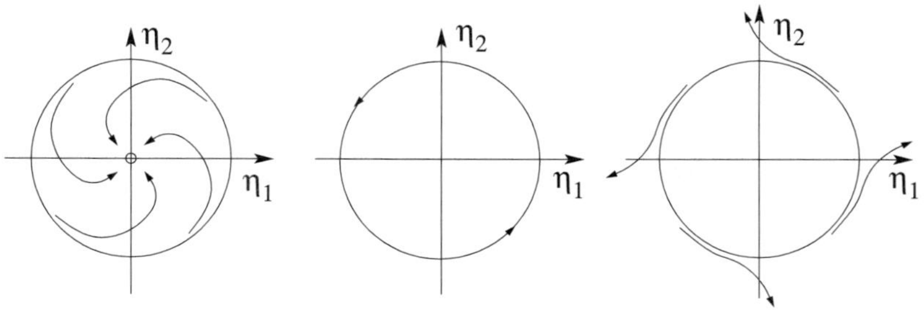

Fig. 3.2. Stable fixed point for Im $c < 0$, limit circle for Im $c = 0$ and unstable behaviour for Im $c > 0$

(3.10) is no longer determined by the simple linear system[2] $\dot{\eta}_1 = \mathrm{i}\Omega\eta_1$ and $\dot{\eta}_2 = -\mathrm{i}\Omega\eta_2$, but by the normal system

$$\dot{\eta}_1 = \mathrm{i}\Omega\eta_1 + c_1\eta_1^2\eta_2 \quad \text{and} \quad \dot{\eta}_2 = -\mathrm{i}\Omega\eta_2 - c_2\eta_1\eta_2^2 \ . \tag{3.19}$$

The substitutions $x_1 = \eta_1 + \mathrm{i}\eta_2$ and $x_2 = \mathrm{i}\,(\eta_1 - \mathrm{i}\eta_2)$ and the agreement $x^2 = x_1^2 + x_2^2$ lead to the real normal form

$$\dot{x}_1 = \Omega x_2 + \frac{x^2}{4}\left[x_1 \mathrm{Im}\ c - x_2 \mathrm{Re}\ c\right] \tag{3.20}$$

and

$$\dot{x}_2 = -\Omega x_1 + \frac{x^2}{4}\left[x_1 \mathrm{Re}\ c + x_2 \mathrm{Im}\ c\right] \ , \tag{3.21}$$

where the real structure of the differential equations requires $c_1 = c$ and $c_2 = \bar{c}$. Such a structure is already expected after the first step of the Poincaré algorithm applied onto (3.10). Only the two parameters Re c and Im c are still open. All other nonlinear terms disappear step by step during the repeated application of the reduction formalism. However, it is not necessary to execute these steps because the resonance terms remain unchanged after their appearance. The stability behavior follows directly from the dynamics of x^2. We obtain from (3.21)

$$\frac{\partial x^2}{\partial t} = \frac{\mathrm{Im}\ c}{2} x^4 \ . \tag{3.22}$$

Thus, the system is stable for Im $c < 0$ and unstable for Im $c > 0$, see Fig. 3.2. Obviously, we need only an estimation about the sign of the quantity Im c, which is usually obtainable after a few iterations of the above-introduced Poincaré algorithm.

[2] The linear system is written in the standard form considering the representation in terms of the eigen vectors of the matrix A.

Ljapunov Theorems

Now we come back to the more general non-autonomous differential equation (3.6). Let us assume that we may construct a scalar function $V(Y,t)$ with $V(0,t) = 0$ which is positive definite and whose total derivative along the solutions of (3.6) is not positive. A function with these properties is called a Ljapunov function, which means if $Y(t, Y_0, t_0)$ is the solution of (3.6) with the initial condition $Y(t_0, Y_0, t_0) = Y_0$ we expect from our construction

$$\frac{\mathrm{d}}{\mathrm{d}t} V(Y(t,Y_0,t_0),t) = \left.\frac{\partial V}{\partial Y}\right|_{Y=Y(t,Y_0,t_0)} H(Y(t,Y_0,t_0),t) + \left.\frac{\partial V}{\partial t}\right|_{Y=Y(t,Y_0,t_0)} \leq 0 \,. \tag{3.23}$$

Because $V(Y,t) > 0$ and the fact that the derivatives along each solution of (3.6) are negative, we get immediately the result

$$V(Y(t,Y_0,t_0),t) \leq V(Y_0,t) \,. \tag{3.24}$$

Since the Ljapunov function $V(Y,t)$ is positive definite, we always find a strictly monotone increasing continuous function Ψ_- with $\Psi_-(0) = 0$ which satisfies

$$V(Y,t) \geq \Psi_-(\|Y\|) \tag{3.25}$$

for all Y and t. The function Ψ_- is also called a conical function (Fig. 3.3). Then, we can always determine an $\epsilon > 0$ so that

$$V(Y_0,t) < \Psi_-(\epsilon) \tag{3.26}$$

for all $\|Y_0\| < \delta$. We take $\delta' = \min(\delta, \epsilon)$. Then the relation $\|Y_0\| < \delta'$ implies

$$\|Y_0\| < \epsilon \quad \text{as well as} \quad V(Y_0,t) < \Psi_-(\epsilon) \,. \tag{3.27}$$

In principle, we may find for each δ' a corresponding ϵ so that the relation (3.27) is satisfied for all $\|Y_0\| < \delta'$. Let us now ask, if a solution with the initial condition $\|Y_0\| < \delta'$ can become

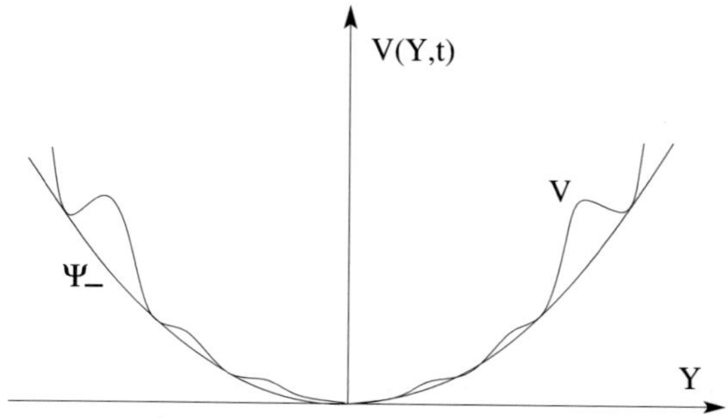

Fig. 3.3. A positive definite function V and the corresponding conical function Ψ_-

$$\|Y(t, Y_0, t_0)\| > \epsilon , \tag{3.28}$$

for all t above a critical time $t_{\rm cr} > t_0$ if the total derivative of $V(Y,t)$ along the solution $Y(t, Y_0, t_0)$ is not positive. If this were the case, we would expect

$$\|Y(t_{\rm cr}, Y_0, t_0)\| = \epsilon \tag{3.29}$$

due to the continuity of the solution $Y(t, Y_0, t_0)$. But this is because of the conical bound property (3.25),

$$V(Y(t_{\rm cr}, Y_0, t_0), t_{\rm cr}) \geq \Psi_- (\|Y(t_{\rm cr}, Y_0, t_0)\|) = \Psi_- (\epsilon) , \tag{3.30}$$

in contradiction to (3.27). Thus we conclude that the existence of a positive definite function $V(Y,t)$ with $V(0,t) = 0$ and whose total derivative along the solutions of (3.6) is not positive is a sufficient condition for the stability, namely that each solution of (3.6) with sufficiently small initial conditions $\|Y_0\| < \delta'$ is always localized in the phase space onto the region $\|Y(t, Y_0, t_0)\| < \epsilon$. This is the content of Ljapunov's first stability theorem. But this theorem gives no information about the convergence behavior of $Y(t, Y_0, t_0)$ for $t \to \infty$. For this problem we need an additional requirement, namely the decrescent property of $V(Y,t)$. The function $V(Y,t)$ is called a decrescent if there exists another conical function Ψ_+ so that for all Y and t

$$V(Y,t) \leq \Psi_+(\|Y\|) \tag{3.31}$$

holds (Fig. 3.4). Since $V(Y,t) \geq 0$ and the derivatives along the solutions of (3.6) are negative, we expect

$$\lim_{t \to \infty} V(Y(t, Y_0, t_0), t) = V_\infty \geq 0 . \tag{3.32}$$

The claim is now to show that $V_\infty = 0$ for each decrescent Ljapunov function. Obviously, the functions Ψ_- and Ψ_+ cover $V(Y,t)$. Furthermore, since the total derivative $\mathrm{d}V/\mathrm{d}t$ along the solution is negative definite, we always find a conical function Φ such that

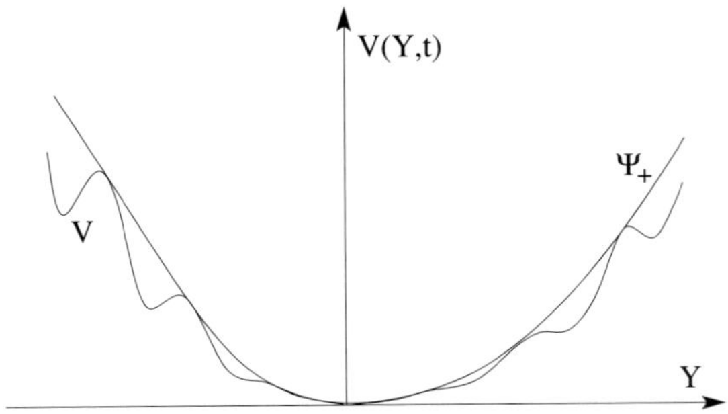

Fig. 3.4. A positive definite decrescent function V bounded by the conical function Ψ_+

$$\frac{d}{dt}V(Y(t,Y_0,t_0),t) \le -\Phi\left(\|Y(t,Y_0,t_0)\|\right) . \tag{3.33}$$

Hence, if $V_\infty > 0$, we conclude that for all t the inequality $V(Y(t,Y_0,t_0),t) \ge V_\infty$ and therefore $\Psi_+(\|Y(t,Y_0,t_0)\|) \ge V_\infty$ hold. The last inequality requires that there exists a finite ε such that

$$\|Y(t,Y_0,t_0)\| \ge \varepsilon \tag{3.34}$$

for all t. But then we have because of (3.33) the inequality

$$\frac{d}{dt}V(Y(t,Y_0,t_0),t) \le -\Phi(\varepsilon) \tag{3.35}$$

and therefore

$$\begin{aligned}\int_0^t \frac{d}{dt'}V(Y(t',Y_0,t_0),t') &= V(Y(t,Y_0,t_0),t) - V(Y_0,t_0)\\ &\le -\Phi(\varepsilon)(t-t_0)\end{aligned} \tag{3.36}$$

i.e., we get for $t \to \infty$ always $V_\infty < 0$ if $\Phi(\varepsilon) \neq 0$ in contradiction to (3.32). The only possible way to avoid contradictions is that $\Phi(\varepsilon) = 0$. Because of the conical character of Φ, we then have $\varepsilon = 0$ and therefore, because of the required decrescent character of the Ljapunov function, $V_\infty = 0$. Hence, we obtain the second Ljapunov theorem: the successful construction of one decrescent Ljapunov function is sufficient for the convergence

$$\lim_{t\to\infty} Y(t,Y_0,t_0) = 0 \tag{3.37}$$

and consequently for the stability of the fixed point $Y = 0$.

We illustrate this behavior with the simple example of a particle in the potential $v(x) \ge 0$ under a Newtonian friction with the coefficient γ. The potential may monotonously increase if $|x|$ increases. Let Y be a two component vector (x,p). Then the evolution equations read

$$\dot{x} = p/m \quad \text{and} \quad \dot{p} = -v'(x) - \gamma p . \tag{3.38}$$

A possible decrescent Ljapunov function is then

$$V(x,p,t) = \frac{p^2}{2m} + v(x) \tag{3.39}$$

because its derivatives along the trajectories are given by

$$\frac{dV}{dt} = \frac{p\dot{p}}{m} + v'(x)\dot{x} = -\frac{\gamma}{m}p^2 \le 0 . \tag{3.40}$$

Thus we get the well-known result that the fixed point $Y = 0$ is stable.

If we come back to our original control problem, we may summarize that the stability analysis is a first step to decide if a certain system requires a control in order to stabilize the optimum trajectory against possible perturbations. If the system is unstable, such a control is absolutely necessary. On the other hand, a stable system does not necessarily need a control. However, in cases where the initially slightly disturbed system relaxes very slowly back to the optimum trajectory, an additional control may support this process in order to make this convergence faster.

3.1.4 The General Solution of Linear Quadratic Problems

Following the above-introduced concept, the linear quadratic problem consists in the determination of the optimum control w^*, of the optimum trajectory Y^* which solve the evolution equations of the linear system

$$\dot{Y}(t) = A(t)Y(t) + B(t)w(t) \tag{3.41}$$

with the initial condition

$$Y(0) = Y_0 \tag{3.42}$$

and which minimizes the performance functional

$$J[Y,w] = \frac{1}{2}\int_0^T \mathrm{d}t\,[Y(t)Q(t)Y(t) + w(t)R(t)w(t)] + \frac{1}{2}Y(T)\Omega Y(T)\,. \tag{3.43}$$

Here, we have used a generalized representation of the version of (3.5) consisting of an integral and an endpoint function. As mentioned in Sect. 2.2.2, the minimization of this mixed performance is called a Bolza problem. The additional consideration of the endpoint is a real extension against (3.5) in the framework of linear quadratic problems. This is not in contrast to the general statement[3] that each endpoint functional can be transformed in an integral representation. This is also possible in the present case, but then we obtain an additional evolution equation which is not linear.

In principle, the problem is only a special case of the large class of deterministic control problems. This can be solved by the techniques discussed in Chap. 2. Here, we use the Hamilton approach. To this aim we rewrite the performance integral

$$J[Y,w] = \frac{1}{2}\int_0^T \mathrm{d}t\left[Y(t)\widetilde{Q}(t)Y(t) + w(t)R(t)w(t)\right] \tag{3.44}$$

with $\widetilde{Q}(t) = Q(t) + \Omega\delta(t-T)$ and construct the Hamiltonian

$$H = P[AY + Bw] - \frac{1}{2}Y\widetilde{Q}Y - \frac{1}{2}wRw \tag{3.45}$$

with the generalized momentum $P(t)$. Because the small control is not assumed to be restricted, we obtain from $\partial H/\partial w = 0$ the pre-optimal control

$$w^{(*)} = R^{-1}B^TP\,, \tag{3.46}$$

[3] See the discussion in Sect. 2.2.2.

and the preoptimized Hamiltonian now reads

$$H^{(*)} = PAY - \frac{1}{2}Y\widetilde{Q}Y + \frac{1}{2}PBR^{-1}B^{T}P\,. \tag{3.47}$$

From here, we obtain the canonical system of evolution equations for the optimal control

$$\dot{Y}^{*} = AY^{*} + BR^{-1}B^{T}P^{*} \tag{3.48}$$

and

$$\dot{P}^{*} = -A^{T}P^{*} + \widetilde{Q}Y^{*}\,. \tag{3.49}$$

Now, we introduce the time-dependent transformation matrix $G(t)$ connecting momenta $P^{*}(t)$ and the state vector $Y^{*}(t)$ via $P^{*}(t) = -G(t)Y^{*}(t)$ and substitute this expression in (3.49)

$$-\dot{G}Y^{*} - G\dot{Y}^{*} = A^{T}GY^{*} + \widetilde{Q}Y^{*}\,. \tag{3.50}$$

From here, we obtain with (3.48)

$$\begin{aligned}\dot{G}Y^{*} &= -A^{T}GY^{*} - \widetilde{Q}Y^{*} - GAY^{*} - GBR^{-1}B^{T}P^{*}\\ &= -A^{T}GY^{*} - \widetilde{Q}Y^{*} - GAY^{*} + GBR^{-1}B^{T}GY^{*}\,,\end{aligned} \tag{3.51}$$

which means the problem is solved if we find a matrix $G(t)$ which satisfies the equation

$$\dot{G} + A^{T}G + GA - GBR^{-1}B^{T}G = -\widetilde{Q}\,. \tag{3.52}$$

Because of $\widetilde{Q}(t) = Q(t) + \Omega\delta\left(t - T\right)$ we conclude that for all $t \neq T$, the matrix $G(t)$ is a solution of

$$\dot{G} + A^{T}G + GA - GBR^{-1}B^{T}G = -Q\,. \tag{3.53}$$

The equation is called the differential Ricatti equation with the boundary condition

$$G(T) = \Omega\,. \tag{3.54}$$

which follows immediately from (3.52) by an integration over the time interval $[T - \varepsilon, T + \varepsilon]$. The symmetry of (3.53) and (3.54) requires the symmetry of $G(t) = G^{T}(t)$. Of course, (3.53) is a nonlinear system of $N \times N$ ordinary coupled differential equations. Although a detailed analysis of (3.53) often requires the application of numerical tools [4,5,7,8], the differential Ricatti equation is usually considered to be the complete solution of the linear quadratic problem.

Finally, we get the expression for the optimal control from (3.46),

$$w^{*} = -R^{-1}B^{T}GY^{*} \tag{3.55}$$

while the optimal trajectory follows from the homogeneous linear system of differential equations

$$\dot{Y}^{*} = \left(A - BR^{-1}B^{T}G\right)Y^{*} \tag{3.56}$$

with the initial condition $Y^{*}(0) = Y_{0}$. The linear relation between the current state and control (3.55) is often called the control law. This law indicatesagain the above-suggested closed-loop character of the control mechanism because the coupling between the control and state, $R^{-1}B^TG$, depends only on quantities characterizing the dynamics of the system or the performance of the control [5, 9].

3.2 Extensions and Applications

3.2.1 Modifications of the Performance

Generalized Quadratic Forms

We may extend the performance integral by adding a mixed bilinear term $Y(t)W(t)w(t)$. In principle, this idea corresponds to the intermediate stage (3.3) of our heuristic derivation of quadratic linear problem. In the case of empirically chosen matrices $Q(t)$, $R(t)$, and $W(t)$, we must be aware that this change can essentially modify the problem. In fact, the addition of such bilinear terms can change the necessary positive definite structure of the performance integral

$$J[Y,w] = \frac{1}{2}\int_0^T \mathrm{d}t \left[Y(t)\widetilde{Q}(t)Y(t) + w(t)R(t)w(t) + 2Y(t)W(t)w(t)\right] . \quad (3.57)$$

Therefore, these extension requires a further check of the composed matrix

$$\begin{pmatrix} Q & W \\ W^T & R \end{pmatrix}, \quad (3.58)$$

which must be positive definite for all times $t \in [0, T]$.

Linear Quadratic Performance

The quadratic performance functional may be extended to a linear quadratic functional by adding linear functions of the control functions and the state variables into functional (3.43)

$$\begin{aligned} J[Y,w] = \frac{1}{2}\int_0^T \mathrm{d}t \left[Y(t)Q(t)Y(t) + w(t)R(t)w(t)\right] \\ + \int_0^T \mathrm{d}t \left[\alpha(t)Y(t) + \beta(t)w(t)\right] \\ + \frac{1}{2}Y(T)\Omega Y(T) + \omega Y(T) . \end{aligned} \quad (3.59)$$

It is easy to check that in this case the optimum control is given by

$$w^* = -R^{-1}\left\{B^T\left[GY^* + \xi\right] + \beta\right\} , \tag{3.60}$$

where $G(t)$ solves the differential Ricatti equation (3.53) with the boundary condition (3.54) while the newly introduced vector function $\xi(t)$ solves the following system of linear differential equations

$$\dot{\xi} = -\left(A - BR^{-1}B^T G\right)^T \xi + GBR^{-1}\beta - \alpha , \tag{3.61}$$

and the boundary condition

$$\xi(T) = \omega . \tag{3.62}$$

The optimal trajectory follows from a modified version of (3.56), namely

$$\dot{Y}^* = \left(A - BR^{-1}B^T G\right) Y^* + BR^{-1}B^T \xi . \tag{3.63}$$

In principle, the derivation of (3.60) follows the same scheme as the derivation of (3.55) in Sect. 3.1.4. The only difference is the application of the generalized relation $P^*(t) = -G(t)Y^*(t) + \xi(t)$ instead of $P^*(t) = -G(t)Y^*(t)$.

Tracking Problems

Let us assume that we wish a certain, but small modification $\psi(t)$ of the optimum trajectory $X^*(t)$, i.e., the desired ideal evolution of the system under control is now given by $X_{\text{ideal}}(t) = X^*(t) + \psi(t)$, which means, we have to ask for a small modification $w(t)$ of the control such that the actually realized trajectory $X(t) = X^*(t) + Y(t)$ is close to the ideal trajectory $X_{\text{ideal}}(t)$. In other words, the control aim is to find a trajectory which follows a given external signal $\psi(t)$. This can be done by considering the performance functional

$$\begin{aligned} J[Y, w] = \frac{1}{2} \int_0^T \mathrm{d}t \left[(Y(t) - \psi(t))\, Q(t)\, (Y(t) - \psi(t)) + w(t) R(t) w(t) \right] \\ + \frac{1}{2} (Y(T) - \psi(T))\, S\, (Y(T) - \psi(T)) . \end{aligned} \tag{3.64}$$

This problem is a special case of a linear quadratic problem with a linear quadratic performance with $\beta(t) = 0$ and $\alpha(t) = -Q\psi$. Therefore, we can employ the results presented above. In particular, the optimal control of the tracking problem is given by

$$w^* = -R^{-1}B^T\left[GY^* + \xi\right], \tag{3.65}$$

where $G(t)$ again solves the differential Ricatti equation (3.53) while the function $\xi(t)$ is a solution of

$$\dot{\xi} = -\left(A - BR^{-1}B^T G\right)^T \xi + Q\psi \tag{3.66}$$

with the boundary condition $\xi(T) = -S$. The optimal trajectory is given by

$$\dot{Y}^* = \left(A - BR^{-1}B^T G\right) Y^* + BR^{-1}B^T \xi . \tag{3.67}$$

Tracking problems occur in several scientific problems. Typical examples are electronic or hydraulic amplifiers, where an incoming signal is transformed into a response signal with another amplitude and phase.

3.2.2 Inhomogeneous Linear Evolution Equations

It may be possible that the linear evolution equations have an inhomogeneous structure

$$\dot{Y} = AY + Bw + F \, , \tag{3.68}$$

where $F(t)$ is an additional generalized force. This problem can be solved by a transformation of the state vector $Y \to Y' = Y - \theta$, where θ satisfies the equation

$$\dot{\theta} = A\theta + F \tag{3.69}$$

so that the new evolution equation for Y'

$$\dot{Y}' = AY' + Bw \tag{3.70}$$

remains. Furthermore, the transformation modifies the original performance functional (3.43) in

$$J[Y, w] = \frac{1}{2} \int_0^T \mathrm{d}t \left[(Y'(t) + \theta(t)) \, Q(t) \, (Y'(t) + \theta(t)) + w(t)R(t)w(t) \right] \\ + (Y'(t) + \theta(T)) \, S \, (Y'(t) + \theta(T)) \, . \tag{3.71}$$

This result suggests that the class of linear quadratic control problems with inhomogeneous linear evolution equations can be mapped onto the class of tracking problems.

3.2.3 Scalar Problems

A special class of linear quadratic problems concerns the evolution in a $1d$-dimensional phase space. In this case all vectors and matrices degenerate to simple scalar values. Especially, the differential Ricatti equation is now given by

$$\dot{G} + 2AG - \frac{B^2}{R} G^2 + Q = 0 \quad \text{with} \quad G(T) = \Omega \, . \tag{3.72}$$

This equation is the scalar Ricatti equation, originally introduced by J.F. Ricatti (1676–1754). A general solution of (3.72) is unknown. But if a particular solution $G^{(0)}$ of (3.72) is available, the Ricatti equation can be transformed by the map $G \to G^{(0)} + g$ into a Bernoulli equation

$$\dot{g} + 2\left(A - \frac{B^2}{R} G^{(0)} \right) g - \frac{B^2}{R} g^2 = 0 \, , \tag{3.73}$$

which we can generally solve. This is helpful as far as we have an analytical or numerical solution of (3.72) for a special initial condition.

We remark that some special elementary integrable solutions are available [10–12]. Two simple checks should be done before one starts a numerical solution [13]:

- If $B^2\alpha^2 = 2\alpha\beta AR + \beta^2 QR$ for a suitable pair of constants (α, β) then α/β is a special solution of the Ricatti equation and it can be transformed into a Bernoulli equation.
- If $(QR)'B - 2QRB' + 4ABQR = 0$, the general solution reads

$$G(t) = \sqrt{\frac{QR}{B^2}} \tanh\left(\int_0^t \frac{\sqrt{QR^{-1}}}{|B|} d\tau + C\right) . \tag{3.74}$$

An instructive example of a scalar problem is the temperature control in a homogeneous thermostat. The temperature follows the simple law

$$\dot{\vartheta} = -\kappa\vartheta + u , \tag{3.75}$$

where ϑ is the temperature difference between the system and its environment, u is the external changeable heating rate and κ is the effective heat conductivity. A possible optimal control is a certain stationary state given by $u^* = \kappa\vartheta^*$. Uncertainties in preparation of the initial state lead to a possible initial deviation $Y(0) = \vartheta(0) - \vartheta^*(0)$, which should be gradually suppressed during the time interval $[0, T]$ by a slightly changed control $u = u^* + w$. Thus, we have the linear evolution equation $\dot{Y} = -\kappa Y + w$, i.e., $A = -\kappa$ and $B = 1$. A progressive control means that the accuracy of the current temperature with respect to the desired value ϑ^* should increase with increasing time. This can be modeled by $Q = \alpha t/T$, $R = 1$, and $\Omega = 0$. We obtain the Ricatti equation

$$\dot{G} - 2\kappa G - G^2 + \frac{\alpha t}{T} = 0 \quad \text{with} \quad G(T) = 0 . \tag{3.76}$$

The solution is a rational expression of Ayri functions

$$G(t) = \frac{\widehat{\kappa}\mathcal{A}(x) + \mathcal{A}'(x) - C\widehat{\kappa}\mathcal{B}(x) - C\mathcal{B}'(x)}{\mathcal{A}(x) - C\mathcal{B}(x)} \tag{3.77}$$

with $\mathcal{A}$ and $\mathcal{B}$ the Ayri-A and the Ayri-B function, $\widehat{\kappa} = \kappa(T/\alpha)^{1/3}$ and $x = \widehat{\kappa}^2 + (\alpha/T)^{1/3}t$. The boundary condition defines the constant C

$$C = \frac{\widehat{\kappa}\mathcal{A}(\widehat{\kappa}^2 + \alpha^{1/3}T^{2/3}) + \mathcal{A}'(\widehat{\kappa}^2 + \alpha^{1/3}T^{2/3})}{\widehat{\kappa}\mathcal{B}(\widehat{\kappa}^2 + \alpha^{1/3}T^{2/3}) + \mathcal{B}'(\widehat{\kappa}^2 + \alpha^{1/3}T^{2/3})} . \tag{3.78}$$

In order to understand the corresponding control law $w^* = -GY^*$ and the optimal relaxation behavior of the temperature difference to the nominal state, see Fig. 3.5, we must be aware that the performance integral initially suppresses a strong heating or cooling. In other words, a very fast reaction on an initial disturbance cannot be expected. The first stage of the control regime is dominated by a natural relaxation following approximately $\dot{Y} = -\kappa Y$ because the contributions of the temperature deviations, $QY^2 \sim tY^2$, to the performance are initially small in comparison to the contributions of the control function Rw^2. The dominance of this mechanism increases with increasing heat conductivity κ. The subsequent stage is mainly the result of the control via (3.77). We remark that the final convergence of $G(t)$ to zero is a

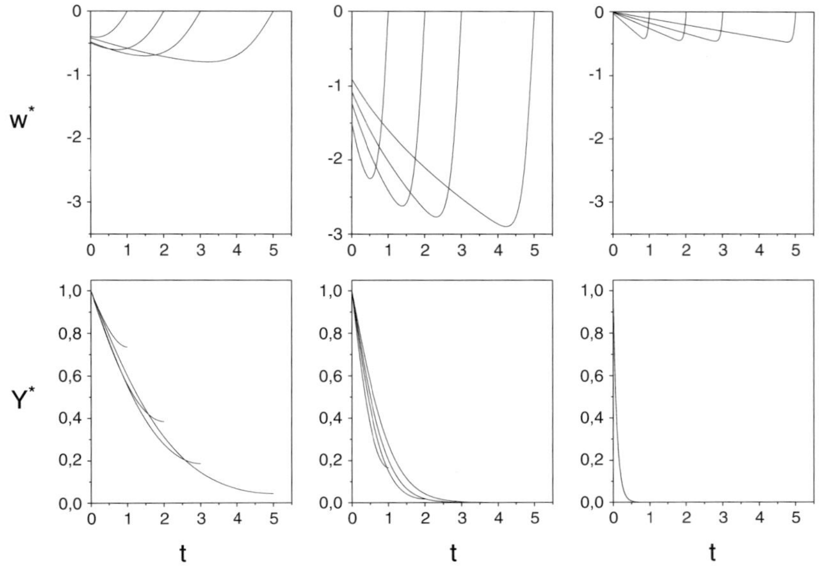

Fig. 3.5. Scalar thermostat: optimal control functions w^* (*top*) and optimal temperature relaxation Y^* (*bottom*) for different time horizons ($T = 1$, 2, 3, and 5. The initial deviation from the nominal temperature is $Y(0) = 1$. The parameters are $\kappa = 0$, $\alpha = 1$ (*left*), $\kappa = 0$, $\alpha = 10$ (*center*) and $\kappa = 10$, $\alpha = 10$ (*right*)

consequence of the corresponding boundary condition. The consideration of a nonvanishing end point contribution to the performance allows also other functional structures.

3.3 The Optimal Regulator

3.3.1 Algebraic Ricatti Equation

A linear quadratic problem with an infinite time horizon and with both the parameters of the linear system and the parameters of the performance functional being time-invariant is called a linear regulator problem [14]. Obviously, the resulting problem is a special case of the previously discussed linear quadratic problems. The independence of the system parameters on time offers a substantial simplification of the required mathematical calculus. Hence, optimal regulator problems are well established in different scientific fields and commercial applications [7, 15].

The mathematical formulation of the optimal regulator problem starts from the performance functional with the infinitely large control horizon

$$J_0[Y, w] = \frac{1}{2} \int_0^\infty \mathrm{d}t \, [Y(t)QY(t) + w(t)Rw(t)] \to \inf \tag{3.79}$$

to be minimized and the linear evolution equations (3.41) with constant coefficients

$$\dot{Y}(t) = AY(t) + Bw(t) \; . \tag{3.80}$$

By no means can the extension of a linear quadratic problem with a finite horizon to the corresponding problem with an infinitely large horizon be interpreted as a special limit case. The lack of a well-defined upper border requires also the lack of an endpoint contribution. To overcome these problems, we consider firstly a general performance

$$J[Y, w, t_0, T] = \frac{1}{2} \int_0^T \mathrm{d}t \, [Y(t)QY(t) + w(t)Rw(t)] + \frac{1}{2} Y(T)\Omega Y(T) \tag{3.81}$$

with finite start and end points t_0 and T instead of functional (3.79). We may follow the same way as in Sect. 3.1.4 in order to obtain the control law (3.55), the evolution equations for the optimum trajectory (3.56), and the differential Ricatti equation (3.53). The value of the performance at the optimum trajectory using (3.55) becomes

$$\begin{aligned} J^* = J[Y^*, w^*, t_0, T] &= \frac{1}{2} \int_{t_0}^T \mathrm{d}t \, [Y^*QY^* + w^*Rw^*] + \frac{1}{2} Y(T)\Omega Y(T) \\ &= \frac{1}{2} \int_{t_0}^T \mathrm{d}t Y^* \left[Q + GBR^{-1}B^T G \right] Y^* \\ &+ \frac{1}{2} Y(T)\Omega Y(T) \; . \end{aligned} \tag{3.82}$$

From here, we obtain with (3.53) and (3.56)

$$\begin{aligned} J^* &= \frac{1}{2} \int_{t_0}^T \mathrm{d}t Y^* \left[-\dot{G} - A^T G - GA + 2GBR^{-1}B^T G \right] Y^* \\ &+ \frac{1}{2} Y(T)\Omega Y(T) \\ &= -\frac{1}{2} \int_{t_0}^T \mathrm{d}t \left[Y^*\dot{G}Y^* + \dot{Y}^*GY^* + Y^*G\dot{Y}^* \right] + \frac{1}{2} Y(T)\Omega Y(T) \\ &= -\frac{1}{2} \int_{t_0}^T \mathrm{d}t \frac{\mathrm{d}}{\mathrm{d}t} [Y^*GY^*] + \frac{1}{2} Y(T)\Omega Y(T) \end{aligned}$$

$$= \frac{1}{2} Y^*(t_0) G(t_0) Y^*(t_0) , \tag{3.83}$$

where the last step follows from the initial condition (3.54). We remark that this result is valid also for the general linear quadratic problem with time-dependent matrices. We need (3.54) for the application of a time-symmetry argument. The performance of the optimal regulator may be written as

$$J_0[Y^*, w^*] = J[Y^*, w^*, 0, \infty] . \tag{3.84}$$

Since the performance of the optimal regulator is invariant against a translation in time, we have

$$J_0[Y^*, w^*] = J[Y^*, w^*, 0, \infty] = J[Y^*, w^*, \tau, \infty] \tag{3.85}$$

for all initial times τ if uniform initial conditions, $Y(\tau) = Y_0$, are considered. Thus we obtain from (3.83) the relation

$$Y_0^* G(\tau) Y_0^* = \text{const} \quad \text{for} \quad -\infty < \tau < \infty . \tag{3.86}$$

Hence, we conclude that the transformation matrix G is time-independent. This requires that the differential Ricatti equation (3.53) degenerates to a so-called algebraic Ricatti equation [6]

$$A^T G + GA - GBR^{-1}B^T G + Q = 0 , \tag{3.87}$$

and the optimal control as well as the optimal trajectory is described by (3.55) and (3.56) with completely time-independent coefficients. Therefore, the optimal regular can be also interpreted as the mathematical realization of a static feedback strategy.

3.3.2 Stability of Optimal Regulators

If the algebraic Ricatti equation is solved, the dynamics of an optimal regulator is completely defined by the control law (3.55) and the dynamics of the state of the system (3.41). These both equations lead to the equation of motion of the optimal trajectory (3.56). An initially disturbed system should converge to its nominal state for sufficiently long times, i.e., we expect $Y^* \to 0$ for $t \to \infty$. This behavior has comprehensive consequences. If we justify a regulator in such a manner that (3.55) holds, the initial deviation as well as any later spontaneous appearing perturbation decreases gradually. The necessary condition for this intrinsic stability of the regulator is that the evolution equation of the optimal trajectory (3.56) is stable. That means the so-called transfer matrix D of the linear differential equation system

$$\dot{Y}^* = \left(A - BR^{-1}B^T G\right) Y^* = DY^* \tag{3.88}$$

must be positive definite.

Let us study the inverted, frictionless pendulum as an instructive example. The pendulum consists of a cart of mass M and a homogeneous rod of mass

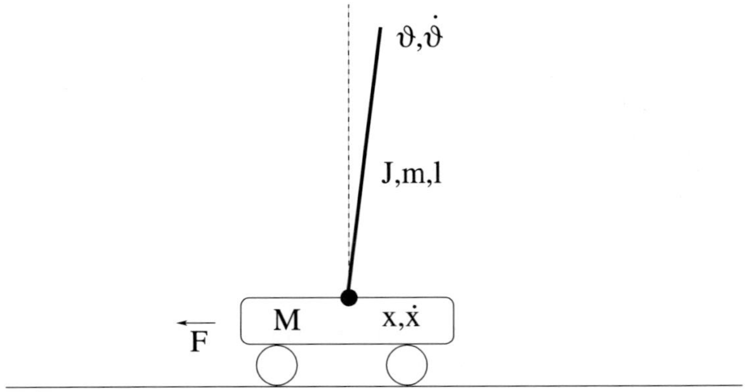

Fig. 3.6. The inverted pendulum problem

m, inertia J and length $2l$ hinged on the cart (Fig. 3.6). The cart may move frictionless under the external control force F along a straight line. Denoting with ϑ the angle between the rod and the vertical axis and with x the position of the cart, the equations of motion are given by

$$(M+m)\ddot{x} = ml(\dot{\vartheta}^2 \sin\vartheta - \ddot{\vartheta}\cos\vartheta) + F \tag{3.89}$$

$$(J+ml^2)\ddot{\vartheta} = mgl\sin\vartheta - ml\ddot{x}\cos\vartheta\ . \tag{3.90}$$

The stationary but instable solution of this problem, $\dot{\vartheta}^* = \dot{x}^* = x^* = F^* = 0$ and $x^* = \text{const.}$, may be our optimum solution. Now, we are interested in the control of small perturbations. To this aim we introduce the dimensionless quantities

$$y = \frac{M+m}{ml}(x - x^*), \quad \tau = \sqrt{\frac{m(M+m)gl}{(J+ml^2)(M+m) - m^2l^2}}\,t\ , \tag{3.91}$$

and

$$w = \frac{F}{mg}\sqrt{\frac{Jl^{-2}+m}{M+m}}\ , \tag{3.92}$$

and the system parameter

$$\varepsilon = \sqrt{\frac{M+m}{m}\frac{J+ml^2}{ml^2}}\ . \tag{3.93}$$

Thus, the linearized equations of motion are now $\ddot{x} = -\vartheta + w\varepsilon$ and $\ddot{\vartheta} = \vartheta - w/\varepsilon$. This leads us to the state vector $Y = (y, v, \vartheta, \omega)$ with $v = \dot{y}$ and $\omega = \dot{\vartheta}$. The control has only one component, namely, w. Hence, we get the matrices

$$A = \begin{pmatrix} 0 & 1 & 0 & 0 \\ 0 & 0 & -1 & 0 \\ 0 & 0 & 0 & 1 \\ 0 & 0 & 1 & 0 \end{pmatrix} \quad \text{and} \quad B = \begin{pmatrix} 0 \\ \varepsilon \\ 0 \\ -\varepsilon^{-1} \end{pmatrix}\ . \tag{3.94}$$

The matrix A is unstable, i.e., there exists some positive eigenvalues. Although this example seems to be very simple, a numerical solution [16] of the algebraic Ricatti equation is required for a reasonable structure of the quadratic performance functional. The main problem is that the nonlinear Ricatti equation has usually more than one real solution. However, the criterion to decide which solution is reasonable follows from the eigenvalues of the transfer matrix $D = A - BR^{-1}B^TG$.

Inverted pendulum systems are classical control test rigs for verification and practice of different control methods with wide ranging applications from chemical engineering to robotics [17]. Of course, the applicability of the linear regulator concept is restricted to small deviations from the nominal behavior. It is a typical feature of linear optimal regulators that they can control the underlying system only in a sufficiently close neighborhood of the equilibrium or of another nominal state. However, the inverted pendulum or several modifications [18,19], e.g., the rotational inverted pendulum, the two-stage inverted pendulum, the triple-stage inverted pendulum or more general a multi-link-pendulum, are also popular candidates for the check of several nonlinear control methods. However, the investigation of such problems in beyond the scope of this book. For more information, we refer the reader to the comprehensive literature [20–25].

In principle, one can also invert the optimal regulator problem, i.e., we ask for the performance which makes a certain controller to an optimum regulator. The first step, of course, is now the creation of a regulator as an abstract or real technological device. We assume that the regulator stabilizes the system. This is not at all a trivial task, but this problem concerns the wide field of modern engineering [1, 26–29]. The knowledge of the regulator is equivalent to the knowledge of the transfer matrix $D = \left(A - BR^{-1}B^TG\right)Y^*$. The remaining problem consists now in finding the performance index to which the control law of the control instrument is optimal. This problem makes sense because the structure of Q allows us to determine the weight of the degrees of freedom involved in the control process [30].

3.4 Control of Linear Oscillations and Relaxations

3.4.1 Integral Representation of State Dynamics

Oscillations

Oscillations are a very frequently observed type of movement. In principle, most physical models with a well-defined ground state can be approximated by the so-called harmonic limit. This is, roughly spoken, the expansion of the potential of the system in terms of the phase space coordinates $X = \{X_1, X_2, \ldots, X_N\}$ up to the second-order around the ground state or equilibrium state[4]. This physically pronounced state can be interpreted as the nominal state of a possible control theory. Without any restriction we may identify the origin of the coordinate system with the ground state, $X^* = 0$. This expansion leads to a linear system of second-order differential equations

$$\ddot{X}_\alpha + \sum_{\beta=1}^{N} \Omega_{\alpha\beta} X_\beta = 0 \quad \text{for} \quad \alpha = 1, \ldots, N \tag{3.95}$$

or in a more compact notation $\ddot{X} + \Omega X = 0$ with the frequency matrix[5] Ω. Of course, this linearization is an idealization of the real object. However, the linearized motion was thoroughly studied because of its wide applications. The harmonic theory is a sufficient and suitable approximation in many scientific fields, e.g., molecular physics and solid state physics or engineering. The influence of external forces $f_{i\alpha}$ requires the consideration of an inhomogeneous term in (3.95). Thus, this equation can be extended to a more generalized case

$$\ddot{X}_\alpha + \sum_{\beta=1}^{N} \Omega_{\alpha\beta} X_\beta = f_\alpha \,. \tag{3.96}$$

The force $f = \{f_1, f_2, \ldots, f_N\}$ can be interpreted as a superposition of driving forces from external, but noncontrollable, sources $\psi_\alpha(t)$ acting on each degree of freedom α and the contributions of N' possible control functions $u = \{u_1, u_2, \ldots, u_{N'}\}$ linearly coupled with the equations of motion

$$f_\alpha(t) = \psi_\alpha(t) + \sum_{\beta=1}^{N'} B_{\alpha\beta} u_\beta \,, \tag{3.97}$$

where B is a matrix of type $N \times N'$ with usually time-independent coefficients (Fig. 3.7). In principal, system (3.96) can be extended to the generalized system of linear differential equations

$$\widehat{D} X(t) = \widehat{M} f(t) \tag{3.98}$$

with the differential operators[6]

$$\widehat{D} = \sum_{k=1}^{n} a_k \frac{\mathrm{d}^k}{\mathrm{d}t^k} \quad \text{and} \quad \widehat{M} = \sum_{k=1}^{n} b_k \frac{\mathrm{d}^k}{\mathrm{d}t^k} \tag{3.99}$$

---

[4] Or another sufficiently strong pronounced stationary state.

[5] The frequency matrix is sometimes also denoted as the dynamical matrix.

[6] Of course, we may reduce the higher derivatives to first-order derivatives but this requires an extension of the phase space by velocities, accelerations, etc. This prolongation method is the standard procedure discussed in the previous chapters. However, in the present case such an extension of the phase space is not desirable.

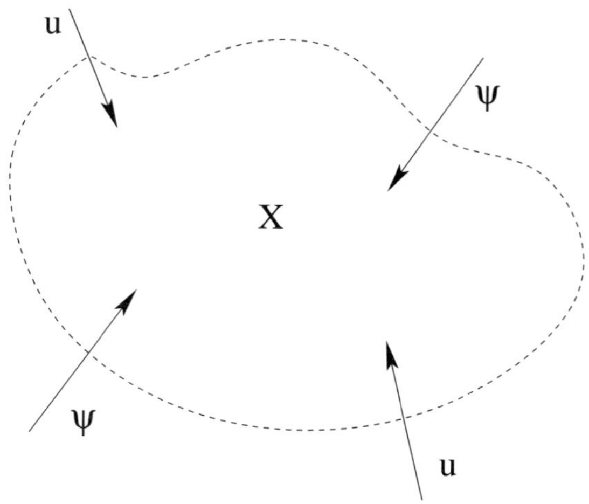

Fig. 3.7. External driving forces and control forces

The time-independent coefficients[7] a_k and b_k are matrices of the order $N \times N$. For instance, a vibrational system with the linear Newtonian friction has the operator

$$\widehat{D} = \frac{\mathrm{d}^2}{\mathrm{d}t^2} + \Lambda \frac{\mathrm{d}}{\mathrm{d}t} + \Omega \tag{3.100}$$

where the matrix Λ contains the friction coefficients. Equations of type (3.98) can be formally solved. The result is a superposition of a solution with zero external forces considering the initial state and a solution with a zero initial state considering the external forces

$$X(t) = \sum_{k=1}^{n} H_k(t) \left. \frac{d^{k-1}X(t)}{dt^{k-1}} \right|_{t=0} + \int_0^t H(t-\tau) f(\tau) d\tau \, . \tag{3.101}$$

The functions $H_k(t)$ and $H(t)$ are called the response functions of the system. These quantities are straightforwardly obtainable for example by application of the Laplace transform

$$\mathcal{A}(p) = \int_0^\infty \mathrm{d}t \exp\{-pt\} A(t) \, , \tag{3.102}$$

which especially yields a polynomial representation of the differential operators

$$\mathcal{D}(p) = \sum_{k=1}^{n} a_k p^k \quad \text{and} \quad \mathcal{M}(p) = \sum_{k=1}^{n} b_k p^k \, . \tag{3.103}$$

[7] A more generalized theory can be obtained for time-dependent coefficients. For the sake of simplicity we focus here only on constant coefficients.

From here we conclude that the Laplace transformed response functions are simple algebraic ratios of two polynomials, e.g., $\mathcal{H}(p) = \mathcal{D}(p)^{-1}\mathcal{M}(p)$.

Relaxation Processes

Obviously, (3.101) can be extended to all processes following generalized kinetic equations of the type

$$\widehat{D}X(t) + \int_0^t K(t-\tau)X(\tau)\mathrm{d}\tau = \widehat{M}f(t) \tag{3.104}$$

with a suitable memory kernel $K(t)$. Physically, the convolution term in (3.104) can be interpreted as a generalized friction indicating the hidden interaction of the relevant degrees of freedom of the system, collected in the state vector X, and other degrees of freedom constituting a thermodynamic bath. The causality of real physical processes requires always the upper limit t of the integral. A general difference between (3.104) and the time-local equation (3.98) is that the latter may be transformed always in a type of structure, but a time-local representation of (3.104) cannot be obtained with the exception of special cases. However, the integral representation of the solution of (3.104) is again (3.101) with the exception that the Laplace transform of the response function $H(t)$ is now given by

$$\mathcal{H}(p) = [\mathcal{D}(p) + \mathcal{K}(p)]^{-1}\,\mathcal{M}(p)\,. \tag{3.105}$$

Evolution equation with memory terms are very popular in several fields of modern physics, for example, condensed matter science, hydrodynamics, and the theory of complex systems. The processes underlying the dynamics of glasses [32–34] or the folding of proteins [35] are typical examples with a pronounced memory. In particular, we can observe a stretched exponential decay

$$K(t) \sim \exp\{-\lambda t^{\gamma}\} \quad \text{and} \quad \gamma < 1 \tag{3.106}$$

close to the glass transition of supercooled liquids [31, 36, 37].

The memory kernel can be determined by several theoretical and experimental methods. Well established theoretical concepts are perturbation techniques in the framework of the linear response theory [38, 40] or the calculus of Green's functions [39], or mode-coupling approaches [31, 33, 36], while various dielectric [42] and mechanical [41] methods as well as x-ray or neutron scattering [43] are available for the experimental detection of memory effects.

Fractional Derivations and Integrals

A very compact representation of a special class of memory kernels is provided by the fractional calculus [44]. Under certain conditions fractional integrals and derivatives can capture parsimoniously long-run structures [45] that decay very slowly. The advantage of fractional derivatives is that these involve only one parameter, the fractional order α of the derivation. This makes fractional derivatives and integrals an interesting candidate for several theoretical approaches.

The fractional integral can be introduced in the following way. First of all we define the nth order integral over a given function as the inverse n-order derivative $\widehat{\mathrm{d}}_n = (\mathrm{d}/\mathrm{d}t)^n$

$$\begin{aligned}\widehat{\mathrm{d}}_{-n} f(t) &= \int_0^t \mathrm{d}\xi_1 \int_0^{\xi_1} \mathrm{d}\xi_2 \ldots \int_0^{\xi_{n-1}} \mathrm{d}\xi_n f(\xi_n) \\ &= \frac{1}{(n-1)!} \int_0^t \mathrm{d}\xi \, (t-\xi)^{n-1} f(\xi) \, . \end{aligned} \tag{3.107}$$

In fact, the application of $\widehat{\mathrm{d}}_n$ on $\widehat{\mathrm{d}}_{-n} f(t)$ yields $\widehat{\mathrm{d}}_n \widehat{\mathrm{d}}_{-n} f(t) = f(t)$, or more general we have the group property $\widehat{\mathrm{d}}_n \widehat{\mathrm{d}}_{-m} = \widehat{\mathrm{d}}_{n-m}$. Now, we generalize the integral to a noninteger n and obtain the Riemann–Liouville fractional integral of order a which corresponds to a derivative of order $-a$

$$\widehat{\mathrm{d}}_{-a} f(t) = \frac{1}{\Gamma(a)} \int_0^t \mathrm{d}\xi \, (t-\xi)^{a-1} f(\xi) \, , \tag{3.108}$$

where $\Gamma(a)$ is the Gamma function. The operator $\widehat{\mathrm{d}}_a$ is called the fractional differential operator. We remark that the classical distinction between integrals and derivatives disappears in the fractional calculus. Obviously, (3.108) is a representation of the memory term with the long-run memory $K(t) \sim t^{a-1}$.

Let us finally point out the role of fractional derivations in the description of the time evolution of dynamical systems. By no means, the fractional calculus applied to physical problems provides other physical basic laws. But it is indeed possible to use the fractional calculus as a completely consistent theory of time evolution of dynamical systems associated with algebraic memory terms.

3.4.2 Optimal Control of Generalized Linear Evolution Equations

As mentioned above, oscillations and linear relaxations correspond to small changes of the state variables X against a stable equilibrium or ground state of the system. Therefore, we may use the same arguments as in Sect. 3.1.2 to introduce a quadratic or linear quadratic performance functional. Then we are able to construct a generalized action

$$S = J[X, u] + S_{\mathrm{constr}} + \sum_{i=1}^{I} \lambda_i G_i + \sum_{i=1}^{K} \overline{\lambda}_i \overline{G}_i \tag{3.109}$$

with the contributions of constraints (3.101)

$$S_{\mathrm{constr}} = \int_0^T \mathrm{d}t P(t) \left[X(t) - X_H(t) - \int_0^t H(t-\tau) f(\tau) \mathrm{d}\tau \right] \tag{3.110}$$

containing the momentum $P(t)$ as a functional Lagrange multiplier. The homogeneous solution $X_H(t)$ is given by

$$X_H(t) = \sum_{k=1}^{n} H_k(t) \left. \frac{\mathrm{d}^{k-1} X(t)}{\mathrm{d}t^{k-1}} \right|_{t=0} . \tag{3.111}$$

The third term of (3.109) may consider I isoperimeter links

$$G_i = \int_0^T \mathrm{d}t g_i(t, X(t)) = 0 \quad i = 1, \ldots, I , \tag{3.112}$$

while the last term is a consequence of the possible point constraints

$$\overline{G}_i = \overline{g}_i(t_i, X(t_i)) = 0 \quad 0 < t_i < T \quad i = 1, \ldots, K , \tag{3.113}$$

which are coupled to the generalized action via the Lagrange multipliers λ_i and $\overline{\lambda}_i$. The structure of the generalized action allows us to determine the Lagrangian L, which is given by

$$\begin{aligned} L = {} & \frac{1}{2}\left[X(t)Q(t)X(t) + u(t)R(t)u(t)\right] + \frac{1}{2} Y(T) \Omega Y(T) \delta(t-T) \\ & + P(t) \left[X(t) - \sum_{k=1}^{n} H_k(t) \left. \frac{\mathrm{d}^{k-1} X(t)}{\mathrm{d}t^{k-1}} \right|_{t=0} - \int_0^t H(t-\tau) f(\tau) \mathrm{d}\tau \right] \\ & + \sum_{i=1}^{I} \lambda_i g_i(t, X(t)) + \sum_{i=1}^{K} \overline{\lambda}_i \overline{g}_i(t, X(t)) \delta(t - t_i) . \end{aligned} \tag{3.114}$$

The main difference with the Lagrangian discussed in Sect. 2.4.1 is the lack of the velocities $\dot{X}(t)$. On the other hand, we have the disadvantage that the Lagrangian is no longer local in time. For the sake of simplicity, we now focus on a problem without constraints of type (3.112) and (3.113) and with vanishing initial conditions for the state vector X[8]. Then, the variation of the generalized action with respect to the momentum P leads again to the dynamic equation

$$X^*(t) = \int_0^t H(t-\tau)\left[\psi(\tau) + Bu^*(\tau)\right] \mathrm{d}\tau \tag{3.115}$$

[8] We remark that the consideration of isoperimetric constraints, point constraints, and nonvanishing initial conditions for the state vector does not change the below-discussed general relation between the state dynamics and the control functions.

for the optimum trajectory. Note that we have substituted the forces by the superposition (3.97) of external sources and control forces. The variation with respect to the state X yields

$$P^*(t) = -Q(t)X^*(t) , \tag{3.116}$$

while the variation with respect to the control functions gives

$$u^*(t) = R^{-1}(t) \int_0^T B^T H^T(\tau - t) P^*(\tau) d\tau . \tag{3.117}$$

Finally, the substitution of (3.116) in (3.117) eliminates the momentum from the control equations, and we obtain

$$u^*(t) = -R^{-1}(t) \int_t^T B^T H^T(\tau - t) Q(\tau) X^*(\tau) \mathrm{d}\tau . \tag{3.118}$$

This relation and (3.115) are a complete set of integral equations determining the corresponding control problem. It is interesting to remark that (3.115) requires the history of all controls in order to determine the current state while the current control follows from the future evolution of the state; see (3.118) and Fig. 3.8. This does not contradict the physically necessary causality because the optimal trajectory and the optimal control are defined by the complete knowledge of the evolution of the environment, given by $\psi(t)$, and the control parameters, $Q(t)$ and $R(t)$, over the whole period $[0, T]$. Finally, the substitution of (3.115) in (3.118) leads to a closed relation between the external forces and the optimum answer via the control functions

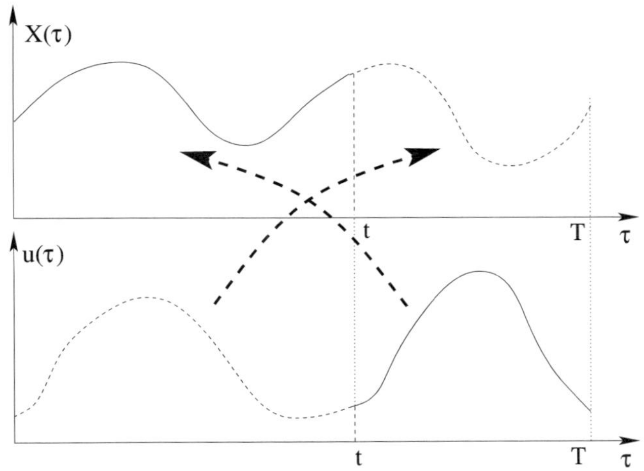

Fig. 3.8. Schematic representation of causality and anticausality in the relations between the optimal trajectory and the optimal control function

$$R(t)u^*(t) + \int_0^T \mathrm{d}\tau B^T U(t,\tau)\, Bu^*(\tau) = -\int_0^T \mathrm{d}\tau B^T U(t,\tau)\, \psi(\tau) \qquad (3.119)$$

with

$$U(t,\tau) = \int_{\max(t,\tau)}^T \mathrm{d}\tau' H^T(\tau' - t) Q(\tau') H(\tau' - \tau) \, . \qquad (3.120)$$

The integral equation (3.119) is of the Fredholm type. Although the theory of the class of integral equations is well established, a quantitative solution of (3.119) usually needs a numerical support.

We end this chapter with an important remark concerning the estimation of the performance. As we have stressed repeatedly, the performance or cost functional has no strict physical meaning, in contrast to the constraints, i.e., the equations of motion for the system state. That means in the case of a quadratic performance, the weight matrices, Q and R, must be chosen empirically with respect to the control aim. But this is, with the exception of the formulation of the classical mechanics as a special control problem, by no means a pure physical task. Economical and technical factors may be as important as physical arguments. An often used performance is the time-averaged energy of the system [46], but the power of heat loss [47] or the injected energy from a certain external source are also reasonable cost functionals.

3.4.3 Perturbation Theory for Weakly Nonlinear Dynamics

If a sufficiently complex system contains nonlinear terms in its equations of motion, then it is, as a rule, often impossible to determine a closed analytical solution. In this case, an approximative solution may be helpful. The standard method is a perturbation expansion, which is also denoted by the concept of successive approximations. Let us add a small nonlinear term to the evolution equation (3.98) of the state X,

$$\widehat{\mathrm{d}} X(t) + \varepsilon \Phi(t, X(t)) = f(t) \, . \qquad (3.121)$$

For the sake of simplicity, we have set $\widehat{M} = 1$. The modification of (3.121) leads to slightly changed integral equations for the optimum control, namely

$$X^*(t) = \int_0^t H(t-\tau) \left[\psi(\tau) - \varepsilon \Phi(\tau, X^*(\tau)) + Bu^*(\tau)\right] \mathrm{d}\tau \qquad (3.122)$$

for the optimum trajectory, and

$$P^*(t) = -Q(t)X^*(t) + \varepsilon \left. \frac{\partial \Phi^T(t,X)}{\partial X} \right|_{X=X^*(t)} \int_t^T H^T(\tau - t) P^*(\tau) d\tau \qquad (3.123)$$

for the momentum. The equation for the control function remains unchanged

$$u^*(t) = R^{-1}(t) \int_{\tau}^{T} B^T H^T(\tau - t) P^*(\tau) \mathrm{d}\tau \,. \tag{3.124}$$

The set of equations (3.122), (3.123), and (3.124) can now be treated in terms of a perturbation theory. The first step is the calculation of the reference solution (P_0^*, X_0^*, u_0^*) corresponding to $\varepsilon = 0$. In a subsequent step, these solutions are substituted into the right-hand side of the three equations considering now the small perturbation. Then, the left-hand side of (3.122), (3.123), and (3.124) yield the first approximation[9] (P_1^*, X_1^*, u_1^*). The successive repetition of this procedure leads eventually to a convergence of the series (P_i^*, X_i^*, u_i^*) $(i = 0, 1, 2, \ldots)$ to the solution of systems (3.122), (3.123), and (3.124). The convergence of this successive approximation method depends on the underlying system. It is still an open problem whether a series converges or not. Some necessary and sufficient conditions can be found in the literature [48–51]. However, the first-order correction is often a sufficient quantitative estimation of the effects due to a small nonlinearity in the evolution equations of the system.

Finally, we remark that a large class of weak nonlinear control problems is related to bilinear systems. They represent large number of real world phenomena [53–55]. Bilinear control systems are described by the following evolution equation:

$$\sum_{\beta=1}^{N} \widehat{\mathrm{d}}_{\alpha\beta} X_\beta + \sum_{\beta=1}^{N} \sum_{\gamma=1}^{N'} \Gamma_{\alpha\beta}^{\gamma} X_\beta u_\gamma = \psi_\alpha(t) + \sum_{\gamma=1}^{N'} B_{\alpha\gamma} u_\gamma \,. \tag{3.125}$$

The bilinear contributions, $X_\beta u_\gamma$, distinguish (3.125) from the above-discussed linear ones. Although bilinear systems are also a special case of nonlinear interacting systems, bilinear systems were brought to attention. A first comprehensive study was given by Wiener [52], who believed that such equations are the essence of understanding the behavior of neural networks. Very import problems, such as the control of nuclear reactors [56], of the dynamics of heat exchange [57], or of the induction motor systems [58], and rotary multi motor systems [59] were successfully described by bilinear systems.

Although the structure of (3.125), does not seem to be very complicated, the control equations of bilinear systems are very hard to handle. In general, there exists no analytical solution. However, it has been shown [60] that the method of successive approximations converges to a stable solution under relatively weak conditions.

[9] That means the reference state plus the first-order correction in the language of the perturbation theory.

References

1. M. Caccamo, L. Y. Zhang, L. Sha and G. Buttazzo: 'An Implicit Prioritized Access Protocol for Wireless Sensor Networks' in *Proceedings of the 23rd IEEE Real-Time Systems Symposium (RTSS'02), Austin, Texas, USA, December 3-5, 2002.* (IEEE Computer Society Press, Los Alamitos CA, 2002) page 39–48.
2. V.I. Arnold: *Geometrische Methoden in der Theorie gewöhnlicher Differentialgleichungen* (Deutscher Verlag der Wissenschaften, Berlin, 1987)
3. C.L. Siegel: Ann. Math. **46**, 423 (1945)
4. S. Bittani, A.J. Laub, J.C. Willems (eds): *The Ricatti Equation* (Springer, Berlin Heidelberg New York, 1991)
5. B.A. Francis, W.M. Wonham: Automatica **12**, 457 (1976)
6. P. Lancaster, L. Rodman: *Algebraic Ricatti Equations* (Oxford Science Publications, Oxford, 1995)
7. A. Locatelli: *Raccolta di problemi di controllo attimo* (Pitagora, Bologna, 1989)
8. V.M. Mehrmann: *The Autonomous Linear Quadratic Control Problem: Theory and Numerical Solutions* (Springer, Berlin Heidelberg New York, 1991)
9. A. Saberi, B.M. Chen, P. Sannuti: *Loop Transfer Recovery Analysis and Design* (Springer, Berlin Heidelberg New York, 1993)
10. M. Kourensky: Proc. London Math. Soc. **24**, 202 (1926)
11. R. Lagrange: Bull. Soc. Math. France **66**, 155 (1938)
12. L. Tchacaloff: Giornale Mat. **63**, 139 (1925)
13. E. Kamke: *Differentialgleichungen, Lösungsmethoden und Lösungen* (Akademische Verlagsgesellschaft, Leipzig, 1951)
14. M.G. Safonov, M. Athans: Gain and phase margin for multiloop LQG regulators. IEEE Trans. Automat. Control **AC 22**, 173 (1977)
15. A. Locatelli: *Elementi di controllo ottimo* (CLUP, Milano, 1987)
16. P. Rudra: *Getting Started with Matlab 5: A Quick Introduction for Scientists and Engineers* (Oxford University Press, New York, 1999)
17. M.H. Raibert: *Legged Robots that Balance* (MIT Press, Cambridge, MA, 1997)
18. S. Yurkovich, M. Widjaja: IFAC Control Eng. Pract. **4**, 445 (1996)
19. M. Widjaja, S. Yurkovich: Intelligent control for swing up and balancing of an inverted pendulum systems. In: *Proceedings of the IEEE International Conference on Control Applications*, Albany, NY, September 28–29, 1995, p. 534
20. C.C. Chung, J. Hauser: Automatica **31**, 851 (1995)
21. K.J. Astrom, K. Furuta: Automatica **36**, 287 (2000)
22. H. Nijmeijer, A.J. van der Schaft: *Nonlinear Dynamical Control Systems* (Springer, Berlin Heidelberg New York, 1990)
23. A. Isidori: *Nonlinear Control Systems* (Springer, Berlin Heidelberg New York, 1989)
24. H.-F. Chen, D.-Z. Cheng, J.-F. Zhang, A. Isidori, C.V. Hollot: *Nonlinear System II, Optimal Control Proceedings of the 14th World Congress of IFAC* (Pergamon Press, New York, 1999)
25. S.H. Strogatz: *Nonlinear Dynamics and Chaos* (Addison-Wesley, Reading, MA, 1994)
26. M. Caccamo, L.Y. Zhang: J. Embedded Comput. **1**, 2 (2004)
27. E. Cox: *The Fuzzy Systems Handbook: A Practitioner's Guide to Building, Using, and Maintaining Fuzzy Systems* (Academic, New York, 1993)

28. T.L. Crenshaw, A. Tirumala, S. Hoke, M. Caccamo: A robust implicit access protocol for real-time wireless collaboration. In: *Proceedings of the IEEE Euromicro Conference on Real-Time Systems*, Palma de Mallorca (2005)
29. J. Yen, R. Langari, L. A. Zadeh: *Industrial Applications of Fuzzy Logic and Intelligent Systems* (IEEE Press, Los Alamitos CA, 1995).
30. A. Locatelli: *Optimal Control* (Birkhäuser, Basel, 2001)
31. M. Einax, M. Schulz: J. Chem. Phys. **115**, 2282 (2001)
32. W. Götze, L. Sjögren: J. Non-Cryst. Solids **131–133**, 161 (1991)
33. W. Götze, L. Sjögren: Rep. Prog. Phys. **55**, 241 (1992)
34. U. Balucanu, M. Zoppi: *Dynamics of the Liquid State* (Clarendon Press, Oxford, 1994)
35. U.H.E. Hansmann, L.T. Wille: Phys. Rev. Lett. **88**, 068105 (2002)
36. E. Leutheusser, Phys. Rev. A **29**, 2765 (1984)
37. S.A. Brawer: *Relaxation in Viscous Liquids and Glasses* (American Ceramic Society, New York, 1983)
38. R. Balescu: *Equilibrium and Nonequilibrium Statistical Mechanics* (Wiley, New York, 1975)
39. S. Doniach, E.H. Sondheimer: *Green's Functions for Solid State Physicists* (Imperial College Press, London, 1998)
40. D. Zubarev, D. Zubarev, G. Röpke: *Statistical Mechanics of Nonequilibrium Processes: Basic Concepts, Kinetic Theory* (Akademie-Verlag, Berlin, 1996)
41. E. Donth: *Relaxation and Thermodynamics in Polymers* (Akademie-Verlag, Berlin, 1992)
42. N.G. McCrum, B.E. Read, G. Williams: *Anelastic and Dielectric Effects in Polymeric Solids* (Wiley, New York, 1967)
43. E. Bartsch, O. Debus, F. Fujara, M. Kiebel, W. Petry, H. Sillescu, J.H. Magill: Physica B **180–181**, 808 (1992)
44. R. Hilfer: Fractional Derivatives in static and dynamics scaling. In: *Scale Invariance and Beyond* ed by B. Dubrulle, F. Graner, D. Sornette (EDP Sciences and Springer, Berlin Heidelberg New York, 1997), pp. 54–62
45. J. D. Hamilton *Time Series Analysis* (Princeton University Press, Princeton, New Jersey, 1994).
46. V.L. Veitz, M.Z. Kolovsky, A.E. Kotchura: *Dynamics of Guided Machine Units* (Nauka, Moscow, 1984)
47. A.M. Ashavskii, A.Y. Sheinbaum: *Power Impuls Systems* (Mashinostroenie, Moscow, 1978)
48. A. Kovaleva: *Optimal Control of Mechanical Oscillations* (Springer, Berlin Heidelberg New York, 1999)
49. M.Z. Kolovsky: Prikl. Mat. Mekh. **4**, 738 (1960)
50. H. Kwakernaak, R. Sivan: *Linear Optimal Control Systems* (Whiley-Interscience, New York, 1972)
51. J.G. Malkin: *Theorie der Stabilität einer Bewegung* (Oldenbourg, München, 1959)
52. N. Wiener: *Cybernetics* (MIT Press, Cambridge, 1948)
53. R. Mohler: *Nonlinear Systems—Applications to Bilinear Control* (Prentice-Hall, Englewood Cliffs, NJ, 1991)
54. M. Sundareshan, R. Fundkowski: IEEE Trans. Automat. Control **31**, 1022 (1986)
55. M. Espana, I. Landau: Automatica **14**, 345 (1978)

56. R. Mohler, W. Koludziej: IEEE Trans. Syst., Man Cybernet. **10**, 683 (1980)
57. C. Bruni, D. DiPillo, G. Koch: Richerche di Automatica **2**, 11 (1971)
58. G. Figalli, M. Cava, L. Tomasi: Int. J. Control **39**, 1007 (1984)
59. L. Guo, A. Schone, X. Ding: Automatica **30**, 1445 (1994)
60. W. Cebuhar, V. Constanza: J. Optim. Theory Appl. **43**, 615 (1984)

4

Control of Fields

4.1 Field Equations

4.1.1 Classical Field Theory

Euler–Lagrange Equations

The theoretical representation of classical mechanics and field theory is very similar in some ways. This also applies to the axiomatic structure. It is well known that essential aspects of the classical mechanics can be translated into the language of field theory. Especially, there exist equivalent relations between the canonical formalism of the field theory and of the classical mechanics with respect to the Hamilton principle and the Euler–Lagrange equations. Let us briefly study the basic ideas of the variational principle for classical fields, in particular, with respect to some specific features occurring in field theoretical equations. This is not the place to discuss the mathematical and physical details of this well-established theoretical concept. The interested reader may find more information in the literature [1–4]. But this very short overview should give at least a guideline for the following chapter, namely for the translation of the previously demonstrated variational method for the control of systems with a finite set of degrees of freedom to the variational method for the control of fields.

Of course, we must extend the terms of classical mechanics in such a way that they become suitable for a field theory. The basic quantities of the Newtonian mechanics are the positions of the particles, or more general the set of degrees of freedom which are collected in the state vector $X(t)$. All admissible states are embedded in a finite-dimensional phase space $\mathbb{P}$.

The state vector is now generalized to a field function

$$X(t) \to \Psi(\mathbf{r}, t) , \tag{4.1}$$

which is defined for each time t and each site $\mathbf{r}$ of a given d-dimensional space $\mathbb{R}^d$. The knowledge of the geometry and topology of this space and any

M. Schulz: *Control Theory in Physics and other Fields of Science*
STMP **215**, 93–121 (2006)

information about a possible fusion with the time coordinate to a continuous space–time structure as they are known from all modern field theories are not important for the variational principle. These questions become relevant if we are interested in solving a special field theoretical problem. Formally, the field function $\Psi(\mathbf{r}, t)$ is an N-dimensional vector describing the N components of the field at the spatial position $\mathbf{r}$ at time t. The physical meaning of the field components Ψ_i $(i = 1, \ldots, N)$ and their relation to the space $\mathbb{R}^d$ or the space–time continuum depends on the theory in mind. Scalar field theories have only one field component[1], while the Maxwell theory requires a four-dimensional vector with well-defined symmetry relations and transformation rules between the field components considering the geometry of the underlying space–time continuum.

The main difference between the classical mechanics and a field theory is the dependence of the fields on the spatial coordinates. That means, a field theory consists of an uncountable infinitely large set of degrees of freedom[2], in contrast to the mechanics with a finite number of degrees of freedom.

The field theoretical action is defined as the integral

$$S = \int_{\Omega} \mathrm{d}^d r \mathrm{d}t \mathcal{L}\left(t, \mathbf{r}, \Psi, \nabla\Psi, \Psi_t\right) \tag{4.2}$$

over a so-called Lagrange density $\mathcal{L}$ with $\Psi_t = \partial\Psi/\partial t$. Here, Ω is a certain region of the whole space–time continuum which is defined by possibly time-dependent boundary conditions. The appearance of gradient terms $\nabla\Psi$ is a typical feature of a physical field theory reflecting the Nahwirkungsprinzip[3]. Furthermore, we may interpret the integral

$$L = \int_{\Omega} \mathrm{d}^d r \mathcal{L}\left(t, \mathbf{r}, \Psi, \nabla\Psi, \Psi_t\right) \tag{4.3}$$

as Lagrangian in order to obtain another analogy to the classical mechanics. However, in the framework of a field theory it is much more conventional to denote the Lagrange density as the Lagrangian.

In the classical mechanics, the Euler–Lagrange equations are the equations of motion of the underlying system. This statement is also true for a field theory. Usually, the Euler–Lagrange equations are called the field equations. The derivation of these equations follows from Hamiltonian's variational principle. Just as we have pointed out for the classical mechanics in

---

[1] Or two, in the case of a complex field.

[2] In the case of a lattice field theory, the degrees of freedom form a countable infinite large set or sometimes a finite set (for example on a lattice with a spherical topology). In this sense, lattice field theories may be interpreted as the link between mechanics and field theory.

[3] We remark that several field theories also contain higher derivatives in space and time. As an example we refer to Einstein's general relativity theory with a Lagrangian containing second-order derivatives of the field components.

Chap. 2.3, this profound principle requires that action (4.2) arrives the extremum $S[X,T] \to \text{extr.}$ for the solution of the field equations. The necessary condition for this demand is that the first variation with respect to all degrees of freedom, i.e., for all field components, vanishes:

$$\begin{aligned}\delta S &= \int_{\Omega} \mathrm{d}^d r \mathrm{d}t \left[\frac{\partial \mathcal{L}}{\partial \Psi}\delta\Psi + \frac{\partial \mathcal{L}}{\partial \nabla\Psi}\delta\nabla\Psi + \frac{\partial \mathcal{L}}{\partial \Psi_t}\delta\Psi_t\right] \\ &= \int_{\Omega} \mathrm{d}^d r \mathrm{d}t \left[\frac{\partial \mathcal{L}}{\partial \Psi}\delta\Psi + \frac{\partial \mathcal{L}}{\partial \nabla\Psi}\nabla\delta\Psi + \frac{\partial \mathcal{L}}{\partial \Psi_t}\frac{\partial}{\partial t}\delta\Psi\right] \\ &= 0 \end{aligned} \tag{4.4}$$

or

$$\begin{aligned} 0 = &\int_{\Omega} \mathrm{d}^d r \mathrm{d}t \left[\frac{\partial \mathcal{L}}{\partial \Psi} - \nabla\left(\frac{\partial \mathcal{L}}{\partial \nabla\Psi}\right) - \frac{\partial}{\partial t}\left(\frac{\partial \mathcal{L}}{\partial \Psi_t}\right)\right]\delta\Psi \\ &+ \int_{\Omega} \mathrm{d}^d r \mathrm{d}t \left[\nabla\left(\frac{\partial \mathcal{L}}{\partial \nabla\Psi}\delta\Psi\right) + \frac{\partial}{\partial t}\left(\frac{\partial \mathcal{L}}{\partial \Psi_t}\delta\Psi\right)\right] . \end{aligned} \tag{4.5}$$

The last term is a generalized divergence with respect to the space–time continuum. Therefore, this contribution can be transformed into a generalized surface integral. We assume again that the variations of the field vanish at the boundaries. But in contrast to the mechanical variations, where the boundary conditions are formulated with respect to the initial and final times, we now have to consider also spatial boundary conditions.

The treatment of the first term of (4.5) remains. Because this contribution must vanish for all admissible variations $\delta\Psi$ of the field components, one obtains the Euler–Lagrange equations

$$\frac{\partial}{\partial t}\left(\frac{\partial \mathcal{L}}{\partial \Psi_t}\right) + \nabla\left(\frac{\partial \mathcal{L}}{\partial \nabla\Psi}\right) = \frac{\partial \mathcal{L}}{\partial \Psi} \tag{4.6}$$

of the classical field theory. These equations, together with correctly defined boundary conditions, completely describe the evolution of the fields Ψ in the space-time continuum.

Klein–Gordon Field Equation

Let us derive the field equations in an explicit form for some standard fields. The basic assumption that fundamental physical laws take places in the same manner independently of the position and the current time requires that a well formulated Lagrangian of a free field need not depend explicitly on $\mathbf{r}$ and t. The simplest physically motivated example is the Klein–Gordon field. This is a scalar field, i.e., Ψ consists of only one real-valued[4] component $\phi(\mathbf{r}, t)$. The Lagrangian of the free Klein–Gordon field is given by

[4] The extension of the Klein–Gordon field to a complex-valued field is always possible and has a deep physical sense [3] when charged scalar fields are considered.

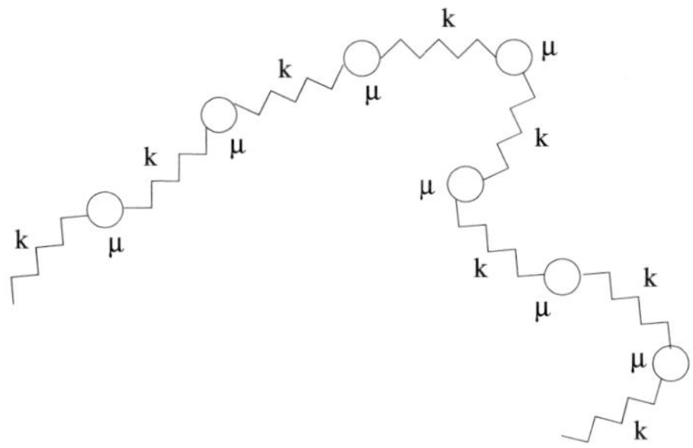

Fig. 4.1. Bead spring model

$$\mathcal{L}(\Psi, \nabla\Psi, \Psi_t) = \frac{1}{2}\left(\phi_t^2 - (\nabla\phi)^2 - m^2\phi^2\right) . \tag{4.7}$$

Inserting this into the Euler–Lagrange equation (4.6) we find the standard Klein–Gordon equation

$$\frac{\partial^2\phi}{\partial t^2} - \nabla^2\phi + m^2\phi = 0 . \tag{4.8}$$

Such a free field is not controllable because there is no coupling with an external changeable source[5]. We may introduce such a coupling by the consideration of the additional term $J\phi$ in the Lagrangian. Thus, we obtain the new Lagrangian

$$\mathcal{L}(t, \mathbf{r}, \Psi, \nabla\Psi, \Psi_t) = \frac{1}{2}\left(\phi_t^2 - (\nabla\phi)^2 - m^2\phi^2\right) + J\phi \tag{4.9}$$

which can now be explicitly dependent on the time and space coordinates, $\mathbf{r}$ and t, thanks to the general functional structure of $J = J(\mathbf{r}, t)$. The presence of the source term changes the homogeneous field equation (4.8) into the inhomogeneous field equation

$$\frac{\partial^2\phi}{\partial t^2} - \nabla^2\phi + m^2\phi = J(\mathbf{r}, t) . \tag{4.10}$$

The Klein–Gordon equation is a very instructive example to study the above-mentioned transition from a finite set to an infinitely large set of degrees of freedom which reflects the crossover from classical mechanics to the field theory. To this aim let us consider a classical harmonic chain (Fig. 4.1), a series of N beads of mass μ arranged in a line of length ℓ, each connected to the next via springs of strength k. The position of the i-th particle is then given by $i\ell/N + x_i$, where x_i is the displacement of the bead located at the ground state position $i\ell/N$. The Lagrangian of the classical system is therefore given by

---

5 The possibility of controling the field evolution by time-dependent boundary conditions remains. However, this is rather a rare case.

$$L = \frac{\mu}{2}\sum_{i=1}^{N}\dot{x}_i^2 - \frac{1}{2}\sum_{i=1}^{N-1} k\left(x_{i+1} - x_i\right)^2 . \tag{4.11}$$

Now let us assume that the number of mass points diverges while the total mass M and the total length ℓ of the chain is conserved, i.e., we write $N = \ell/\epsilon$ and $\mu = \varrho\epsilon$ where ϵ is a small length scale, $\epsilon \to 0$ and ϱ is the mass density. In the limit $\epsilon \to 0$, the Lagrangian becomes

$$\begin{aligned} L &= \frac{\rho}{2}\sum_{i=1}^{N}\epsilon\dot{x}_i^2 - \frac{k\epsilon}{2}\sum_{i=1}^{N-1}\frac{\left(x_{i+1} - x_i\right)^2}{\epsilon^2}\epsilon \\ &\to \frac{1}{2}\int_0^{\ell} dr\left[\rho\left(\frac{\partial\phi(r,t)}{\partial t}\right)^2 - \kappa\left(\frac{\partial\phi(r,t)}{\partial r}\right)^2\right] \end{aligned} \tag{4.12}$$

where we have taken the limit via

$$\begin{aligned} \epsilon &\to dx \\ k\epsilon &\to \kappa \\ i\ell/N = i\epsilon &\to r \\ x_i(t) &\to \phi(r,t) . \end{aligned} \tag{4.13}$$

Here, the field $\phi(r,t)$ is now the displacement of an infinitesimally small particle with respect to its equilibrium position r, and κ is the Young's modulus. If we use the Euler–Lagrange equation, we arrive at

$$\frac{\partial^2\phi(r,t)}{\partial t^2} - \frac{\kappa}{\varrho}\frac{\partial^2\phi(r,t)}{\partial r^2} = 0 \tag{4.14}$$

which is just the familiar wave equation for a one-dimensional system traveling with the velocity $(\kappa/\varrho)^{1/2}$. This equation is a special case of (4.8) with $m = 0$ and a rescaled time scale $t \to t\,(\kappa/\varrho)^{1/2}$.

The physical importance of the Klein–Gordon equation is mainly of quantum field theoretical character which is not the topic of this book. However, the degenerated case of a simple wave equation is especially often used also for modeling several control theoretical problems. Furthermore, the Klein–Gordon equation is a suitable candidate for fundamental studies of the behavior of controlled partial differential equations of a hyperbolic type [5–8].

Schrödinger Equation

Another interesting field equation is the Schrödinger equation. Although the main property of this equation is the foundation of the non-relativistic quantum mechanics, the Schrödinger equation itself is a field equation for the complex wave function $\psi(r,t)$. Here, we have the Lagrangian

$$\begin{aligned} \mathcal{L}\left(t, \mathbf{r}, \Psi, \nabla\Psi, \Psi_t\right) = &-\frac{\hbar^2}{2m}\left[\left|\nabla\psi\right|^2 + \frac{im}{\hbar}\left(\psi\psi_t^* - \psi^*\psi_t\right)\right] \\ &- V(\mathbf{r},t)\left|\psi\right|^2 \end{aligned} \tag{4.15}$$

with the potential $V(\mathbf{r},t)$. This Lagrangian is one of the few physically reasonable candidates of canonical field theories with an explicite dependence on space and time due to the potential. The application of the Euler–Lagrange equation (4.6) with respect to the two independent field components ψ^* and ψ yield the Schrödinger equation,

$$\hbar i \psi_t = -\frac{\hbar^2}{2m}\nabla^2\psi + V(\mathbf{r},t)\psi\,, \tag{4.16}$$

and the complex conjugated equation. In contrast to the Klein–Gordon equation which is a partial differential equation of a hyperbolic type, the Schrödinger equation is essentially a partial differential equation of the parabolic type. The control of the Schrödinger field takes place via the time dependent potential $V(\mathbf{r},t)$. It should be remarked that the origin of the potential is often the interaction with a classical electromagnetic field. In this case, $V(\mathbf{r},t)$ is defined by the scalar electromagnetic potential, while further terms containing the vector potential additionally occur in the Schrödinger equation. However, in contrast to the Klein–Gordon equation, the coupling between the Schrödinger field and external sources does not change the homogeneous character of the corresponding field equation.

Maxwell Equations

The third example which should be mentioned is the classical field theory of electromagnetism. Here, we have to consider a four-component field $\Psi = (\mathbf{A},\varphi)$ consisting of the scalar potential φ and the vector potential $\mathbf{A}$. The Lagrangian is given by

$$\mathcal{L}(\mathbf{E},\mathbf{B}) = \frac{1}{2}\left(\mathbf{E}^2 - \mathbf{B}^2\right) + \mathbf{jA} - \varphi\varrho\,, \tag{4.17}$$

where the electric field, $\mathbf{E}$, and the magnetic field, $\mathbf{B}$, are given by

$$\mathbf{E} = -\nabla\varphi - \frac{\partial \mathbf{A}}{\partial t} \quad \text{and} \quad \mathbf{B} = \operatorname{curl}\mathbf{A}\,. \tag{4.18}$$

The external sources are given by the arbitrarily changeable charge density $\varrho(\mathbf{r},t)$ and the electric current $\mathbf{j}(\mathbf{r},t)$. We remark that the three-dimensional fields $\mathbf{B}$ and $\mathbf{E}$ are not independent from each other. It is simple to check that (4.18) immediately requires

$$\operatorname{curl}\mathbf{E} + \frac{\partial \mathbf{B}}{\partial t} = 0 \quad \text{and} \quad \operatorname{div}\mathbf{B} = 0 \tag{4.19}$$

i.e., the first group of the Maxwell equations. The second group of field equations follows from the application of the Euler–Lagrange equations (4.6) on the Lagrangian (4.17), namely

$$\operatorname{div}\mathbf{E} = \varrho \quad \text{and} \quad \operatorname{curl}\mathbf{B} - \frac{\partial \mathbf{E}}{\partial t} = \mathbf{j}\,. \tag{4.20}$$

Another important phenomenon belongs to the construction of a control due to the external sources $\varrho(\mathbf{r},t)$ and $\mathbf{j}(\mathbf{r},t)$. These quantities are not completely independent from each other, because the second group of the Maxwell equations automatically requires the balance equation

$$\frac{\partial \varrho}{\partial t} + \operatorname{div} \mathbf{j} = 0 \tag{4.21}$$

as a constraint for the sources.

Maxwell equations are of great importance for the control of transmitting and receiving areals [9, 10], for the field controlled phase transition [10, 11] in magnetic materials, or for the optimization of laser beams [12, 13].

In addition to the few classical examples presented here, the modern physics knows a large set of further canonical field theories, e.g., the general gravitation theory or the theory of Dirac fields. But a detailed discussion of all these interesting theories would lead us too far. Here, we refer to the large fundus of textbooks and monographs [14–18].

4.1.2 Hydrodynamic Field Equations

There exists a large class of field theories besides the canonical ones. A characteristic property of these theories is that they cannot generally be reduced to a physically reasonable Lagrangian. A standard example of these types of field equations is the evolution equations for liquids and gases. The state of a liquid is mathematically described by the three-dimensional velocity field $\mathbf{v}(\mathbf{r},t)$ and two other thermodynamic quantities, for example, the local pressure $p(\mathbf{r},t)$ and the density $\varrho(\mathbf{r},t)$. Thus, the field state is defined by the five components of $\Psi = (\mathbf{v}, p, \varrho)$. Obviously, the use of these state variables as basic elements of a field theory requires a continuously smeared liquid. In other words, such a theory fails for too short length scales of an order of magnitude of the liquid molecules. On the other hand, the concept of a continuous smearing of a discrete structure to a field has a widespread application also for other field theoretic descriptions. Therefore, the transition from a discrete structure of a many-body problem to a continuous description is often called the hydrodynamic limit.

The five components of the hydrodynamic field Ψ require five field equations. The first equation is the mass balance

$$\frac{\partial \varrho}{\partial t} + \operatorname{div} \varrho\mathbf{v} = Q\,, \tag{4.22}$$

where the right-hand side describes the effects of possible positive and negative liquid sources injecting and removing matter. But these sources must be located at the surface of the hydrodynamic system so that the right-hand side contributions can be installed in the boundary conditions. The remaining bulk material balance always requires $Q = 0$.

The second field equation concerns the velocity field. This vector equation represents the balance of the momenta and consists of the components

$$\varrho\left(\frac{\partial v_\alpha}{\partial t}+\sum_{\beta=1}^{3} v_\beta \frac{\partial v_\alpha}{\partial r_\beta}\right) = -\frac{\partial p}{\partial r_\alpha}+f_\alpha + \sum_{\beta=1}^{3}\frac{\partial}{\partial r_\beta}\left[\eta\left(\frac{\partial v_\alpha}{\partial r_\beta}+\frac{\partial v_\beta}{\partial r_\alpha}\right)\right] + \frac{\partial}{\partial r_\alpha}\left[\left(\zeta-\frac{2}{3}\eta\right)\operatorname{div}\mathbf{v}\right] \tag{4.23}$$

These equations are also denoted as Navier–Stokes equations. They contain two material parameters, the shear viscosity η and the bulk viscosity ζ. These are generally dependent on the pressure p and the density ϱ. Furthermore, the force density components[6] f_α represent the influence of external forces, e.g., of the gravity field or also partially of electromagnetic fields. A control of the liquid dynamics via these forces is strongly restricted from a technological point of view.

The last equation which is necessary to complete the set of hydrodynamic equations is the entropy balance

$$\varrho T\left(\frac{\partial s}{\partial t}+\sum_{\beta=1}^{3} v_\beta \frac{\partial s}{\partial r_\beta}\right) = \sum_{\alpha,\beta=1}^{3}\eta\left[\left(\frac{\partial v_\alpha}{\partial r_\beta}\right)^2+\frac{\partial v_\alpha}{\partial r_\beta}\frac{\partial v_\beta}{\partial r_\alpha}\right] + \left(\zeta-\frac{2}{3}\eta\right)(\operatorname{div}\mathbf{v})^2+\operatorname{div}\,(\kappa\nabla T) \tag{4.24}$$

with the temperature T and the temperature-dependent heat conductivity κ. The temperature is not an independent quantity, because the thermodynamic state equations always require a well-defined representation $T = T(s, p)$. The entropy density s is a thermodynamic potential which may be expressed as a function of p and ϱ. Similar to the balance equations of mass and momenta, an additional input or output of entropy is only possible via the boundaries[7].

The hydrodynamic field equations are strongly nonlinear equations. Hence, they are treated almost by numerical techniques, e.g., by finite element methods [19–24]. Furthermore, the viscosity moduli and the heat conductivity indicate that the processes described by the hydrodynamic equations (4.22), (4.23), and (4.24) are irreversible ones. The control of hydrodynamic equations is an important, but largely still open, problem for chemical engineering and several related scientific fields. Some important problems are the control of turbulent effects during the process of combustion [25–28], the optimal

---

6 The force density is defined as force per unit volume.

7 A creation or annihilation of entropy inside the bulk may be possible by chemical reactions. But in this case, the hydrodynamic system must be extended by transport and balance equations for the reacting components. In other words, the equation for entropy now contains new source terms, but these contributions are controlled by further field equations with no bulk source terms but a possible input at the output via the surface of the hydrodynamic system.

control of mixing procedures, and reaction processes [29] in chemical plants or the optimal control of flow-transport [30,31] in contaminated natural and artificial lakes and rivers.

4.1.3 Other Field Equations

Vibrational Fields

The classical elasticity theory can be interpreted as the hydrodynamic limit of solids. An initial deformation of a given solid is followed by vibrations spreading the whole body. In contrast to the hydrodynamics, the basic equations describing this phenomena are almost linear equations. The basic quantity is the three-dimensional local deviation $\mathbf{h}(\mathbf{r}, t)$. This is simply the difference between the time-dependent position of an observed point of the solid from its equilibrium position $\mathbf{r}$. In other words, the general field function Ψ is now given by $\Psi = \mathbf{h}$. The corresponding field equations depend on the symmetry of the underlying solid. Therefore, the complete field equations for the vibrational problem contain a set of different moduli. We restrict, for the sake of simplicity, our brief overview on an isotropic solid. In this case, the field equations are given by

$$\varrho \frac{\partial^2 \mathbf{h}}{\partial t^2} - \frac{E}{2(1+\nu)} \left[\nabla^2 \mathbf{h} + \frac{1}{(1-2\nu)} \nabla \mathrm{div}\, \mathbf{h} \right] = \mathbf{f} \,, \tag{4.25}$$

where the Young modulus, E, and the Poisson modulus, ν, are material constants. Although a real vibrating system can be usually influenced only via its surface, it is, from a technical point of view, reasonable to introduce these effects as an external force field $\mathbf{f} = \mathbf{f}(\mathbf{r}, t)$. The field equation is a partial differential equation of the hyperbolic type. In fact, simple transverse and longitudinal plane waves are the basic solutions of (4.25). We remark that the consideration of additional viscosity terms in (4.25) leads to damped vibrations. Such generalizations are considered in several model equations, e.g., the Kelvin model, the Maxwell model, or the Maxwell–Kelvin model [32–34].

The optimal control of undamped or damped vibrations is very important for the stabilization of buildings and bridges [35–37], but also for the reduction of energy losses of machines, motors, and generators [38].

Continuous Reaction–Diffusion Processes

Another very important class of field equations belongs to the theoretical description of diffusion and reaction processes. Usually, systems in which reactants are transported by diffusion [39,40] are denoted as reaction–diffusion systems. Two fundamental time scales characterize these systems, namely the diffusion time as a typical time between collisions of reacting particles and the reaction time defining the inverse reaction rate of neighboring particles. When the reaction time is much larger then the diffusion time, the process follows approximately the classical kinetic equations. Such a process is reaction-limited while the opposite case is called a diffusion-limited processes [41–46]. But slow reaction processes may also be diffusion-limited. For instance, heterogeneous reaction systems form increasing product layers between the educt regions, so that a particle has to overcome this barrier by diffusion processes in order to find an appropriate reaction partner. Such a system shows initially an reaction-limited behavior. But with increasing thickness of the product layer the further evolution is dominated by a diffusion limited law. In particular, the structure of the educt-product interfaces at this late stage of the heterogeneous chemical reaction is strongly determined by local diffusion coefficients [47].

The corresponding fields of an N-component diffusion-reaction system are the local concentrations $c_\alpha(\mathbf{r},t)$ which form the field vector $\Psi = (c_1, c_2, \ldots, c_N)$. Diffusion and reaction processes are considered in continuous mean field equation for the particle concentrations which are of the type

$$\frac{\partial c_\alpha(\mathbf{r},t)}{\partial t} = \sum_{\beta=1}^{N} \nabla D_{\alpha\beta}(\Psi)\nabla c_\beta(\mathbf{r},t) + \sum_{\beta,\gamma=1}^{N} K_{\beta\gamma} c_\beta(\mathbf{r},t) c_\gamma(\mathbf{r},t) - \sum_{\delta=1}^{N} \overline{K}_{\delta\alpha} c_\delta(\mathbf{r},t) c_\alpha(\mathbf{r},t) \tag{4.26}$$

with the concentration-dependent diffusion coefficients $D_{\alpha\beta}(\Psi)$ and the kinetic coefficients $K_{\beta\gamma}$ and $\overline{K}_{\delta\alpha}$ characterizing the reaction rate of creation processes $[\beta]+[\gamma] \rightarrow [\alpha]+[\delta]$ and annihilation processes $[\delta]+[\alpha] \rightarrow [\beta]+[\gamma]$ with respect to the component α. Obviously, (4.26) considers only pair reactions which may be straightforwardly extended to multicomponent reactions.

Such diffusion–reaction equations are very important for understanding and modeling of an optimal control of contaminated soils by venting procedures [48–50], but also for the maximum suppression of undesirable interface reactions between chemically active solids [47].

A very-often analyzed special case is diffusion–reaction systems in which the concentrations of all components with the exception of only one are fixed or eliminated due to the presence of fast processes restoring the chemical equilibrium for the involved components[8]. In this case, one get the parabolic partial differential equation

$$\frac{\partial c(\mathbf{r},t)}{\partial t} = D\nabla^2 c(\mathbf{r},t) + ac(\mathbf{r},t) - \sum_{k\geq 2} b_k c^k(\mathbf{r},t) , \tag{4.27}$$

where the coefficients a and b_k depend on the underlying chemical processes. This equation is the starting point of a broad class of theoretical investigations concerning the time evolution of reaction–diffusion processes [51–54].

[8] This approximative procedure is also known as adiabatic elemination.

Nonlocal Field Theories

Field theories does not necessarily correspond to local partial differential equations. There are several physical reasons to formulate a physical field theory with nonlocal terms. The basic idea is always the finite velocity of propagation of excitations in materials. That means the response on a disturbance at a certain point $\mathbf{r}'$ and at time t' reaches another point $\mathbf{r}$ at a later time t. Evolution equations of the following type are well known:

$$\frac{\partial\varphi(\mathbf{r},t)}{\partial t} = F(\mathbf{r},t,\varphi(\mathbf{r},t)) + \int_0^t \mathrm{d}t' \int_G \mathrm{d}^d r' K(\mathbf{r}-\mathbf{r}',t-t')\varphi(\mathbf{r},t) \tag{4.28}$$

with a space–time-dependent memory term. Such equations are used for the modeling of active walker motions [55–58] of diffusion processes in glasses or supercooled liquids [60] or of the evolution of biological fields and related quantities [61–63]. A typical control problem analyzed repeatedly by several authors [64–66] is the heat conductivity in materials with memory terms. However, a systematic discussion of field equations with memory is still open.

4.2 Control by External Sources

4.2.1 General Aspects

Before we start the discussion of field control problems, we should briefly summarize the general concept of formulation of a control problem. The basic consideration is the existence of a set of evolution equations or equivalent relations describing the dynamical state of the system. These equations have a clearly physical origin independent wether these equations describe the motion of particles or the movable components of a machine. But it is important that the evolution equations are controllable, i.e., they must contain terms which couple the system with its external environment. From a philosophical point of view, these equations characterize the objective part of the control problem.

The second important quantity is the performance or cost functional. This quantity defines the control aim and is primarily not of a physical origin. The performance is more or less a measure of penalty for the deviation from a desired ideal result of an experiment or from the desired ideal output of a process. Therefore, the cost functional is the subjective part of the control problem.

The procedures determining the optimal solution for a system under control generally yield three types of equations. The first group is the originally formulated evolution equations of the system state. These equations are identically reproduced by all the techniques discussed in the previous chapters. The second group is the set of the adjoint evolution equations describing the evolution of the generalized momenta[9] of the system. The third group is the set of control equations which connect the control functions with the dynamics of the state variables and the generalized momenta.

We expect for the control of fields an analogous situation. The control problem extends the field equations by the dynamics of the adjoint fields and the control fields. The main difference, except that the evolution equations are now partial differential equations, with the control of systems with a finite degree of freedom is the importance of the boundary conditions. While these conditions are defined as initial and final conditions in the mechanical approach to a control problem, the field theoretical approach also allows spatial boundary conditions. These conditions fix a general field solution to a certain condition at the surface of the considered volume.

This situation opens a second type of field control. While the control of a field by the source terms is the counterpart to the control of a system with a finite number of degrees of freedom by external forces, fields may also be controlled by an appropriate change of the boundary conditions.

4.2.2 Control Without Spatial Boundaries

Optimal Control Field Equations

We assume, that the field equations for a given N-component field $\Psi(\mathbf{r},t)$ can be written as a set of partial differential equations of the form

$$\frac{\partial \Psi}{\partial t} = F\left(t, \mathbf{r}, \Psi, \nabla\Psi, \nabla^2\Psi, \ldots, u, \nabla u, \nabla^2 u, \ldots, J\right) . \tag{4.29}$$

Obviously, there is no symmetry between the time scale and the spatial coordinates. Each evolution equation contains only first-order derivatives with respect to the time while higher-order spatial derivatives are allowed. All above-discussed local field theories can be transformed into this representation[10]. The external sources are explicitely separated into the n-component control field $u(\mathbf{r},t)$, see Fig. 4.2, which may be used to change the dynamical behavior of $\Psi(\mathbf{r},t)$, and further external, but unchangeable and therefore noncontrollable sources $J(\mathbf{r},t)$ with n' components. The field equations (4.29) are declared for all sites $\mathbf{r} \in \mathbb{R}^d$ of a d-dimensional space and for all times $t \in (-\infty, \infty)$.

The performance of the field theoretical control problem may be a functional of the type

[9] In order to avoid confusion, we stress again that the generalized momenta are not the physical momenta. The latter are usually a part of the state vector.

[10] It may be possible that these representation requires an extension of the number of field components. For example, the inhomogeneous wave equation, $\ddot{\varphi} - \nabla^2\varphi = u$ may be rewritten as a system of two partial differential equations with first-order derivatives with respect to the time. In other words, we obtain the two-component field $\Psi = (\varphi, \psi)$ and the components satisfy the equations $\dot{\varphi} = \psi$ and $\dot{\psi} = \nabla^2\varphi + u$.

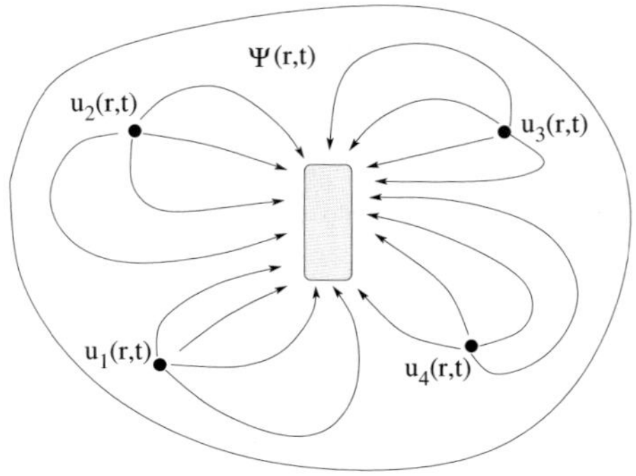

Fig. 4.2. A typical example for a control via external sources: The changes of the sources $u(\mathbf{r}, \mathbf{t})$ propagate due to the field $\Psi(\mathbf{r}, \mathbf{t})$ in the region under control

$$R[\Psi, u] = \int_{-\infty}^{\infty} \mathrm{d}t \int \mathrm{d}^d r \phi(t, \mathbf{r}, \Psi(\mathbf{r}, t), \nabla\Psi(\mathbf{r}, t), \ldots, u(\mathbf{r}, t), \nabla u(\mathbf{r}, t), \ldots) \, . \tag{4.30}$$

The integration is taken over all points of the d-dimensional space $\mathbb{R}^d$ and over all times so that no boundary conditions occur. Therefore, this problem is called a field control problem without boundaries.

As in Sect. 2.4.1, (4.29) and (4.30) can be combined to a common generalized action

$$S[\Psi, \Pi, u] = \int_{-\infty}^{\infty} \mathrm{d}t \int \mathrm{d}^d r \mathcal{L}\left(t, \mathbf{r}, \Psi, \Pi, u\right) \tag{4.31}$$

with the Lagrangian

$$\begin{aligned} \mathcal{L} &= \phi(t, \mathbf{r}, \Psi, \nabla\Psi, \nabla^2\Psi, \ldots, u, \nabla u, \nabla^2 u, ...) \\ &+ \Pi \left[\frac{\partial \Psi}{\partial t} - F\left(t, \mathbf{r}, \Psi, \nabla\Psi, \nabla^2\Psi, \ldots, u, \nabla u, \nabla^2 u, \ldots, J\right) \right] \end{aligned} \tag{4.32}$$

and the generalized momentum field or adjoint field $\Pi = \Pi(\mathbf{r}, t)$. The derivation of the Euler–Lagrange equations follows the same procedure as in Sect. 4.1.1. The variation with respect to the generalized momentum field reproduces the field equations while the variation with respect to the control fields leads to

$$\left[\frac{\partial}{\partial u} - \sum_{i=1}^{d} \nabla_i \frac{\partial}{\partial \nabla_i u} + \sum_{i,j=1}^{d} \nabla_i \nabla_j \frac{\partial}{\partial \nabla_i \nabla_j u} \pm \cdots \right] [\phi - (\Pi \mid F)] = 0 \, , \tag{4.33}$$

where we have used, as in Chap. 2.4.1 the vector scalar product $(\Pi \mid F)$ in order to avoid confusions with respect to the presence of more than two vectors[11].

[11] In the present case F, Π, and $\partial/\partial u$.

The last group of equations is the set of the adjoint evolution equations following from the variation with respect to Ψ. Here, we obtain

$$\frac{\partial \Pi}{\partial t} = \left[\frac{\partial}{\partial \Psi} - \sum_{i=1}^{d} \nabla_i \frac{\partial}{\partial \nabla_i \Psi} + \sum_{i,j=1}^{d} \nabla_i \nabla_j \frac{\partial}{\partial \nabla_i \nabla_j \Psi} \pm \cdots \right] \times [\phi - (\Pi \mid F)] \tag{4.34}$$

with $\nabla_i = \partial / \partial r_i$. The set of equations (4.29), (4.33), and (4.34) is a system of partial differential equations solving the optimal control problem[12]. A general solution of these more or less formal equations cannot be expected.

From this point of view, a specification of the class of field theories seems to be necessary. Fortunately, most of the known and physically realistic field equations are linear expressions in the spatial derivatives[13] of the field functions. Furthermore, these equations contain no derivatives ∇u, $\nabla^2 u$, ... and they are only linearly coupled with the control field u. In this case, we may formally rewrite the field equations (4.29) as

$$\frac{\partial \Psi(\mathbf{r},t)}{\partial t} = \widehat{\mathcal{F}} \Psi(\mathbf{r},t) + V(\mathbf{r},t,\Psi) + A(\mathbf{r},t,\Psi) u(\mathbf{r},t) + B(\mathbf{r},t,\Psi) J(\mathbf{r},t) \tag{4.35}$$

with the N component vector V, the $N \times n$ matrix A and the $N \times n'$ matrix B. The linear operator $\widehat{\mathcal{F}}$ is defined by

$$\widehat{\mathcal{F}} = \sum_{i=1}^{d} \mathcal{F}_i(\mathbf{r},t,\Psi) \nabla_i + \frac{1}{2} \sum_{i,j=1}^{d} \mathcal{F}_{ij}(\mathbf{r},t,\Psi) \nabla_i \nabla_j + \ldots , \tag{4.36}$$

and it contains the differentiable $N \times N$ matrices $\mathcal{F}_i(\mathbf{r},t,\Psi)$, $\mathcal{F}_{ij}(\mathbf{r},t,\Psi), \ldots$. The second specification belongs to the performance integral. In many applications it is sufficient to assume a performance function of the type

$$\phi = Q(\mathbf{r},t,\Psi) + \frac{1}{2} \sum_{\alpha,\beta=1}^{N} \sum_{i=1}^{d} (\nabla_i \Psi_\alpha) \Omega_{\alpha\beta}(\mathbf{r},t) (\nabla_i \Psi_\beta) + R(\mathbf{r},t,u) \tag{4.37}$$

with the symmetric matrix Ω of rank $N \times N$. The component representation is chosen to obtain a clear representation. The performance (4.37) is sufficient for many problems. The insertion of (4.37) and (4.35) in (4.34) yields the adjoint field equations (now in the component representation with $\gamma = 1, \ldots, N$)

$$\frac{\partial \Pi_\gamma}{\partial t} = \frac{\partial Q}{\partial \Psi_\gamma} - \sum_{\beta=1}^{N} \nabla (\Omega_{\gamma\beta} \nabla \Psi_\beta) - \sum_{\alpha=1}^{N} \Pi_\alpha \frac{\partial V_\alpha}{\partial \Psi_\gamma} - \sum_{\alpha=1}^{N} \sum_{\beta=1}^{n} \Pi_\alpha \frac{\partial A_{\alpha\beta}}{\partial \Psi_\gamma} u_\beta$$

[12] In this chapter we refrain from an additional indication of the optimal solution, i.e., we continuously use Ψ, Π, and u instead of the correct style Ψ^*, Π^*, and u^*.

[13] The only exception in the above-introduced field theories is the equation for the entropy balance (4.24).

$$-\sum_{\alpha=1}^{N}\sum_{\beta=1}^{n'}\Pi_\alpha\frac{\partial B_{\alpha\beta}}{\partial\Psi_\gamma}J_\beta-\sum_{\alpha,\beta=1}^{N}\Pi_\alpha\widehat{\mathcal{F}}^\gamma_{\alpha\beta}\Psi_\beta$$
$$-\sum_{\alpha=1}^{N}\widehat{\mathcal{F}}^\dagger_{\gamma\alpha}\Pi_\alpha \tag{4.38}$$

with the adjoint operator

$$\widehat{\mathcal{F}}^\dagger=-\sum_{i=1}^{d}\nabla_i\mathcal{F}_i^T(\mathbf{r},t,\Psi)+\frac{1}{2}\sum_{i=1}^{d}\nabla_i\nabla_j\mathcal{F}_{ij}^T(\mathbf{r},t,\Psi)+\cdots \tag{4.39}$$

and the N operators

$$\widehat{\mathcal{F}}^\gamma=\sum_{i=1}^{d}\frac{\partial\mathcal{F}_i}{\partial\Psi_\gamma}\nabla_i+\frac{1}{2}\sum_{i,j=1}^{d}\frac{\partial\mathcal{F}_{ij}}{\partial\Psi_\gamma}\nabla_i\nabla_j+\ldots\,. \tag{4.40}$$

Furthermore, the control laws (4.33) are reduced to a simple algebraic law

$$\frac{\partial R(u)}{\partial u_\mu}=\sum_{\alpha=1}^{N}A^T_{\mu\alpha}(\mathbf{r},t,\Psi)\Pi_\alpha \tag{4.41}$$

with $\mu=1,\ldots,n$.

Example: Economic Field Theories

An interesting, but not directly physically motivating example of field control without boundaries are so-called economic field theories [67]. Such theories have a more or less economic origin. A natural way to handle an economic system at large scales consists in embedding these into a two-dimensional geographic space and, subsequently, a continuous description. Such a procedure reduces the detailed economic relations to several economic and social fields satisfying suitable field equations. Various publications [68–72, 78] deal with such economic field theoretical concepts. We remark that other names, such as urban fields, continuous flow models, or spatial economy are also popular.

Historically, the first problems of this kind were analyzed by Thünen in 1826 [74]. He found that agricultural production would be distributed among a set of concentric rings around the central town, i.e., a singular consumption region, according to the cost of transportation. Heavy or bulky goods, such as wood for energy production and building trade would be produced closer to the city, while goods more easily transportable farther away. This special theory is also known as land use model.

Other early publications [75,76], so-called location theories, consider the location of production plants instead of consumers. The standard theory of spatial economics was developed in the early 1950s [70,77] using Euler-Langrange variation principles and hydrodynamic concepts. As such the model is extremely elegant and versatile, being able to represent all previously known continuous models as special cases. Here, we will give a simple example [78] in order to demonstrate the basic ideas and the relation to the control of physical field theories.

In a two-dimensional space, the trade flow can be represented by a vector field $\mathbf{v}(\mathbf{r},t) = (v_1(\mathbf{r},t), v_2(\mathbf{r},t))$ with $\mathbf{r} = (r_1, r_2)$. The absolute value of the trade flow, $|\mathbf{v}(\mathbf{r},t)|$, represents the quantity of goods traded, whereas the unit direction field $\mathbf{n} = \mathbf{v}/|\mathbf{v}|$ defines the local direction of the flow. Just as the flow of a liquid satisfies various balance equations controlling the local conservation of mass, momentum, or energy, the flow of traded commodities is defined by a balance equation

$$\frac{\partial c(\mathbf{r},t)}{\partial t} + \operatorname{div} \mathbf{v}(\mathbf{r},t) = u(\mathbf{r},t) . \tag{4.42}$$

The source term $u(\mathbf{r},t)$ is related to the local excess supply of production over consumption and can be interpreted as control field. Positive values of $u(\mathbf{r},t)$ correspond to local sources of commodities due to production centers while negative values of $u(\mathbf{r},t)$ represent a local excess of consumers. The quantity $c(\mathbf{r},t)$ defines the local stock on hand. In general, we can expect that $c(\mathbf{r},t)$ depends on the local number of traded goods. Hence we get a relation of the type

$$c(\mathbf{r},t) = c(|\mathbf{v}(\mathbf{r},t)|) . \tag{4.43}$$

For example, a possible assumption is a power law $c = c_0 |\mathbf{v}(\mathbf{r},t)|^\beta$ with the reserve exponent β.

In order to determine the trade flow we need a further equation. This equation follows from the economic principle of minimum transportation costs. We assume a cost field $\kappa(\mathbf{r})$ which is determined by the local state of the infrastructure and the structure of the ground. Then, the total transportation costs $K(t)$ at a given time t are given by [70, 71, 79]

$$K(t) = \int d^2r\, |\mathbf{v}(\mathbf{r},t)|\, \kappa(\mathbf{r}) , \tag{4.44}$$

where we have assumed an infinitely large area, and the total transportation costs over the whole period[14] are given by

$$\widetilde{K} = \int_{-\infty}^{\infty} \mathrm{d}t K(t) . \tag{4.45}$$

Furthermore, the costs also depend on the control field $u(\mathbf{r},t)$. We must be aware that the local excess depends on the geographical landscape. Obviously, there exist regions which are favored for the settlement of production plants while other regions are more appropriate for residential areas. Therefore, without any reference to the transportation costs, there exists a natural excess $u_0(\mathbf{r})$ which is, of course, defined by the landscape. A large positive deviation

[14] We assume here a infinite large period, $-\infty < t < \infty$.

$u(\mathbf{r},t) - u_0(\mathbf{r})$ means a large local production and consequently a strong contamination of the environment which requires additional high-cost measures in order to avoid these unwanted side-effects. On the other hand, strongly negative values of $u(\mathbf{r},t) - u_0(\mathbf{r})$ represent a strong density of consumers and therefore higher costs for the maintenance of buildings and infrastructure. Thus, we should complete (4.45) by an additional contribution in order to obtain the performance functional

$$R = \int_{-\infty}^{\infty} \mathrm{d}t\mathrm{d}^2 r \left[|\mathbf{v}(\mathbf{r},t)| \, \kappa(\mathbf{r}) \; + \frac{1}{2} \left(u(\mathbf{r},t) - u_0(\mathbf{r}) \right)^2 \right] . \tag{4.46}$$

These relation and the field equation (4.42) build the complete field control problem. First of all, we extract the Lagrangian

$$\mathcal{L} = |\mathbf{v}(\mathbf{x},t)| \, \kappa(\mathbf{r}) + \frac{1}{2} \left(u(\mathbf{r},t) - u_0(\mathbf{r}) \right)^2 \\ + \Pi(\mathbf{r},t) \left[\frac{\partial c(|\mathbf{v}(\mathbf{r},t)|)}{\partial t} + \operatorname{div} \mathbf{v}(\mathbf{r},t) - u(\mathbf{r},t) \right] . \tag{4.47}$$

This representation requires two important remarks. First of all, we must consider that (4.42) is a one-component field equation. That means only one degree of freedom of the vector $\mathbf{v} = (v_1, v_2)$ can be interpreted as a state field in the sense of the control theory, the other degree of freedom is simply another control field. Therefore, the generalized momentum $\Pi(\mathbf{r},t)$ is only a scalar field. The second remark concerns the field equation itself. Obviously, (4.42) is a field equation in an implicit representation. However, the structure of (4.47) may be changed by an appropriate transformation of the generalized momentum in such a manner that the last contribution contains a field equation in the explicit form.

The variation of the action corresponding to (4.47) with respect to the control field, the state field, and the momentum field yields the Euler–Lagrange equations. Especially, the variation with respect to $\mathbf{v}(\mathbf{r},t)$ leads to

$$\left[\kappa(\mathbf{r}) - c'(|\mathbf{v}(\mathbf{r},t)|) \frac{\partial \Pi(\mathbf{r},t)}{\partial t} \right] \frac{\mathbf{v}(\mathbf{r},t)}{|\mathbf{v}(\mathbf{r},t)|} = \nabla \Pi(\mathbf{r},t) . \tag{4.48}$$

The compact form of (4.48) partially hides the fact that these equations are essentially one equation for the state field and one for the second control field. The variation with respect to $\Pi(\mathbf{r},t)$ leads again to (4.42), while the variation with respect to the first control field gives

$$u(\mathbf{r},t) - u_0(\mathbf{r}) = \Pi(\mathbf{r},t) . \tag{4.49}$$

Equations (4.42), (4.48), and (4.49) form a complete field-theoretical problem. Unfortunately, these equations are dominated by strong nonlinearities. This problem also remains if we focus in particular on the steady state. This special case reduces the field equations to

$$\kappa(\mathbf{r})\frac{\mathbf{v}(\mathbf{r})}{|\mathbf{v}(\mathbf{r})|} = \nabla\Pi(\mathbf{r}) \quad \text{and} \quad \operatorname{div}\mathbf{v}(\mathbf{r}) = \Pi(\mathbf{r}) + u_0(\mathbf{r})\,. \tag{4.50}$$

The decomposition of the transport field $\mathbf{v}(\mathbf{r})$ into a vortex-free part $\nabla\varphi$ and a divergent-free part $\mathbf{w}(\mathbf{r})$ with div $\mathbf{w}(\mathbf{r}) = 0$ transforms the second equation into the two-dimensional Poisson equation

$$\nabla^2\varphi(\mathbf{r}) = \Pi(\mathbf{r}) + u_0(\mathbf{r})\,. \tag{4.51}$$

The solution of this equation is possible with well-known standard methods if $\Pi(\mathbf{r})$ is known. The second step is the determination of the scalar momentum field $\Pi(\mathbf{x})$. If we take squares of both sides of the first equation of (4.50), we obtain the closed nonlinear field equation

$$(\nabla\Pi)^2 = \left(\frac{\partial\Pi}{\partial r_1}\right)^2 + \left(\frac{\partial\Pi}{\partial r_2}\right)^2 = \kappa(\mathbf{r})\,. \tag{4.52}$$

The local transportation costs $\kappa(\mathbf{r})$ are, just as the quantity $u_0(\mathbf{r})$, an external field which may be empirically determined by suitable observations or estimations. Finally, inserting the solution of (4.52) into the first equation of (4.50) we obtain a nonlinear, algebraic equation for the field $\mathbf{w}(\mathbf{r})$ which depends on the local structure of the fields $\varphi(\mathbf{x})$ and $\Pi(\mathbf{r})$.

The present theory allows the construction of optimal roads. Let us assume that we know the scalar field of the transportation costs $\kappa(\mathbf{r})$. As discussed above, the vector field $\mathbf{v}(\mathbf{r})$ defines the direction of the local flow. A road may be defined by the curve $\mathbf{y}(s)$ where s is an arbitrary curve parameter. Then the tangent of an optimal road

$$\mathbf{t}(s) = \frac{\mathrm{d}\mathbf{y}(s)}{\mathrm{d}s} \tag{4.53}$$

always shows in the direction of the local flow. That means we have the relation

$$\mathbf{t}(s) = \frac{\mathbf{v}(\mathbf{y}(s))}{|\mathbf{v}(\mathbf{y}(s))|} \tag{4.54}$$

along the road. Therefore, we expect that an optimal road fulfils the equation

$$\kappa(\mathbf{y}(s))\mathbf{t}(s) = \kappa(\mathbf{y}(s))\frac{\mathbf{v}(\mathbf{y}(s))}{|\mathbf{v}(\mathbf{y}(s))|} = \nabla\Pi(\mathbf{y}(s))\,. \tag{4.55}$$

Let us try to eliminate the momentum field Π from (4.55). To this aim we differentiate with respect to the curve parameter

$$\frac{\mathrm{d}}{\mathrm{d}s}\left[\kappa(\mathbf{y}(s))\frac{\mathrm{d}\mathbf{y}(s)}{\mathrm{d}s}\right] = \frac{\mathrm{d}}{\mathrm{d}s}\nabla\Pi(\mathbf{y}(s))\,. \tag{4.56}$$

The vector components on the right-hand side can be written as

$$\begin{aligned}\frac{\mathrm{d}}{\mathrm{d}s}\left.\frac{\partial\Pi(\mathbf{r})}{\partial r_i}\right|_{\mathbf{r}=\mathbf{y}(s)} &= \sum_{j=1}^{2}\left.\frac{\partial^2\Pi(\mathbf{r})}{\partial r_i\partial r_j}\right|_{\mathbf{r}=\mathbf{y}(s)}\frac{\partial y_j(s)}{\partial s}\\ &= \sum_{j=1}^{2}\left.\frac{\partial^2\Pi(\mathbf{r})}{\partial r_i\partial r_j}\right|_{\mathbf{r}=\mathbf{y}(s)} t_j(s)\,.\end{aligned} \tag{4.57}$$

We multiply this expression with $\kappa(\mathbf{y}(s))$ and apply (4.55) in order to obtain

$$\kappa(\mathbf{y}(s))\frac{d}{ds}\left.\frac{\partial \Pi(\mathbf{r})}{\partial r_i}\right|_{\mathbf{r}=\mathbf{y}(s)} = \sum_{j=1}^{2}\left.\frac{\partial^2 \Pi(\mathbf{r})}{\partial r_i \partial r_j}\right|_{\mathbf{r}=\mathbf{y}(s)} \left.\frac{\partial \Pi(\mathbf{r})}{\partial r_j}\right|_{\mathbf{r}=\mathbf{y}(s)} \tag{4.58}$$

$$= \frac{1}{2}\frac{\partial}{\partial r_i}\sum_{j=1}^{2}\left.\left(\frac{\partial \Pi(\mathbf{r})}{\partial r_j}\right)^2\right|_{\mathbf{r}=\mathbf{y}(s)} \tag{4.59}$$

or with (4.52)

$$\kappa(\mathbf{y}(s))\frac{\mathrm{d}}{\mathrm{d}s}\nabla \Pi(\mathbf{y}(s)) = \frac{1}{2}\left.\nabla \kappa(\mathbf{r})\right|_{\mathbf{r}=\mathbf{y}(s)} . \tag{4.60}$$

Considering (4.55), we obtain the optimal road equation

$$\kappa(\mathbf{y}(s))\frac{\mathrm{d}}{\mathrm{d}s}\left[\kappa(\mathbf{y}(s))\frac{\mathrm{d}\mathbf{y}(s)}{\mathrm{d}s}\right] = \frac{1}{2}\nabla \kappa(\mathbf{y}(s)) . \tag{4.61}$$

We remark that the last equations are close to Fermat's law in optics. In particular, roads are equivalent to light rays, the momentum field corresponds to the eikonal function, and the transportation cost may be interpreted as the refraction index. For example, if we have spatially separated types of transportation, for example transportation over sea and over land, the trading routes are straight lines, broken at the coastline via the well-known refraction law [80, 81]. A similar phenomenon belongs to roads through high mountain regions. For instance, the highways from Rome to Milan pass the Appenines in a similar way as light rays pass a glassy plate; see Fig. 4.3.

Pontryagin's Maximum Principle

If the Lagrangian depends only on the control field u but not on the spatial derivatives $\nabla u, \nabla^2 u, \ldots$, we are able to apply Pontryagin's maximum principle. In other words, we search for the control field $u(\mathbf{r}, t)$ which minimizes the Lagrangian (4.32) at each point $(\mathbf{r}, t)$ and for each field configuration $(\Pi, \Psi, \nabla\Psi, \nabla^2\Psi, ...)$. As in the case of systems with a finite degree of freedom, we may define the preoptimized field, $u^{(*)}\left(\mathbf{r}, t, \Pi, \Psi, \nabla\Psi, \nabla^2\Psi, ...\right)$, which satisfies the inequality

$$\begin{aligned} &\mathcal{L}\left(t, \mathbf{r}, \Pi, \Psi, \nabla\Psi, \nabla^2\Psi, \ldots, u^{(*)}(t, \mathbf{r}, \Pi, \Psi, \nabla\Psi, \nabla^2\Psi, \ldots)\right) \\ &\leq \mathcal{L}\left(t, \mathbf{r}, \Pi, \Psi, \nabla\Psi, \nabla^2\Psi, \ldots, u\right) \end{aligned} \tag{4.62}$$

for all allowed control fields u. The Pontryagin maximum principle allows the extension of the field control problem to control fields $u(t, \mathbf{r})$ restricted to an arbitrary, possibly time- and space-dependent region $\mathbb{U}'(\mathbf{r}, t) \subset \mathbb{U}$. In other words, the preoptimized Lagrangian is then given by

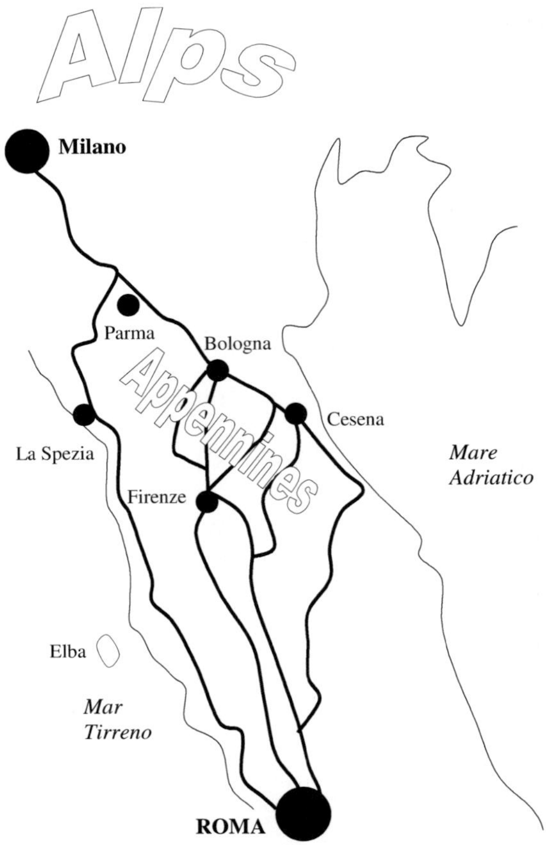

Fig. 4.3. Schematic representation of the highways from Rome to Milan

$$\begin{aligned}
&\mathcal{L}^{(*)}\left(t, \mathbf{r}, \Pi, \Psi, \nabla\Psi, \nabla^2\Psi, \ldots\right) \\
&= \mathcal{L}\left(t, \mathbf{r}, \Pi, \Psi, \nabla\Psi, \nabla^2\Psi, \ldots, u^{(*)}(t, \mathbf{r}, \Pi, \Psi, \nabla\Psi, \nabla^2\Psi, \ldots)\right) \\
&= \min_{u \in \mathbb{U}'(\mathbf{r},t) \subset U} \mathcal{L}\left(t, \mathbf{r}, \Pi, \Psi, \nabla\Psi, \nabla^2\Psi, \ldots, u\right) . \qquad (4.63)
\end{aligned}$$

The Pontryagin maximum principle becomes important if the control law, $\partial\mathcal{L}/\partial u = 0$, see for example (4.41), as a specialized version has no global minimum solution inside the allowed set $\mathbb{U}'(\mathbf{r}, t)$.

Linear–Quadratic Problems

The system of partial differential equations consisting of the field equations (4.35), the adjoint field equations (4.38), and the control law (4.41) is again a strongly nonlinear problem. Most of such problems[15] have to be solved numerically and it is often uncertain whether the applied numerical algorithm has actually found the correct solution.

However, a systematic procedure is possible for linear–quadratic problems. To this aim we consider linear field equations. This may be physically motivated by the circumstance that a large class of classical field theories automatically generates linear field equations because linearity is a necessary condition for fields satisfying the superposition principle. Linear field equations are a special class of (4.35) and may be written in the form

$$\frac{\partial \Psi(\mathbf{r},t)}{\partial t} = \widehat{\mathcal{F}}\Psi(\mathbf{r},t) + Au(\mathbf{r},t) + BJ(\mathbf{r},t) \tag{4.64}$$

with the matrices $A = A(\mathbf{r},t)$ and $B = B(\mathbf{r},t)$ and the linear operator

$$\widehat{\mathcal{F}} = \mathcal{F}_0 + \sum_{i=1}^{d} \mathcal{F}_i \nabla_i + \frac{1}{2} \sum_{i,j=1}^{d} \mathcal{F}_{ij} \nabla_i \nabla_j + \ldots \tag{4.65}$$

with the matrices $\mathcal{F}_0(\mathbf{r},t)$, $\mathcal{F}_i(\mathbf{r},t)$, $\mathcal{F}_{ij}(\mathbf{r},t)$, ... We remark that without any restrictions the matrix B can be combined with the external sources J to a common quantity, $BJ \to J$. This is not generally allowed for the matrix A and the control fields u. If we replace $Au \to u$ we must simultaneously substitute $u \to A^{-1}u$ in the performance functional. But this is possible only if the inverse matrix A^{-1} exists for all points $(\mathbf{r},t)$ of the space–time continuum.

Furthermore, the performance may be characterized by the quadratic form

$$\begin{aligned} \phi = {} & \frac{1}{2} \sum_{\alpha,\beta=1}^{N} \left[\Psi_\alpha Q_{\alpha\beta}(\mathbf{r},t) \Psi_\beta + \sum_{i=1}^{d} (\nabla_i \Psi_\alpha) \Omega_{\alpha\beta}(\mathbf{r},t) (\nabla_i \Psi_\beta) \right] \\ & + \frac{1}{2} \sum_{\alpha,\beta=1}^{n} u_\alpha R_{\alpha\beta}(\mathbf{r},t) u_\beta \end{aligned} \tag{4.66}$$

which is a specialized version of (4.37). The symmetric matrix Q is of order $N \times N$, while the symmetric matrix R has the rank $n \times n$. Considering the above-introduced linear structure of the field equations and the quadratic form of the performance, we obtain the adjoint field equations

$$\frac{\partial \Pi_\gamma}{\partial t} = \sum_{\beta=1}^{N} \left[Q_{\gamma\beta}(\mathbf{r},t) \Psi_\beta - \nabla (\Omega_{\gamma\beta} \nabla \Psi_\beta) \right] - \sum_{\alpha=1}^{N} \mathcal{F}^\dagger_{\gamma\alpha} \Pi_\alpha , \tag{4.67}$$

which depend no longer on the control fields and other external sources. The adjoint operator $\mathcal{F}^\dagger$ is defined by

$$\mathcal{F}^\dagger = \mathcal{F}_0^T(\mathbf{r},t) - \sum_{i=1}^{d} \nabla_i \mathcal{F}_i^T(\mathbf{r},t) + \frac{1}{2} \sum_{i,j=1}^{d} \nabla_i \nabla_j \mathcal{F}_{ij}^T(\mathbf{r},t) + \cdots . \tag{4.68}$$

---

[15] The control of hydrodynamic equations yields several standard examples for such a nonlinear coupled system of partial differential equations.

Thus, the adjoint field equation is also linear. Finally, we get the linear control law

$$\sum_{\alpha=1}^{n} R_{\gamma\alpha}(\mathbf{r},t)\,u_\alpha = \sum_{\alpha=1}^{N} A_{\mu\alpha}^T(\mathbf{r},t)\Pi_\alpha \ . \tag{4.69}$$

If R^{-1} exists, (4.69) may be used for the elimination of the control fields from the field equations (4.64). Thus, the linear–quadratic control problem leads to a system of $2N$ coupled linear partial differential equations for the N components of the field Ψ and the N components of the momentum field Π. Such systems can be solved by various approximative numerical techniques. A standard method is the expansion of the pair of field components in terms of a complete set of orthogonal functions or basic functions. This reduces the system of partial differential equations to an infinitely large linear algebraic system. Such a system can be solved numerically if only a finite subset of orthogonal functions is taken into account. In principel, this approximative solution should converge to the true solution with an increasing number of considered basic functions. The convergence velocity depends strongly on the type of the more or less empirically chosen set of orthogonal functions. The choice of an appropriate set of basic functions is often a question of the numerical and scientific experience with respect to the underlying problem.

4.2.3 Passive Boundary Conditions

Passive boundary conditions occur if we have a field under control in a certain spatial region with unchangeable boundary conditions. In other words, the boundaries cannot be used for a control of the fields. The above-introduced techniques can also be extended to those boundary problems. However, just as in the case of optimal field control without boundaries, it remains the general situation that systematic concepts for solving such problems exist only for linear field equations. A very elegant method is the application of Green's functions. The knowledge of these functions allows the solution of linear field equation for arbitrary sources, but with well-defined boundary conditions.

Let us assume that the linear equation (4.64) is declared over a well-defined region $\mathbb{G}$ of the space $\mathbb{R}^d$ and that the field has at the border $\partial\mathbb{G}$ well-defined boundary conditions. Then, the formal solution of (4.64) is given by

$$\Psi(\mathbf{r},t) = \Psi_{\text{hom}}(\mathbf{r},t) + \int_0^t \mathrm{d}\tau \int_G \mathrm{d}^d x \Gamma(\mathbf{r},t \mid \mathbf{x},\tau)\left[Au(\mathbf{r},\tau) + BJ(\mathbf{r},\tau)\right] \tag{4.70}$$

with the so-called Green function $\Gamma(\mathbf{r},t \mid \mathbf{x},\tau)$ and the homogeneous solution $\Psi_{\text{hom}}(\mathbf{r},t)$ satisfying

$$\frac{\partial \Psi_{\text{hom}}(\mathbf{r},t)}{\partial t} = \widehat{\mathcal{F}}\Psi_{\text{hom}}(\mathbf{r},t) \ . \tag{4.71}$$

This formal solution may be interpreted as an alternative version of the field equations. We remark that the homogeneous solution is mainly determined by the initial conditions. We use again the quadratic performance function (4.66) in order to construct the generalized action

$$
\begin{aligned}
S = & \int_0^T \mathrm{d}t \int_G \mathrm{d}^d r \phi\left(\mathbf{r}, t, \Psi, \nabla\Psi, u\right) \\
& + \int_0^T \mathrm{d}t \int_G \mathrm{d}^d r \Pi\left(\mathbf{r}, t\right)\left[\Psi\left(\mathbf{r}, t\right) - \Psi_{\mathrm{hom}}\left(\mathbf{r}, t\right)\right. \\
& \left. - \int_0^t \mathrm{d}\tau \int_G \mathrm{d}^d x \Gamma\left(\mathbf{r}, t \mid \mathbf{x}, \tau\right)\left[Au(\mathbf{r}, t) + BJ(\mathbf{r}, t)\right]\right] , \qquad (4.72)
\end{aligned}
$$

where we have considered a final control horizon T. The variation of the generalized action with respect to the field momentum Π again leads to the dynamic equation, while the variation with respect to the field leads to

$$
\Pi\left(\mathbf{r}, t\right) = \nabla(\Omega\left(\mathbf{r}, t\right)\nabla\Psi\left(\mathbf{r}, t\right)) - Q\left(\mathbf{r}, t\right)\Psi\left(\mathbf{r}, t\right) \qquad (4.73)
$$

and the variation with respect to the control field gives

$$
u\left(\mathbf{r}, t\right) = R^{-1}\left(\mathbf{r}, t\right) \int_t^T \mathrm{d}\tau \int_G \mathrm{d}^d x A^T\left(\mathbf{r}, t\right) \Gamma^T\left(\mathbf{x}, \tau \mid \mathbf{r}, t\right) \Pi\left(\mathbf{x}, \tau\right) . \qquad (4.74)
$$

The substitution of (4.73) in (4.74) eliminates the momentum field from the control equations and we obtain

$$
\begin{aligned}
u\left(\mathbf{r}, t\right) = & R^{-1}\left(\mathbf{r}, t\right) \int_t^T d\tau \int_G d^d x A^T\left(\mathbf{r}, t\right) \Gamma^T\left(\mathbf{x}, \tau \mid \mathbf{r}, t\right) \\
& \times \left[\nabla(\Omega\left(\mathbf{x}, \tau\right)\nabla - Q\left(\mathbf{x}, \tau\right)\right]\Psi\left(\mathbf{x}, \tau\right) . \qquad (4.75)
\end{aligned}
$$

This integral relation and the formal solution (4.70) are a complete set of integral equations determining the linear field control problem with passive boundaries. The procedure is applicable to all field equations if the Green function is available. But this is, by no means, a trivial problem, and it requires often a large amount of numerical and analytical tools [82–84].

A practical problem with passive boundary conditions is the optimal control of soil venting [85] used for remediation of soil contaminations with volatile organic components, such as chlorinated organic hydrocarbon. Soil venting basically means that the soil air is removed from the contaminated soil region by pumping while the volatile contaminants in the extracted air are collected and annihilated [86, 87]. In principle, the corresponding physical model is a diffusion-reaction process (Fig 4.4). It considers (i) the diffusion of soil air from several boreholes filled with fresh air under pressure (sources) to other

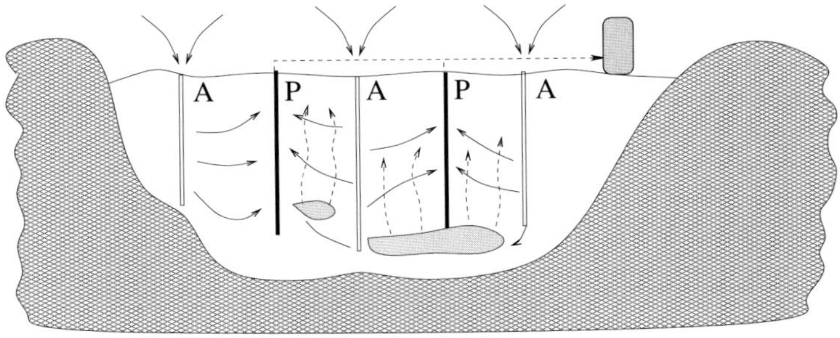

Fig. 4.4. The principle of soil venting: fresh air is pumped into the active boreholes (A) and moves by diffusion processes to the passive boreholes (P). Volatile organic materials follow the diffusion flow to the passive holes. The contaminated air is finally filtered and the organic components are collected in special tanks

boreholes sucking off the contaminated air (sinks) (ii) the transport of organic components following the diffusion gradient of the air, and (iii) a spontaneous reduction of the organic components by the presence of microorganisms. Adsorbing boundary conditions are assumed for the interface between the affected region and the atmosphere as well as the ground water or surface water while reflecting boundary conditions are taken into account for the interface between the soil and rocky grounds.

The optimal field control problem consists in an optimal choice of the distribution of sources and sinks in order to clean the soil in a minimal time with as small as possible costs. The practical solution requires various numerical methods, for example, genetic simulation methods [88] and modified versions of the Powell algorithm [89].

4.3 Control via Boundary Conditions

A very important method controlling the dynamics of fields is the so-called boundary control. This concept is applied in practice very often. For example, the complicated aerodynamic and chemical evolution of a mixture of air and natural gas in a combustion chamber of power station is controlled via the boundaries, e.g., by the change of the current injection rates of the air or the amount of outgoing waste air.

Unfortunately, the control of nonlinear field equations via boundaries is more or less an empirical and essential field for a broad experimental research. Another situation occurs for linear field equations. Here, boundary control can be mapped onto the source control. To this aim, let us assume that the field equation is given by the homogeneous version of (4.64)

$$\frac{\partial \Psi(\mathbf{r},t)}{\partial t} = \widehat{\mathcal{F}}\Psi(\mathbf{r},t) \tag{4.76}$$

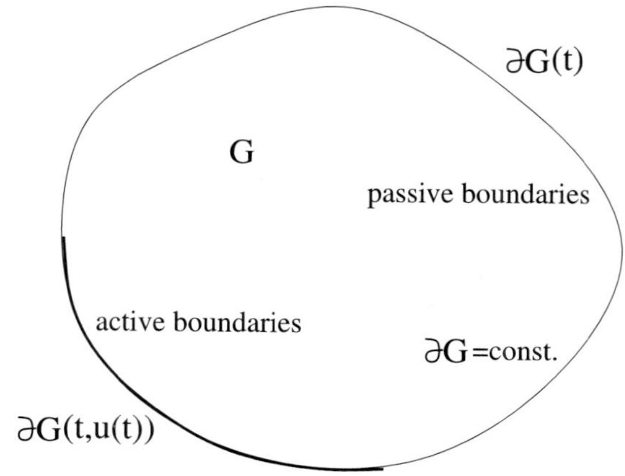

Fig. 4.5. Passive boundaries are defined by time-independent or uncontrolled time-dependent functions, active boundary conditions depend on free changeable control functions $u(t)$

defined over the region $\mathbb{G}$ of the space $\mathbb{R}^d$ while the boundary conditions are given by some constraints at the border $\partial\mathbb{G}$, see Fig. 4.5. Furthermore, the boundary conditions should be changeable by several external control fields $w(\mathbf{x},t)$ which are declared only at the border, $\mathbf{x} \in \partial\mathbb{G}$. Therefore, we assume that at the boundaries the values of the field $\Psi(\mathbf{x},t)$ and possibly its derivatives $\nabla\Psi(\mathbf{x},t)$, ... are well defined functions of the control fields, $\Psi(\mathbf{x},t) = \Psi(\mathbf{x},t,w(\mathbf{x},t))$, $\nabla\Psi(\mathbf{x},t) = \nabla\Psi(\mathbf{x},t,w(\mathbf{x},t)),\ldots$. We split the field $\Psi(\mathbf{r},t)$ in two parts,

$$\Psi(\mathbf{r},t) = \Psi_0(\mathbf{r},t) + \widetilde{\Psi}(\mathbf{r},t) \tag{4.77}$$

with passive boundary conditions for $\Psi_0(\mathbf{r},t)$, $\Psi_0(\mathbf{x},t) = 0$ for $\mathbf{x} \in \partial\mathbb{G}$. The field $\widetilde{\Psi}(\mathbf{r},t)$ satisfies the equation

$$\widehat{\mathcal{F}}\widetilde{\Psi}(\mathbf{r},t) = 0 \tag{4.78}$$

and have the same boundary conditions as the original field $\Psi(\mathbf{r},t)$, i.e., $\widetilde{\Psi}(\mathbf{x},t) = \Psi(\mathbf{x},t)$, $\nabla\widetilde{\Psi}(\mathbf{x},t) = \nabla\Psi(\mathbf{x},t)$, ... Obviously, (4.78) is no evolution equation with respect to time. That means the time t is here only a simple parameter. Now we assume that we are able to solve these equations for arbitrary boundary conditions. That is not a trivial task. The structure of the solution is mainly defined by the structure of the operator $\widehat{\mathcal{F}}$ and the boundary conditions. For example, the one-component equation $\Delta\widetilde{\varphi} = 0$ requires for a given region $\mathbb{G}$ of $\mathbb{R}^3$ the solution

$$\widetilde{\varphi}(\mathbf{r},t) = \int\limits_{\partial\mathbb{G}} \frac{d\mathbf{A}}{4\pi}\left[\frac{1}{|\mathbf{x}-\mathbf{r}|}\nabla\widetilde{\varphi}(\mathbf{x},t,w(\mathbf{x},t)) + \frac{\mathbf{x}-\mathbf{r}}{|\mathbf{x}-\mathbf{r}|^3}\widetilde{\varphi}(\mathbf{x},t,w(\mathbf{x},t))\right], \tag{4.79}$$

where $\mathrm{d}\mathbf{A}$ is the oriented infinitesimal surface element of the border $\partial\mathbb{G}$. However, the explicit solution of (4.78) considering the boundary conditions correctly is a functional of the boundary conditions and therefore of the control functions, $\widetilde{\Psi}(\mathbf{r},t) = \widetilde{\Psi}[\mathbf{r},t,w]$. We substitute this solution and (4.77) in (4.76) and obtain because of (4.78)

$$\frac{\partial \Psi_0(\mathbf{r},t)}{\partial t} = \widehat{\mathcal{F}}\Psi_0(\mathbf{r},t) - \dot{\widetilde{\Psi}}[\mathbf{r},t,w] \,. \tag{4.80}$$

This is an inhomogeneous field equation for the function $\Psi_0(\mathbf{r},t)$ which satisfies passive boundary conditions and which is controlled by an apparently external source $\dot{\widetilde{\Psi}}[\mathbf{r},t,w]$. Thus, the above-discussed formalism can now be used for the control problem.

The most currently intensive discussed field control problems with active boundary conditions are related to the scalar wave equation [5,90], $\ddot{\varphi} - \nabla^2\varphi = 0$, generalized Klein–Gordon equations [91], $\ddot{\varphi} - \nabla a(\mathbf{r})\nabla\varphi + m(\mathbf{r})\varphi = 0$, or vibrating plates [91, 92] $\ddot{\varphi} + \nabla^4\varphi = 0$. It is surprising that these apparently simple problems also offer a lot of mathematical problems related to the controllability and stability of the control.

References

1. W. Thirring: *A Course in Mathematical Physics II. Classical Field Theory* (Springer, Berlin Heidelberg New York, 1998)
2. E. Schmutzer: *Grundprinzipien der klassischen Mechanik und der klassischen Feldtheorie* (Verlag der Wissenschaften, Berlin, 1973)
3. M. Kaku: *Quantum Field Theory* (Oxford University Press, Oxford, 1993)
4. N. Sanchez: *Non-Linear Equations in Classical and Quantum Field Theory* (Springer, Berlin Heidelberg New York, 1985)
5. W. Krabs: Math. Meth. Appl. Sci. **1**, 322 (1979)
6. R. Triggiani: Appl. Math. Optim. 18, 241 (1988)
7. W. Krabs: On time-optimal boundary control of vibrating beams. In: *Control Theory for Distributed Parameter Systems and Applications*, Lecuture Notes in control and Information Sciences, vol. 54, ed by F. Kappel, K. Kunisch, W. Schappacher (Springer, Berlin Heidelberg New York, 1983), p. 127
8. V. Komornik: *Exact Controllability and Stabilization: The Multiplier Method* (Wiley, Chichester, 1994)
9. N.J. Lynch-Aird: IEEE J. Select. Areas Commun. **9**, 830 (1991)
10. A. Asamitsu, Y. Moritomo, Y. Tomioka, T. Arima, Y. Tokura: Nature **373**, 407 (1995)
11. Y. Tomioka, A. Asamitsu, Y. Moritomo, H. Kuwahara, Y. Tokura: Phys. Rev. Lett. **74**, 5108 (1995)
12. H. Risken: Z. Phys. **186**, 85 (1965)
13. H. Haken: *Laser Theory* (Springer, Berlin Heidelberg New York, 1985)
14. W. Thirring: *Classical Dynamical Systems and Classical Field Theory* (Springer, Berlin Heidelberg New York, 1992)
15. F.E. Low: *Classical Field Theory: Electromagnetism and Gravitation* (Wiley-Interscience, New York, 1997)

16. A.O. Barut: *Electrodynamics and Classical Theory of Fields and Particles* (Dover Publications, New York, 1981)
17. E. Binz, J. Sniatycki, H. Fischer: *Geometry of Classical Fields* (Elsevier, North-Holland, 1988)
18. D.E. Soper: *Classical Field Theoryn* (Wiley, Chichester, 1976)
19. E.F. Kaasschieter, A.J.M. Huijben: Report PN 90-92-A, TU Delft (1990)
20. P.G. Ciarlet, J.-L. Lions: *Finite Element Methods, Vol. II of Handbook of Numerical Analysis* (North-Holland, Amsterdam, 1991)
21. F. Brezzi, M. Fortin: *Mixed and Hybrid Finite Element Methods* (Springer, Berlin Heidelberg New York, 1991)
22. H.-C. Huang: *Finite Element Analysis of Non-Newtonian Flow* (Springer, Berlin Heidelberg New York, 1999)
23. G. Dhondt: *The Finite Element Method for Three-Dimensional Thermomechanical Applications* (Wiley, Chichester, 2004)
24. A. Munjiza: *The Combined Finite-Discrete Element Method* (Wiley, Chichester, 2004)
25. T.A. Nooren, H.A. Wouters, T.W.J. Peeters, D. Roekarts, U. Maas, D. Schmidt: Combust. Theory Model. **1**, 79 (1997)
26. J.O. Hirschfelder, C.F. Curtiss: *Theory of Propagation of Flames, Part I: General Equations* (Williams and Wilkins, Baltimore, 1949)
27. N. Peters, K. Seshadri: Combust. Flame **187**, 197 (1988)
28. T. Schmidt, T. Blasenbrey, U. Maas: Combust. Theory Model. **2**, 135 (1998)
29. B. Ruf, O. Deutschmann, F. Behrendt, J. Warnatz: J. Appl. Phys. **79**, 7256 (1996)
30. D.G. Zeitoun, G.F. Pinder: Water Resour. Res. **29**, 217 (1993)
31. O.M. Aamo: *Flow Control by Feedback* (Springer, Berlin Heidelberg New York, 2003)
32. R.M. Christensen: *Theory of Viscoelasticity* (Dover Publications, New York, 1982)
33. J.D. Ferry: *Viscoelastic Properties of Polymers*, 3rd edn (Wiley, Chichester, 1980)
34. K.L. Dorrington: Symp. Soc. Exp. Biol. **34**, 289 (1980)
35. K. Kawashima, S. Unjoh: J. Struct. Eng. **120**, 2583–2600 (1994)
36. R.M. Mutobe, T.R. Cooper: Comput. Struct. **72**, 279 (1999)
37. L.L. Chung, L.Y. Wu, T.G. Jin: Eng. Struct. **20**, 62 (1998)
38. I.I. Blekhman: *Vibrational Mechanics* (World Scientific Publishing, Singapore, 1999)
39. K.J. Laidler: *Chemical Kinetics* (McGraw-Hill, New York, 1965)
40. S.W. Benson: *The Foundations of Chemical Kinetics* (McGraw-Hill, New York, 1960)
41. N.G. van Kampen: *Stochastic Processes in Physics and Chemistry* (North-Holland, Amsterdam, 1981)
42. H. Haken: *Synergetics* (Springer, Berlin Heidelberg New York, 1978)
43. G. Nicolis, I. Prigonine: *Self-Organization in Non-Equilibrium Systems* (Wiley, New York 1980)
44. T.M. Liggett: *Interacting Particle Systems* (Springer, Berlin Heidelberg New York, 1985)
45. V. Kuzovkov, E. Kotomin: Rep. Prog. Phys. **51**, 1479 (1988)
46. K. Kang, S. Redner: Phys. Rev. A **32**, 435 (1985)

47. V.I. Yudson, M. Schulz, S. Stepanow: Phys. Rev. E **57**, 5063 (1998)
48. B. Bock, H. Hötzel, M. Nahold: *Untergrundsanierung mittels Bodenluftabsaugung und In-Situ-Strippen*, vol. 9 (Schriftenreihe Angewandte Geologie, Karlsruhe, 1990)
49. U. Fischer: Experimental and numerical investigation of soil vapor extraction. Thesis no. 11277, ETH Zürich (1995)
50. J.S. Gierke, N.J. Hutzler, D.B. McKenzie: Water Resour. Res. **28**, 323 (1992)
51. V. Privman: *Nonequilibrium Statistical Mechanics in One Dimension* (Cambridge University Press, Cambridge, 1997)
52. Y. Smagina, O. Nekhamkina, M. Sheintuch: Ind. Eng. Chem. Res. **41**, 2023 (2002)
53. A.A. Alonso, B.E. Ydstie: Automatica **37**, 1739 (2001)
54. H. Motz, H. Wise: J. Chem. Phys. **32**, 1893 (1960)
55. B.M. Schulz, S. Trimper, M. Schulz: Eur. Phys. J. B **15**, 499 (2000)
56. B.M. Schulz, M. Schulz, S. Trimper: Phys. Lett. A **291**, 87 (2001)
57. B.M. Schulz, P.Reineker, M.Schulz: Phys. Lett. A **299**, 337 (2002)
58. M. Schulz, P. Reineker, B.M. Schulz, S. Trimper: Phys. Chem. **282**, 379 (2002)
59. B.M. Schulz, S. Trimper, M. Schulz: Phys. Rev. E **66**, 031106 (2002)
60. M. Schulz, S. Stepanow: Phys. Rev. B **59**, 13528 (1999)
61. M. Schulz, K. Zabrocki, S. Trimper: Phys. Rev. E **65**, 056106 (2002)
62. S. Trimper, K. Zabrocki, M. Schulz: Phys. Rev. E **70**, 056133 (2004)
63. M. Schulz, S. Trimper: Phys. Rev. B **64**, 233101 (2001)
64. J.W. Nunziato: Quart. Appl. Math. **29**, 187 (1971)
65. W. Jung: Über ein pseudoparabolisches Rand-Kontroll-Problem aus der Wärmeleitung. Thesis, Universität Frankfurt/M. (1982)
66. W. Scondo: Randsteuerung einer evolutionsgleichung mit Gedächtnis. Thesis, Universität Frankfurt/M. (1982)
67. M. Schulz: *Statistical Physics and Economics* (Springer, Berlin Heidelberg New York, 2003)
68. S. Angel, G.M. Hyman: *A Geometry of Movement for Regional Science* (Pion, London, 1976)
69. T. Puu: Reg. Sci. Urban Econ. **11**, 317 (1981)
70. M.J. Beckmann: Econometrica **20**, 643 (1952)
71. M.J. Beckmann, T. Puu: *Spatial Structures* (Springer, Berlin Heidelberg New York, 1990)
72. T. Puu: Reg. Sci. Urban Econ. **8**, 225 (1978)
73. T. Puu: Chaos, Solitons and Fractals **5**, 35 (1995)
74. J.H.V. Thünen: *Der isolierte Staat in Beziehung auf Nationaleinkommen und Landwirtschaft, reprint of the 1826 edition* (Gustav Fischer, Stuttgart, 1966)
75. W. Launhardt: *Mathematische Begründung der Volkswirtschaftslehre* (B.G. Teubner, Leipzig, 1885)
76. A. Weber: *Über der Standort der Industrien* (Tübingen, 1909)
77. T. Puu: *The Allocation of Road Capital in Two-Dimensional Space: A Continuous Approach* (North-Holland, Amsterdam, 1979)
78. T. Puu: Chaos, Solitons and Fractals **3**, 99 (1993)
79. M.J. Beckmann, T. Puu: *Spatial Economics: Potential, Density and Flow* (North-Holland, Amsterdam, 1979)
80. T.F. Palander: *Beiträge zur Staatstheorie* (Almquist and Wiksells, Uppsala, 1935)

81. H.V. Stackelberg: Jahrb. Nationalökon. Stat. **148**, 680 (1938)
82. E. Zauderer: *Partial Differential Equations of Applied Mathematics* (Wiley-Interscience, Chichester, 1998)
83. Y. Pinchover, J. Rubinstein: *An Introduction to Partial Differential Equations* (Cambridge University Press, Cambridge, 2005)
84. A. Broman: *Introduction to Partial Differential Equations: From Fourier Series to Boundary-Value Problems* (Dover Publications, New York, 1990)
85. H.H. Gerke, U. Hornung, Y. Kelanemer, M. Slodiĉka, S. Schumacher: *Optimal Control of Soil Venting: Mathematical Modeling and Applications* (Birkhäuser, Basel, 1999)
86. F. Schwille: Besond. Mitt. Deutsch. Gewässerkundl. Jahrb. **46**, 72+XII (1984)
87. F. Schwille: *Dense Chlorinated Solvents in Porous and Fractured Media* (Lewis Publishers, Chelson, 1988)
88. D.E. Goldberg: *Genetic Algorithms in Search, Optimizationand Machine Learning* (Addison-Wesley Publishing Co., New York, 1989)
89. R.P. Brent: *Algorithms for Minimization without Derivatives* (Prentice-Hall, New York, 1973)
90. W. Krabs: Z. Operations Res. **26**, 63 (1982)
91. W. Krabs: *Optimal Control of Undamped Linear Vibrations* (Heldermann-Verlag, Lemgo, 1995)
92. M. Niezgődka: Stability of a class of nonlinear evolution free boundary problems with respect to domain variations. In: *Optimal Control of Partial Differential Equations*, ed by K.-H. Hoffmann, W. Krabs (Birkhäuser, Basel, 1984), p. 173

5

Chaos Control

5.1 Characterization of Trajectories in the Phase Space

5.1.1 General Problems

In the previous chapters we have discussed the control theory from a mechanical and field theoretical point of view. This control theory is, of course, applicable to all deterministic evolution equations and leads, under correct formulated initial and boundary conditions and under a reasonable coupling between the control functions and the system, to an optimal control. The scope of these control methods is very large; however these methods require three important considerations: (i) the complete knowledge about the dynamics of the system (ii) accurately determined initial conditions, and (iii) a permanent and nondelayed control by a control function $u(t)$ continuously varying over the whole control period.

In the following chapters we will give up successively these conditions. This necessitates a certain adjustment of the control concepts, since all three points are related to a more or less strong lack of information. In this chapter we focus on a discontinuously and delayed control of unstable system states. Simultaneously we take into account a complete knowledge about the characteristic properties of the system dynamics. But we will consider a weak inaccuracy of the initial conditions and several system parameters so that the uncontrolled system becomes more or less unpredictable at least for large time scales.

A sensitive analysis of the nature of this type of control requires a more detailed knowledge about possible general scenarios concerning the motion of system with a nontrivial degree of complexity. Especially, it is necessary to know some general aspects about the structure of a trajectory through the phase space of a system under a constant or zero control.

M. Schulz: *Control Theory in Physics and other Fields of Science*
STMP **215**, 123–148 (2006)

5.1.2 Conservative Hamiltonian Systems

To this aim we assume that the motion of the mechanical system may be confined to a finite volume of the phase space. That is, for example, the case if a system of finite energy is trapped in a sufficiently deep potential or if we observe a system from its center of mass while all degrees of freedom of the system are coupled by such interaction terms that no component can escape the system. In principle, such bounded systems are the standard situation observed in the majority of physical problems. Furthermore, we take into account that the uncontrolled Hamiltonian is explicitly time-independent.

We know from classical mechanics that the definition of spatial coordinates q_i $(i = 1, \ldots, N)$ and momenta p_i and $(i = 1, \ldots, N)$ are not fixed[1]. It is always possible to find another set of degrees of freedom

$$Q_i = Q_i(q_1, \ldots, q_N, p_1, \ldots, p_N) \quad \text{and} \quad P_i = P_i(q_1, \ldots, q_N, p_1, \ldots, p_N) \tag{5.1}$$

by a canonical transformation

$$p_i = \frac{\partial F(q_1, \ldots, q_N, P_1, \ldots, P_N)}{\partial q_i} \qquad Q_i = \frac{\partial F(q_1, \ldots, q_N, P_1, \ldots, P_N)}{\partial P_i} \tag{5.2}$$

with an arbitrary generating function F which allows the unique representation of the new coordinates Q and momenta P as functions of the old degrees of freedom (q, p) via (5.1) and vice versa. Each transformation of this type conserves the structure of the canonical equations of motion (1.1) for the new Hamiltonian[2]

$$H'(Q, P) = H(q(Q, P), p(Q, P)) \,. \tag{5.3}$$

One special choice of the generating function is the identification of F with the explicitly time-independent part of a complete integral I of the mechanical Hamilton–Jacobi equation

$$\frac{\partial I}{\partial t} + H\left(q, \frac{\partial I}{\partial q}\right) = 0 \,, \tag{5.4}$$

where the momenta of the original Hamiltonian $H = H(q, p)$ are replaced by $\partial I / \partial q$. A complete integral is a special solution of the Hamilton–Jacobi equation consisting of the N coordinates q and $N+1$ independent constants which are denoted as C_0 and $J_1, \ldots, J_N$; see also Sect. 2.5. The structure of (5.4)

[1] In this chapter we follow the traditional mechanical notation and consider a $2N$-dimensional phase space.

[2] If the generating function is explicitly time dependent, the Hamiltonian must be transformed via

$$H(q, p) \to H'(Q, P, t) = H(q(Q, P, t), p(Q, P, t), t) + \left.\frac{\partial F(q, P, t)}{\partial t}\right|_{q = q(Q, P, t)} .$$

This case is not considered here.

suggests the general form $I = C_0 + \overline{F}(q, J) - E(J)t$. Identifying the invariant quantities of the motion, J, with the new momenta P and $\overline{F}(q, J)$ with the generating function $F(q, P)$ of the canonical transformation, we obtain the new Hamiltonian from (5.4) and (5.3)

$$H'(P, Q) = H(q, p) = E(J) = E(P) \tag{5.5}$$

i.e., the transformed Hamiltonian H' depends only on the new momenta P but not on the coordinates. Thus, we obtain the new equations of motion

$$\dot{P}_i = \frac{\partial H'}{\partial Q_i} = 0 \quad \text{and} \quad \dot{Q}_i = \frac{\partial H'}{\partial P_i} = \Omega_i(P) \tag{5.6}$$

with the frequencies Ω_i depending only on the new momenta P. Consequently, the trajectories are given by

$$P_i = \text{const.} \quad \text{and} \quad Q_i = \Omega_i(P)\, t + Q_i^{(0)} \tag{5.7}$$

in the new coordinate system spanned by the new coordinates Q and the new momenta P. Since, on the one hand, the motion of the system was assumed to be constrained on a finite domain of the phase space, but, on the other hand, Q_i increases monotonously with increasing time, the new coordinates must be interpreted as angles. That means the phase space is structured in a set of imbedded N-dimensional tori. Any possible trajectory lies on one of these tori. Obviously, all tori must be covered by closed trajectories or by dense ergodic curves. The decision wether a trajectory forms closed loops or not is determined by the frequencies $\Omega_i(P)$ and therefore by the momenta P. If the equation

$$\sum_{i=1}^{N} \Omega_i(P)\, m_i = 0 \tag{5.8}$$

can be satisfied by nonzero integers m_i, the trajectories form closed loops, and the motion of the system has a periodic character. Such a solution is an exception which requires that all frequency ratios are rational numbers. In other words, a closed trajectory may be expected only for special momenta P and therefore only for special initial conditions of the system. All other initial conditions should lead to conditionally periodic motion on trajectories covering the corresponding torus completely.

However, this nice geometrical picture is too simple for a general discussion of the motion of a system through the phase space. The Hamilton–Jacobi equation is, in general, an extremely complicated mathematical object. Thus, the actual structure of the complete integrals and therefore of the generating function is open. In particular, it is not clear if the conservation laws $P_i = \text{const.}$ leads to isolated or nonisolated tori. If they are not isolated, the trajectories are not necessary confined to one torus. Thus, we would expect that the system moves through a domain of higher dimension than that of the simple torus by successive “jumps” at certain contact points between tori which belong together.

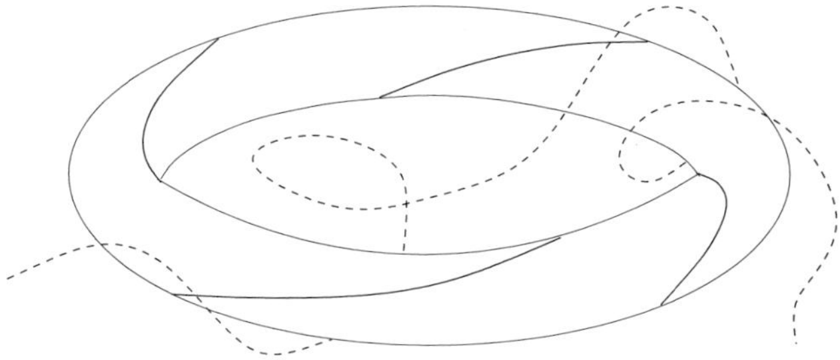

Fig. 5.1. Schematic representation of a closed orbit covering a torus and a chaotic trajectory as typical elements of the phase space

The first guess is that because of this argument most trajectories are wildly erratic and run quite far from initially neighbored trajectories. Surprisingly, it can be shown that, in contrast to this presumption, a finite volume of the admissible domain has the familiar structure of imbedded tori covered with dense trajectories. This is the Kolmogorov–Arnold–Moser theorem [1,52,53]. Although the main statement of this theorem is the result of a perturbation theory with an integrable Hamiltonian as reference model, it should be also applicable to strongly nonlinear coupled systems.

In particular, the Kolmogorov–Arnold–Moser theorem is important for the characterization of mechanical systems with a small number of degrees of freedom. Several early [54,55] as well as recent numerical investigations show that the relative part of the region containing apparently random trajectories increases rapidly with the complexity of the system, and we arrive the mixing behavior discussed in Sect. 1.2.

This is also an important remark for the closed trajectories discussed above. Under the consideration that the system allows isolated tori with closed trajectories, these orbits are strongly unstable. A small change of the initial conditions or the system parameters transfers the system from its unstable period motion either to another torus with a conditionally periodic motion or to an apparently chaotic trajectory (Fig. 5.1).

5.1.3 Nonconservative Systems

Conservative mechanical systems are either in a permanent motion or they are fixed for all times. In contrast to this behavior, the evolution equations of nonconservative systems can have stable and/or unstable fixed points and/or attractors in the phase space. That means such systems can gradually reduce or increase their dynamical mobility. However, this behavior is only possible because such systems exchange permanently energy and matter with external sources and sinks or their initial state is far from the thermodynamical equilibrium. In many cases (see below), the equations of motion reflect this exchange only by several constant parameters, so that one have the first impression that these equations are on the same level as mechanical equations.

In fact, nonconservative evolution equations usually describe only the time-dependence of a few relevant variables which seems to be representative for the characterization of the whole system whereas all other degrees of freedom are hidden in several parameters and some fluctuation terms which are not considered at the deterministic level; see Sect. 1.3.2. As an example, let us consider a chemical reaction model, the Lotka model [56], which is described by the reaction scheme

$$\begin{aligned} A + \mathcal{X} &\longrightarrow 2\mathcal{X} \\ \mathcal{X} + \mathcal{Y} &\longrightarrow 2\mathcal{Y} \\ \mathcal{Y} &\longrightarrow B\,. \end{aligned} \tag{5.9}$$

The variables of the system are the concentrations of the components $\mathcal{X}$ and $\mathcal{Y}$ while the complicated intrinsic microscopic dynamics of the molecules is described by three kinetic coefficients k_1, k_2, and k_3 defining the velocity of the first, second, and third reaction. The kinetic equations of this reaction are simply given by

$$\begin{aligned} \dot{x} &= k_1 c_A x - k_2 x y \\ \dot{y} &= k_2 x y - k_3 y \end{aligned} \tag{5.10}$$

with x being the concentration of the component $\mathcal{X}$, y is the concentration of $\mathcal{Y}$, and c_A is the concentration of A. We keep the concentration of A constant. This is always possible by controlled supply from external sources. In this sense, the component A can be interpreted as the source of the process, while the component B represents a chemical sink.

The concentrations x and y form the phase space of the Lotka model. The evolution equations have a partially unstable fixed point for $x = y = 0$, which means that all initial conditions $x_0 = 0$, $y_0 \neq 0$ converge to the fixed point for sufficiently long times while the initial conditions $x_0 \neq 0$ and $y_0 = 0$ lead to divergent behavior $x \to \infty$ for $t \to \infty$. All other, nonzero initial conditions yield closed cycles. These orbits can be calculated in the case of the Lotka model analytically. We obtain from (5.10)

$$\frac{\mathrm{d}y}{\mathrm{d}x} = \frac{(k_2 x - k_3)\, y}{(k_1 c_A - k_2 y)\, x}\,, \tag{5.11}$$

which can be integrated, and we arrive at

$$x + y - x_0 - y_0 = \frac{k_1 c_A}{k_2} \ln \frac{y}{y_0} - \frac{k_3}{k_2} \ln \frac{x_0}{x}\,. \tag{5.12}$$

Figure 5.2 shows a representation of these closed orbits in the phase space. The Lotka model is a candidate of a large class of similar evolution processes. In general, all these equations may be formally written in the form $\dot{X} = F(X)$. Possible fixed points X^0 follow from the requirement $F(X^0) = 0$. As we have discussed in Sect. 3.1.3, the stability of fixed points is mainly determined by the eigenvalues of the matrix $\partial F(X^0)/\partial X^0$. The type of different fixed points increases rapidly with increasing dimension of the phase space. So we have for $N = 1$ only two standard types of fixed points (stable and unstable), while

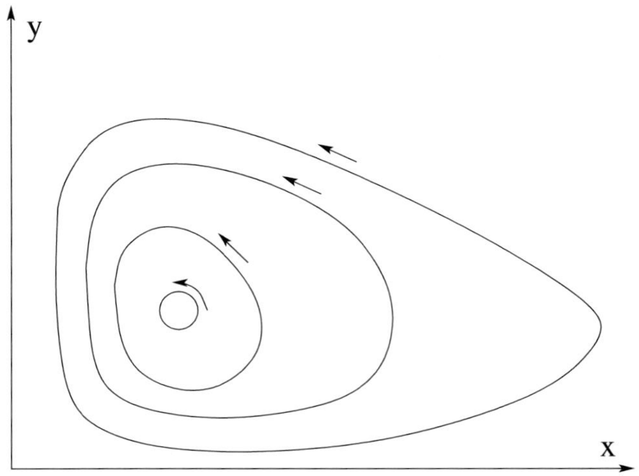

Fig. 5.2. The motion of the Lotka model through its two-dimensional phase space

the classification scheme for $N = 2$ shows six standard cases; see Fig. 5.3. Furthermore, the evolution of a two-component system may converge into a stable limit cycle (attractor). Finally, we remark that chaotic trajectories can be observed also for nonconservative systems. A standard example is the Belousov–Zhabotinskii reaction [57]. Although the behavior of nonconservative evolution equations offers an apparently bewildering variety of different possible scenarios, both conservative and nonconservative systems may be controlled with similar techniques. In several cases, especially close to stable fixed points or attractors, the control of nonconservative evolution equations becomes simpler as the control of conservative equations of motion.

5.2 Time-Discrete Chaos Control

5.2.1 Time Continuous Control Versus Time Discrete Control

Let us now assume that the motion of a mechanical system is strictly confined to a finite domain, $\mathbb{G}$, of the N-dimensional phase space $\mathbb{P}$ and that the control of a system takes place at well-defined times t_i with $t_0 < t_1 < \cdots < t_\mu < \cdots$. In other words, the vector control function u can be replaced by a set of parameters

$$u \to u^{(\mu)} = \left(u_1^{(\mu)}, u_2^{(\mu)}, \ldots, u_n^{(\mu)}\right) , \tag{5.13}$$

which are constant for each period $[t_\mu, t_{\mu+1}]$ with ($\mu = 0, 1, \ldots$) and which change their values only at the control times t_μ, see Fig. 5.4. Let us further assume that the state vector X has reached the value $X^{(\mu)}$ at time t_μ in course of the controlled motion of the system through the phase space $\mathbb{P}$. Then we can,

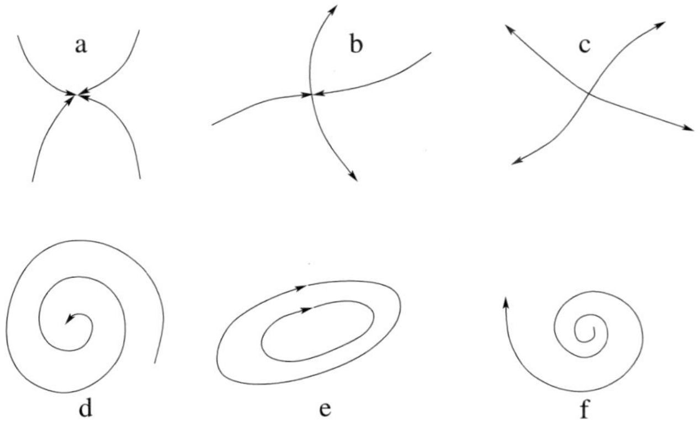

Fig. 5.3. Six standard fixed points in a two-dimensional phase space. Both the eigenvalues λ_1 and λ_2 of the matrix $\partial F(X^0)/\partial X^0$ can be (**a**) Re$\lambda_1 < 0$, Re$\lambda_1 < 0$, Im$\lambda_1 = 0$ and Im$\lambda_2 = 0$ (stable regime), (**b**) Re$\lambda_1 > 0$, Re$\lambda_2 < 0$, Im$\lambda_1 = 0$ and Im$\lambda_2 = 0$ (instable regime), (**c**) Re$\lambda_1 > 0$, Re$\lambda_2 > 0$, Im$\lambda_1 = 0$ and Im$\lambda_2 = 0$ (instable regime), (**d**) Re$\lambda_1 > 0$ and/or Re$\lambda_2 > 0$, Im$\lambda_1 \neq 0$ and/or Im$\lambda_2 \neq 0$ (instable regime), (**e**) Re$\lambda_1 = 0$, Re$\lambda_2 = 0$, Im$\lambda_1 \neq 0$ or Im$\lambda_2 \neq 0$ (metastable regime), and (**f**) Re$\lambda_1 < 0$, Re$\lambda_2 < 0$, Im$\lambda_1 \neq 0$ and/or Im$\lambda_2 \neq 0$ (nstable regime)

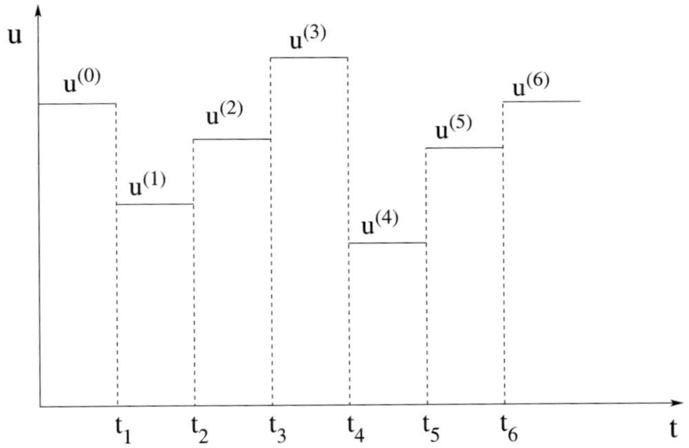

Fig. 5.4. Time discrete control on a continuous time scale

at least formally, solve the evolution equation (2.53) for the period $[t_\mu, t_{\mu+1}]$ considering the initial conditions $X(t_\mu) = X^{(\mu)}$, and we obtain

$$X(t) = \Phi(t, t_\mu, X^{(\mu)}, u^{(\mu)}) \quad \text{for} \quad t_\mu < t < t_{\mu+1} \, . \tag{5.14}$$

Obviously, the current state is a function of the initial state $X^{(\mu)}$, the control $u^{(\mu)}$ chosen at the time t_μ and the current time t. In particular, we obtain for $t = t_{\mu+1}$

$$X^{(\mu+1)} = \Phi(t_{\mu+1}, t_\mu, X^{(\mu)}, u^{(\mu)}) \, , \tag{5.15}$$

which is a functional relation between the states of the system for two control times. If the underlying evolution equations are an autonomous system of differential equations, then (5.15) reduces to

$$X^{(\mu+1)} = \Phi(t_{\mu+1} - t_\mu, X^{(\mu)}, u^{(\mu)}) \tag{5.16}$$

and further, if we focus on equidistant control times $t_{\mu+1} = t_\mu + \Delta t$

$$X^{(\mu+1)} = \Phi(X^{(\mu)}, u^{(\mu)}) , \tag{5.17}$$

where the time difference is now a simple system parameter which is no longer explicitly considered. This is a discrete equation of motion. Because of the facts that (5.17) is the solution of the original evolution equation (2.53) for the time period $[t_\mu, t_{\mu+1}]$ and that the dynamics of the system is confined to the region $\mathbb{G}$, we automatically get the mapping $\Phi : \mathbb{G} \to \mathbb{G}$. The discrete series $\{X^{(1)}, X^{(2)}, \ldots, X^{(\mu)}, \ldots\}$ can be interpreted as a trace of the complete trajectory of the system, which means in the case of a constant control we should expect a discrete realization of a periodic motion, a conditionally periodic motion or a chaotic trajectory similar to the scenarios discussed in Sect. 5.1.2.

Another very instructive method leading to time-discrete controls are related to Poincaré plots. To this aim we imagine that we set up a N_{screen}-dimensional ($N_{\text{screen}} < N$) screen to the phase space defined by the $(N - N_{\text{screen}})$-component screen equation $s(X) = 0$. Every time the system crosses this screen, we make a record of the position and we may change the control function. The set of positions at the screen is called a Poincaré plot; see Fig. 5.5. The solution of the complete equation of motion allows us again

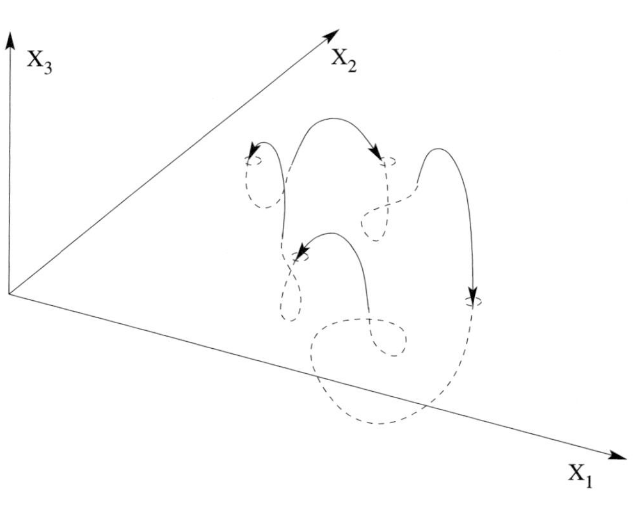

Fig. 5.5. Two-dimensional Poincaré plot generated from a closed orbit in a three-dimensional phase space

to find a mapping of type (5.17). The only difference to the first type of discrete equations is that the control no longer takes place for equidistant times but for all times t_μ satisfying the screen condition $s(X(t_\mu)) = 0$ for a given trajectory.

Discrete evolution equations are not only derivable from mechanical equations[3], they are also characteristic for modelling several processes in physics, chemistry, biology or economics where the internal dynamics of the system is more or less unknown, but the input $\left(X^{(\mu)}, u^{(\mu)}\right)$ and the output $X^{(\mu+1)}$ are connected by deterministic rules. In this sense, mapping (5.17) is not necessarily the result of a well-defined time-continuous evolution equation, as we have argued above, it is also possible that such laws are defined empirically on the basis of a sufficiently large experience. A typical example concerns the exploitation of a Fishery. This example is a very prominent idealized model for understanding the management of renewable resources [58, 59]. The resource stock, i.e., the biomass of the fishes, at the end of a period $[t_{\mu-1}, t_\mu]$ is denoted by $x^{(\mu)}$. With external intervention by fishing, the biological reproduction yields the biomass $x'^{(\mu+1)}$ at time $t_{\mu+1}$ given by $x'^{(\mu+1)} = f(x^{(\mu)})$. It is usual to assume that $f(x) \sim x$ for small x (corresponding to a dominant reproduction of the species in a suitable environment) and $f(x) \to 1$ for $x \to \infty$ (corresponding to saturation effects due to the limited sources of foods and due to the presence of predatory fishes). However, if $y^{(\mu+1)}$ is the harvested in the period $[t_\mu, t_{\mu+1}]$ then the stock at the end of this period is given by $x^{(\mu+1)} = f(x^{(\mu)}) - y^{(\mu+1)}$. Fish harvesting is not costless. In particular, the harvest depends on the biomass resource available for exploitation, $x'^{(\mu+1)}$, and the necessary labor effort, $z^{(\mu+1)}$, for these period, $y^{(\mu+1)} = g(x'^{(\mu+1)}, z^{(\mu+1)})$ with g an increasing, usually concave function in both arguments. Finally, we express the effort, $z^{(\mu+1)}$, as a function of the harvest and the available biomass of fishes by $z^{(\mu+1)} = h\left(y^{(\mu+1)}, x'^{(\mu+1)}\right)$ with h increasing in its first argument and decreasing in its second. Hence, the tree recursion relations

$$\begin{aligned} x^{(\mu+1)} &= f(x^{(\mu)}) - y^{(\mu+1)} \\ y^{(\mu+1)} &= g(f(x^{(\mu)}), z^{(\mu+1)}) \\ z^{(\mu+1)} &= h\left(y^{(\mu+1)}, f(x^{(\mu)})\right) \end{aligned} \tag{5.18}$$

can be solved in order to obtain the closed recursive relation for the evolution of current resource stock $x^{(\mu)} \to x^{(\mu+1)}$.

This example is only one candidate of a large class of empirically constructed discrete evolution equations. Other examples are logistic map discussed below, economic growth models [21, 22], forest management models [19, 20], or the price policy of monopolist [18, 23], but also the mathematical process of generating random numbers by pseudorandom number generators [10].

[3] In contrast that is a rather rare situation and it is only reasonable if the discrete time points play a special role, e.g., in the present case of a discrete control.

A very broad class of physical applications of discrete evolution equations is cellular automata models. Here, relatively simple interaction rules allow the description of complex phenomena [24, 46–51] such as self-organized criticality [29,37], evolution of chemically induced spiral waves [38], oscillations and chaotic behavior of states [38, 39, 51], forest fires [40], earthquakes [41–43], discrete mechanics [44], statistical mechanics [45], the dynamics of granular matter [29, 32], soliton excitations [33], and fluid dynamics [34, 35]. Other applications belong to various domains in biology [36], including neuroscience [30, 31], and the dynamics of traffic systems [25–28].

5.2.2 Chaotic Behavior of Time Discrete Systems

First of all, we will discuss the dynamic behavior of discrete equations of motion for the trivial control $u^{(\mu)} = \lambda = \text{const}$. In this case, we obtain

$$X^{(\mu+1)} = \Phi(X^{(\mu)}, \lambda) \,, \tag{5.19}$$

and the time evolution of the series $\{X^{(0)}, X^{(1)}, X^{(2)}, \ldots.\}$. follows from the recursive application of the function Φ on a given initial state $X^{(0)}$. Let us study the standard example of a logistic map in order to get an impression about the time behavior of discrete evolution equations. The logistic map has a one-component state and is defined by the recursion law

$$x^{(\mu+1)} = \lambda x^{(\mu)} \left(1 - x^{(\mu)}\right) = \phi_{\log}(x^{(\mu)}) \tag{5.20}$$

with the one-component control parameter λ. This model has already been introduced by Verhulst 1845 to simulate the growth of a population in a closed area [11]. Other applications related to economic problems are used to explain the growth of a deposit under progressive rates of interest [12].

As found by several authors [13, 15–17] the iterates $x^{(\mu)}$ $(\mu = 0, 1, \ldots)$ display, as a function of the parameter λ, rather complicated behavior that becomes chaotic at large λ, see Fig. 5.6. The chaotic behavior is not tied to the special form of the logistic map (5.20). Thus, the following results are also characteristic of other functions Φ in (5.19). In particular, the transition from regular (but not necessary simple) behavior to the chaotic regime during the change of an appropriate control parameter is universal behavior for all one-component discrete equations, $x^{(\mu+1)} = f(x^{(\mu)})$, in which the function ϕ has only a single maximum in the properly rescaled unit interval $0 \leq x^{(\mu)} \leq 1$.

It should be remarked that other discrete equations with chaotic properties, for instance, several types of sometimes so-called second-order[4] discrete

[4] We remark that second-order equations with one component can be rewritten into the standard form (5.19) with two components

$$\begin{pmatrix} x^{(\mu+1)} \\ y^{(\mu+1)} \end{pmatrix} = \begin{pmatrix} \phi\left(x^{(\mu)}, y^{(\mu)}\right) \\ x^{(\mu)} \end{pmatrix} .$$

From this point of view, it is not necessary to consider the order of the difference equation as a relevant property.

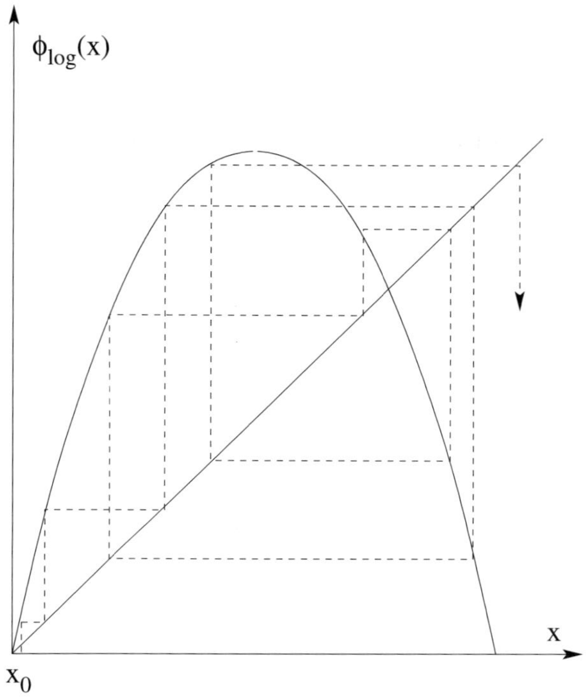

Fig. 5.6. Geometrical construction of the sequence $x^{(1)}, \ldots, x^{(\mu)}, \ldots$: The special schematic representation corresponds to a chaotic dynamics, starting from the initial value x_0 close to the unstable fixed point $x^0 = 0$. Note that $\phi = x$ is the reflection line defining the successive mapping $x^{\mu} \to x^{\mu+1}$

equations $x^{(\mu+1)} = f(x^{(\mu)}, x^{(\mu-1)})$ may belong to other universality classes[5]. However, most of the properties which are valid for the logistic map (5.20) hold at least qualitatively also for other difference equations.

Let us briefly discuss the main properties of the logistic map. The logistic map $x^{(\mu)} \to x^{(\mu+1)}$ has two fixed points, $x^0 = 0$ and $x^0 = 1 - \lambda^{-1}$ satisfying $x^0 = \phi_{\log}(x^0)$. The stability analysis of discrete evolution equations considers, just as in the case of differential equations, see Sect. 3.1.3, weak perturbations of the fixed points, $x^{(\mu)} = x^0 + \varepsilon_\mu$. Thus we obtain $\varepsilon_{\mu+1} = \Lambda\varepsilon_\mu + o(\varepsilon_\mu)$ where the so-called multiplier Λ is given by $\Lambda = \phi'_{\log}(x^0)$. Obviously, for $|\Lambda| < 1$ the fixed point is linearly stable. Conversely, if $|\Lambda| > 1$ the fixed point is unstable. The stability of the marginal case $|\Lambda| = 1$ cannot be decided in the framework of the linear stability analysis, and it requires a more detailed analysis.

For small control parameters, $\lambda < 1$, the quantity $x^{(\mu)}$ develop toward the stable fixed point $x^0 = 0$ because $\phi'_{\log}(0) = \lambda < 1$. For $1 < \lambda < 3$, we get the

[5] That means the number of components considered in (5.19) defines essentially the behavior of the time series $\left\{X^{(0)}, X^{(1)}, \ldots\right\}$.

multiplier $\phi'_{\log}(1-\lambda^{-1}) = 2-\lambda$ so that here the fixed point $x^0 = 1-\lambda^{-1}$ becomes stable.

Both fixed points are unstable for $\lambda > 3$. We call $\lambda_1 = 3$ the first critical value of the control parameter. Now we observe a stable oscillation of period two with the two alternating values $\overline{x}_1^0$ and $\overline{x}_2^0$ which are related together via the equations $\overline{x}_1^0 = \phi_{\log}(\overline{x}_2^0)$ and $\overline{x}_2^0 = \phi_{\log}(\overline{x}_1^0)$. Of course, both values $\overline{x}_1^0$ and $\overline{x}_2^0$ are stable fixed points of the second-iterate map $\phi_{\log}^{(2)}(x) = \phi_{\log}(\phi_{\log}(x))$. In fact, we obtain

$$\overline{x}_{1/2}^0 = \frac{1}{2\lambda}\left[\lambda + 1 \pm \sqrt{(\lambda-3)(\lambda+1)}\right] \tag{5.21}$$

and the corresponding multiplier of the 2-cycle, $\Lambda = 4 + 2\lambda - \lambda^2$, satisfies $|\Lambda| < 1$ for $3 < \lambda < 1 + \sqrt{6}$.

Thus, the unique asymptotic solution $x^{(\mu)} \propto x^0 = 1 - \lambda^{-1}$ for $\mu \to \infty$ splits into alternating solutions $\overline{x}_1^0$ and $\overline{x}_2^0$ while crossing the border $\lambda = 3$. At $\lambda = 3$ the values of $\overline{x}_1^0$ and $\overline{x}_2^0$ coincide and equal $x^0 \doteq 1 - \lambda^{-1} = 2/3$ which shows that the 2-cycle bifurcates continuously from x^0. This bifurcation is sometimes denoted as pitchfork bifurcation.

Above the second critical value of the control parameter, $\lambda_2 = 1 + \sqrt{6}$, the 2-cycle splits into a 4-cycle. Further period-doublings to cycles of period 8, 16, ..., 2^m,... occur as λ increases, Fig. 5.7. The critical values λ_m where the number of fixed points changes from 2^{m-1} to 2^m scale like $\lambda_m = \lambda_\infty - C\delta^{-m}$. Here, C and λ_∞ are specific parameters ($\lambda_\infty = 3.56994\ldots$ for the logistic map), while the Feigenbaum constant $\delta = 4.6692\ldots$ is a universal quantity [9, 14, 15]. All the cycles can be interpreted as attractors of a finite set of points.

For $r > r_\infty$ the asymptotic behavior of the series $x^{(0)}, \ldots, x^{(\mu)}, \ldots$ becomes unpredictable. More precisely, we must say that for certain values of $\lambda > \lambda_\infty$ the sequence never settles down to a fixed point or a periodic cycle. Instead the asymptotic behavior is aperiodic and therefore chaotic. The corresponding attractor changes from a finite to an infinite set of points.

However, the region for $\lambda > \lambda_\infty$ shows a surprising mixture of several periodic p-cycles ($p = 3, 5, 6, \ldots$) and the true chaos. The periodic cycles occur in small λ-windows among other windows with chaotic behavior and also show successive bifurcations p, $2p$, $\ldots 2^n p, \ldots$. The corresponding λ-values scale like the above-mentioned Feigenbaum law, only the nonuniversal constants C and r_∞ are different.

Furthermore, periodic triplings $3^n p$, and quadruplings $4^n p$ occur at $\lambda'_n = \lambda'_\infty - C'\overline{\delta}^{-n}$ with different nonuniversal constants λ'_∞ and C' and different Feigenbaum constants which are again universal for the type of the bifurcation, e.g., $\overline{\delta} = 55.247\ldots$, for the tripling.

Besides the stable fixed points, the stable periodic cycles, and the chaotic sets we also find a set of unstable fixed points and periodic orbits which are partially the continuation of the stable regimes to control parameters above the corresponding critical value. For example, the above-introduced

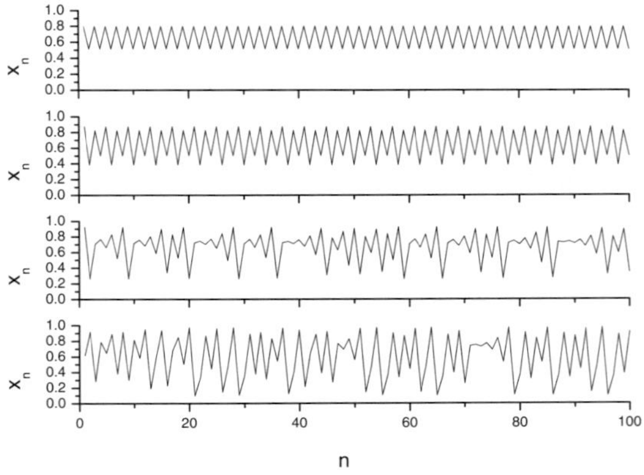

Fig. 5.7. Varios dynamic regimes of the discrete logistic map recursion law: periodic oscillations with period 2, $\lambda = 3.2$ (*top*), periodic oscillations with period 4, $\lambda = 3.5$, chaotic behavior with $\lambda = 3.7$ and chaotic behavior with $\lambda = 3.9$ (*bottom*)

fixed points $x^0 = 0$ and $x^0 = 1 - \lambda^{-1}$ and the 2-cycle (5.21) also exists as unstable orbits for $\lambda > 3$ and $\lambda > 1 + \sqrt{6}$, respectively, and they are also embedded in the chaotic set.

Finally, we remark that the attraction of an arbitrary initial condition $x^{(0)} \in [0,1]$ to a stable orbit or fixed point indicates, from a physical point of view, a hidden dissipative process. In other words, the logistic map has no background in the conservative classical mechanics.

On the other hand, the discrete trajectories obtained from the recursion law (5.17) for mechanical equations of motion can be directly obtained from the continuously trajectories discussed in Sect. 5.1.2 by marking the positions at the discrete times t_μ. Although there are no fixed points and attractive orbits, we again find chaotic sets embedded within it a large number of unstable periodic orbits.

5.2.3 Control of Time Discrete Equations

As we have remarked in the previous chapter, a chaotic set on which the trajectory of the chaotic process lives embeds within it a large number of unstable periodic orbits. Let us assume that the control aim consists in the realization of one of these finite orbits of periodicity m which may be defined by the finite set $\mathbb{C}_m = \left\{Y^{(1)}, Y^{(2)}, \ldots, Y^{(m)}\right\}$ and the periodic boundary condition $Y^{(n+m)} = Y^{(n)}$. Obviously, this control aim can be interpreted as a discrete version of a special tracking problem. The mapping between consecutive values of Y are simply given by the law

$$Y^{(\mu+1)} = \Phi(Y^{(\mu)}, \lambda) \tag{5.22}$$

for a well-defined fixed control λ, i.e., we consider that the set $\mathbb{C}_m$ contains m unstable solutions of

$$Y = \underbrace{\Phi(\ldots.\Phi(\Phi(\Phi(Y,\lambda),\lambda),\lambda),\ldots.,\lambda)}_{m\text{-times}}, \tag{5.23}$$

which are connected[6] by (5.22).

The evolution of the system under control follows the recursion law (5.17). With the special choice $u = \lambda$, the trajectory of the system, given by the set $\{X^{(1)}, X^{(2)}, \ldots, X^{(m)}, \ldots\}$, will diverge from the desired unstable orbit exponentially even if the initial conditions are nearly identical to one of the m states $Y^{(\mu)} \in \mathbb{C}_m$. We know that because of the ergodicity, the chaotic trajectory visits or accesses an arbitrarily small neighborhood of the subsequent value $Y^{(\mu+1)}$ of the desired orbit after a sufficiently long period. But this knowledge is only helpful in obtaining a suitable starting point at which the control starts. Obviously, we need an active, i.e., time dependent, control $u^{(\mu)}$ in order to stabilize the evolution of the system with respect to the control aim. We assume that the admissible values of $u^{(\mu)}$ are located in a small sphere of radius ε and center λ

$$u^{(\mu)} \in \mathbb{U}' \quad \text{if} \quad \left|u^{(\mu)} - \lambda\right| \leq \varepsilon\,, \tag{5.24}$$

which is a subset of the control space, $\mathbb{U}' \subset \mathbb{U}$.

Since the initially chosen value of the control function u is usually not equal to the special value λ defining the control aim, the trajectory of the system will diverge from the desired cycle exponentially even the initial conditions are identical. Furthermore, we assume, that at time t_μ, the trajectory falls into the neighborhood of $Y^{(\eta)}$. Without any restriction, we may define $\eta = \mu \bmod m$. Then the desired next value, $Y^{(\eta+1)}$, is given by (5.22) while the next value of the controlled trajectory, $X^{(\mu+1)}$, is given by (5.17). The difference between both the new values is now

$$X^{(\mu+1)} - Y^{(\eta+1)} = \Phi(X^{(\mu)}, u^{(\mu)}) - \Phi(Y^{(\eta)}, \lambda) \tag{5.25}$$

or with $\delta X^{(\mu)} = X^{(\mu)} - Y^{(\eta)}$ and $\delta u^{(\mu)} = u^{(\mu)} - \lambda$

$$\begin{aligned} \delta X^{(\mu+1)} &= \Phi(Y^{(\eta)} + \delta X^{(\mu)}, \lambda + \delta u^{(\mu)}) - \Phi(Y^{(\eta)}, \lambda) \\ &= \frac{\partial \Phi(Y^{(\eta)}, \lambda)}{\partial Y^{(\eta)}} \delta X^{(\mu)} + \frac{\partial \Phi(Y^{(\eta)}, \lambda)}{\partial \lambda} \delta u^{(\mu)}\,. \end{aligned} \tag{5.26}$$

This equation may be rewritten in the standard form of a linear problem

$$\delta X^{(\mu+1)} = A_\eta \delta X^{(\mu)} + B_\eta \delta u^{(\mu)}\,, \tag{5.27}$$

where the A_η are matrices[7] of type $N \times N$ and the B_η are matrices of type $N \times n$

[6] It is necessary to say that the solutions must be connected by (5.22), because (5.23) is often satisfied for more than one finite sets $\mathbb{C}_m$ of m nonidentical values $Y^{(\mu)}$. For example, the logistic map has two fixed points, $x^* = 0$ and $x^* = 1-\lambda^{-1}$, which are not connected by (5.22), i.e. the fixed points form two seperate sets $\mathbb{C}_1$.

[7] Note that the dimension of the phase space is N while the dimension of the control space is n.

$$A_\eta = \frac{\partial \Phi(Y^{(\eta)}, \lambda)}{\partial Y^{(\eta)}} \quad \text{and} \quad B_\eta = \frac{\partial \Phi(Y^{(\eta)}, \lambda)}{\partial \lambda} \quad (\eta = 1, \ldots, m) \,. \tag{5.28}$$

The aim is now to find a linear control law

$$\delta u^{(\mu)} = K_\eta \delta X^{(\mu)} \tag{5.29}$$

with the $n \times N$ matrix K_η so that $\left|\delta X^{(\mu+1)}\right| < \left|\delta X^{(\mu)}\right|$. In this case, the unstable cycle becomes stable due to the external control. This is by no means a trivial problem and its solution requires some theoretical preparations. To solve this problem, we focus on the special case that the system should be stabilized to an unstable fixed point. Thus we must take into account only three matrices, $A_1 = A$, $B_1 = B$, and $K_1 = K$. A generalization to the stabilization of unstable m-cycles is straightforwardly obtainable [60].

5.2.4 Reachability and Stabilizability

In order to solve the above-introduced problem, we have to explain the term reachability. That is, roughly speaking, the property that each point of the phase space $\mathbb{P}$ can be reached by series of successive control steps. In the case of a linear system

$$X^{(\mu+1)} = AX^{(\mu)} + Bu^{(\mu)} \tag{5.30}$$

characterized by the matrices A and B, reachability is similar to the requirement that there must exist an input $\left\{u^{(1)}, \ldots, u^{(N)}\right\}$ that transfers the state $X^{(\mu)}$ from the origin $X^{(1)} = 0$ to an arbitrary point of the phase space. In this case we say the pair (A, B) is also reachable. To decide if a pair (A, B) is reachable, we consider first a one-dimensional control, $n = 1$. Then we obtain

$$\begin{aligned}
X^{(2)} &= Bu^{(1)} \\
X^{(3)} &= ABu^{(1)} + Bu^{(2)} \\
&\vdots \\
X^{(N+1)} &= A^{N-1}Bu^{(1)} + A^{N-2}Bu^{(2)} + \ldots + ABu^{(N-1)} + Bu^{(N)}
\end{aligned} \tag{5.31}$$

with N free parameters $u^{(1)}, \ldots, u^{(N)}$. Now we have two possibilities: either the N vectors B, $AB, \ldots, A^{N-1}B$ are linearely independent or not. In the first case, these vectors form a complete basis of the phase space, i.e., each point of this space is reachable at the latest after N steps. For the second case we assume that $C = A^\kappa B$ with $(1 < \kappa < N - 1)$ is the first vector which is linearly dependent from the previous vectors, i.e., there exists a relation of the form

$$C = \sum_{i=0}^{\kappa-1} \alpha_i A^i B \,. \tag{5.32}$$

But in this case, the next vector also depends linearly on $A^i B$ $(i = 0, \ldots, \kappa-1)$. In fact, we obtain

$$A^{\kappa+1}B = AC = \sum_{i=0}^{\kappa-1} \alpha_i A^{i+1}B = \alpha_{\kappa-1}A^{\kappa}B + \sum_{i=1}^{\kappa-1} \alpha_{i-1}A^{i}B$$
$$= \alpha_{\kappa-1}C + \sum_{i=1}^{\kappa-1} \alpha_{i-1}A^{i}B$$
$$= \sum_{i=0}^{\kappa-1} \left[\alpha_{\kappa-1}\alpha_i + \alpha_{i-1}\right] A^{i}B \qquad (5.33)$$

with $\alpha_{-1} = 0$. The recursive application of this procedure shows that all subsequent vectors are linearly dependent from the first κ vectors. That means, the infinite large set of vectors $A^{\mu}B$ ($\mu = 0, 1, 2, \ldots$) spans only a κ-dimensional subspace of the phase space and, consequently, not all points of the phase space are reachable after an arbitrary number of control steps. Thus, we conclude that in the case of a one-dimensional control, a pair (A, B) is reachable if

$$\operatorname{rank}\left(A^{N-1}B, A^{N-2}B, \ldots, AB, B\right) = N\,. \qquad (5.34)$$

In the case of more than one control parameter, we may use the same argumentation as before. Especially, we find that the pair (A, B) is reachable if an arbitrary vector $X^{(N+1)}$ must be presented by the superposition

$$X^{(N+1)} = A^{N-1}Bu^{(1)} + A^{N-2}Bu^{(2)} + \cdots + ABu^{(N-1)} + Bu^{(N)} \qquad (5.35)$$

otherwise there exists a subspace of $\mathbb{P}$ which is unreachable after an arbitrarily large number of steps, i.e., the pair (A, B) is not reachable. The necessary condition for the reachability is again (5.34), but now for the $N \times n$ matrix B.

Let us assume that the pair (A, B) is reachable. Then we ask for a control law $u = KX$ so that the mapping

$$X^{(\mu)} \to X^{(\mu+1)} = (A + BK)\, X^{(\mu)} \qquad (5.36)$$

is a contraction. This requires that all eigenvalues ξ_i, $i = 1, \ldots, N$, of the matrix $A+BK$ are smaller than unity, $|\xi_i| < 1$. This problem belongs to the stabilizability of the system under control. We focus again to a one-dimensional input, $n = 1$. For higher dimensions we refer to the literature [61–63]. We get the surprising result that one can arbitrarily choose the spectrum of the matrix $(A + BK)$ if the pair (A, B) is reachable. For such systems there always exists a similarity transformation T yielding a transformed pair

$$A' = T^{-1}AT \quad \text{and} \quad B' = T^{-1}B \qquad (5.37)$$

so that we obtain the controller canonical form [62, 64]

$$A' = \begin{pmatrix} -a_{N-1} & -a_{N-2} & -a_{N-3} & \cdots & -a_1 & -a_0 \\ 1 & 0 & 0 & \cdots & 0 & 0 \\ 0 & 1 & 0 & \cdots & 0 & 0 \\ \vdots & \ddots & \ddots & \ddots & \vdots & \vdots \\ 0 & \cdots & 0 & 1 & 0 & 0 \\ 0 & \cdots & \cdots & 0 & 1 & 0 \end{pmatrix} \quad \text{and} \quad B' = \begin{pmatrix} 1 \\ 0 \\ 0 \\ \vdots \\ 0 \\ 0 \end{pmatrix} . \tag{5.38}$$

It is simple to check that a_i $(i = 1, \ldots, N)$ are the coefficients of the characteristic polynomial of A' and because of (5.37) also of A,

$$\det(z - A') = \det(z - A) = z^N + \sum_{i=0}^{N-1} a_i z^i . \tag{5.39}$$

Transforming the feedback $K \to K' = KT$ and the state vector $X'^{(\mu)} = T^{-1}X^{(\mu)}$ also yields the transformed mapping (5.36)

$$X'^{(\mu+1)} = (A' + B'K')\, X'^{(\mu)} = T^{-1}\,(A + BK)\, TX'^{(\mu)} . \tag{5.40}$$

Thus the transformed matrix $A' + B'K'$ has the same eigenvalues as $A + BK$, since both matrices are related by a similarity transformation. Now we choose the eigenvalues of $A' + B'K'$to be $\{\xi_1, \xi_2, \ldots, \xi_N\}$ with $|\xi_i| < 1$ for $i = 1, \ldots, N$. Then, the characteristic polynomial for this free choice is given by

$$\prod_{i=1}^{N} (z - \xi_i) = z^N + \sum_{i=0}^{N-1} a_i' z^i , \tag{5.41}$$

where the coefficients a_i' follows immediately from the algebraic expansion of the left-hand side. It is clear that by choosing

$$K' = \left(a_{N-1} - a'_{N-1}, a_{N-2} - a'_{N-2}, \ldots, a_0 - a'_0\right) , \tag{5.42}$$

the matrix $A' + B'K'$ has the same form as A', but with the coefficients a_i replaced by a_i'. Obviously, this discussion shows that the eigenvalues of every reachable pair can be arbitrarily assigned. Simultaneously, we have obtained a simple method for constructing the feedback $K = K'T^{-1}$. The transformation T is obtainable in two steps. The first step uses the controllability matrix

$$T_1 = (A^{N-1}B, A^{N-2}B, \ldots, AB, B) , \tag{5.43}$$

which transforms the pair (A, B) to

$$A'' = T_1^{-1} A T_1 \quad \text{and} \quad B' = T^{-1} B \tag{5.44}$$

with

$$A'' = \begin{pmatrix} 0 & \cdots & 0 & -a_0 \\ 1 & \ddots & \vdots & -a_1 \\ & \ddots & 0 & \vdots \\ & & 1 & -a_{N-1} \end{pmatrix} \quad \text{and} \quad B'' = \begin{pmatrix} 1 \\ 0 \\ 0 \\ 0 \end{pmatrix} , \tag{5.45}$$

from where we also obtain the coefficients of the characteristic polynomial. A further transformation,

$$A' = T_2^{-1} A'' T_2 \quad \text{and} \quad B' = T^{-1} B'' \tag{5.46}$$

with

$$T_2 = \begin{pmatrix} 1 & a_{N-1} & \cdots & a_2 & a_1 \\ 0 & 1 & \ddots & \vdots & \vdots \\ & 0 & \ddots & a_{N-1} & a_{N-2} \\ & & \ddots & 1 & a_{N-1} \\ & & & 0 & 1 \end{pmatrix} \tag{5.47}$$

yields the requested form (A', B'). Thus, we have $K' = KT_1T_2$ and therefore the desired matrix K

$$K = K' T_2^{-1} T_1^{-1} , \tag{5.48}$$

where K' is given by (5.42). The components of K' are directly obtainable from the characteristic polynomial of the matrix A (5.39) and from the desired eigenvalues of $A + BK$ via (5.41).

Finally, we remark that these considerations also remain valid for time-continuous linear systems. We refrain from a presentation of this, in comparison to the discrete version, give only slightly modified proof, and refer to the literature [61–63].

5.2.5 Observability

Let us now discuss problems related to a control of system if only a reduced information about the current system state is available. In contrast to what was followed in Chap. 6, we suppose that the equations of motion of the system are still completely known, but the complete state vector X is no longer measurable by the available experimental instruments. In other words, the observability of the system may be restricted to some overall functions of the current state. That means from a mathematical point of view that the state vector $X \in \mathbb{P}$, projected onto an observation vector $Y \in \mathbb{O}$, is an element of the observation space $\mathbb{O} \subset \mathbb{P}$. Since the dimension of the observation space is usually lower than the dimension of the phase space, the projection $X \to Y$ implies a loss of information. This statement applies especially to a control of the system, since the control law consists now in a relation between the control function u and the observation vector Y. It is important to note that observability does not mean the reconstruction of the state X from one measurement Y but the construction of a control u from all previous observations Y. We consider here again the discrete linearized law (5.30) extended by the mapping $Y^{(\mu)} = CX^{(\mu)}$. The initial state may be $X^{(1)}$. Then we obtain from (5.30) for the free system $(u = 0)$

$$
\begin{aligned}
Y^{(1)} &= CX^{(1)} \\
Y^{(2)} &= CAX^{(1)} \\
&\vdots \\
Y^{(N)} &= CA^{N-1}X^{(1)} \,.
\end{aligned}
\tag{5.49}
$$

The space spanned by the vectors $CX^{(1)}$, $CAX^{(1)}, \ldots, CA^{N-1}X^{(1)}$ considering all $X^{(1)} \in \mathbb{P}$ must have the same dimension N as the phase space in order to map an arbitrary initial state $X^{(1)}$ completely to the set of observations. If this is not the case, we may argue in a manner similar to the problem of reachability that then also the consideration of higher vectors $CA^N X^{(1)}$, $CA^{N+1}X^{(1)}$ can never lead to a complete observability. Hence, we can say that the dynamical process is observable if

$$
\operatorname{rank} \begin{pmatrix} C \\ CA \\ \vdots \\ CA^{N-1} \end{pmatrix} = N \,. \tag{5.50}
$$

Again, this statement can be applied in the present form to the characterization of linear evolution equations. We find that a linear system

$$
\dot{X} = AX + Bu \quad \text{and} \quad Y = CX \tag{5.51}
$$

is observable if (5.50) holds. Furthermore, in the case of nonlinear evolution equations we may linearize these equations with respect to a certain reference point. This allows us to introduce locally defined observability (and also reachability) conditions.

5.3 Time-Continuous Chaos Control

5.3.1 Delayed Feedback Control

We have already found that time-continuous trajectories of dynamical systems may form, in particular, unstable periodic orbits. Such an orbit is characterized by cyclic boundary conditions, $\overline{X}(t) = \overline{X}(t + T_{\text{orbit}})$, where T_{orbit} is the period. Now we will present a control which stabilizes this orbit. Furthermore, we consider the above-introduced projection concept. The whole N-dimensional state vector $X(t)$ is not the measurable quantity but the lower dimensional vector,

$$
Y(t) = \beta(X(t)) \tag{5.52}
$$

where $\beta : \mathbb{P} \to \mathbb{O}$ with $\dim \mathbb{O} = \mathrm{N}' < N$, is. From here we construct the control function [65]

$$
u(t) = U(Y(t) - Y(t - T)) \tag{5.53}
$$

with a freely maneuverable $n \times N'$ matrix U defining the coupling between the measurements and the control, where the basic idea of delayed feedback control methods enters. The time T is called the delay time. The coupling to the internal dynamics is given by the equations of motion (2.53) which have now the time-delayed form

$$\dot{X}(t) = F\left(t, X(t), U(\beta(X(t)) - \beta(X(t-T)))\right) . \tag{5.54}$$

For the sake of simplicity, we assume that the explicit time-dependence of the function F is also periodic with T, i.e., $F(t, X, u) = F(t + T, X, u)$. This condition is helpful in defining experimentally the period of the unstable orbits especially in case that the orbit periods T_{orbit} are integer multiples of the external driving period. Let us now fix $T_{\text{orbit}} = T$ for what follows[8]. The existence of the unstable periodic orbit poses the reference equation

$$\dot{\overline{X}}(t) = F\left(t, \overline{X}(t), 0\right) . \tag{5.55}$$

Now we consider a small deviation from the unstable orbit, $Y(t) = X(t) - \overline{X}(t)$, and a small external control. This allows us to linearize the evolution equation for the system under control (5.54) with respect to the unstable periodic orbit constraint

$$\dot{Y}(t) = A(t)Y(t) + B(t)\left[Y(t) - Y(t-T)\right] \tag{5.56}$$

with

$$A(t) = \left.\frac{\partial F(t, X, 0)}{\partial X}\right|_{X=\overline{X}(t)} \tag{5.57}$$

and

$$B(t) = \left.\frac{\partial F(t, \overline{X}(t), u)}{\partial u}\right|_{u=0} U \left.\frac{\partial \beta(X)}{\partial X}\right|_{X=\overline{X}(t)} . \tag{5.58}$$

Obviously, the $N \times N$ matrices $A(t)$ and $B(t)$ are periodic functions with the period T. The periodically linear time dependent control equation (5.56) can be decomposed into eigen-functions according to the Floquet theory [66]. To this aim we use the ansatz

$$Y(t) = \exp\{-\gamma t\} \, y(t) , \tag{5.59}$$

where γ is the characteristic exponent or the Floquet exponent and $y(t)$ is a periodic function, $y(t) = y(t + T)$. The substitution of (5.59) in (5.56) yields

$$\dot{y}(t) = \left[A(t) + B(t)\left(1 - e^{\gamma T}\right) + \gamma\right] y(t) . \tag{5.60}$$

This is an eigenequation with eigenvalues γ and eigenfunctions $y(t)$. The solution of these equation requires special knowledge about the structure of the

[8] The generalization of the problem to different time scales for the external driving period, the delay time, and the orbit period (as integer multiples of the driving period) is, of course, always possible.

matrices A and B and therefore about the dynamics of the system [67]. We avoid the problem and focus on the discussion of some general aspects. To this aim, we consider first the uncontrolled evolution equation which follows from (5.60) setting $B(t) \equiv 0$,

$$\dot{y}(t) = [A(t) + \overline{\gamma}] \, y(t) \; , \tag{5.61}$$

with $\overline{\gamma}$ an eigenvalue of the uncontrolled problem. The geometrical meaning of the Floquet exponents $\overline{\gamma}_i$ $(i = 1, \ldots, N)$ is quite clear from (5.59). The real parts $\overline{\gamma}'_i = \mathrm{Re}\overline{\gamma}_i$ define the radial attraction to the unstable orbit while the imaginary parts $\overline{\gamma}''_i = \mathrm{Im}\overline{\gamma}_i$ determine the convolution of the trajectory around the reference orbit. Since the periodic orbit was assumed to be unstable, at least one eigenvalue has a negative real part.

Now we are ready to consider the complete equation (5.60). For the moment we introduce the quantity $\sigma = 1 - \mathrm{e}^{\gamma T}$ as a free parameter [71]. Thus, the eigenvalues of (5.60) for a well-defined dynamics of the system, given by the matrix $A(t)$, and a control law defined by the matrix U and contained in $B(t)$, are simple functions of the parameter σ, i.e., we have

$$\gamma_i = \varphi_i(U, \sigma) \qquad \text{with} \qquad \varphi_i(U, 0) = \varphi_i(0, \sigma) = \overline{\gamma}_i \; . \tag{5.62}$$

Now we are able to estimate the behavior of the eigenvalues qualitatively. We expand (5.62) in terms of σ and all components of U up to the first non-trivial order and we recall that σ is also a function of the Floquet exponent

$$\begin{aligned} \gamma_i &\approx \varphi_i(0,0) + \sum_{\alpha=1}^{n} \sum_{\beta=1}^{N'} R_i^{\alpha\beta} U_{\alpha\beta} \sigma \\ &= \overline{\gamma}_i + \sum_{\alpha=1}^{n} \sum_{\beta=1}^{N'} R_i^{\alpha\beta} U_{\alpha\beta} \left(1 - \mathrm{e}^{\gamma_i T}\right) \end{aligned} \tag{5.63}$$

with

$$R_i^{\alpha\beta} = \left. \frac{\partial^2 \varphi_i(U, \sigma)}{\partial U_{\alpha\beta} \partial \sigma} \right|_{U=0, \sigma=0} \; . \tag{5.64}$$

The coefficients $R_i^{\alpha\beta}$ contain all the details of the system, i.e., the dynamics of the state X, the coupling between the control and the system, and the projection of the state to the measurable quantities Y. These coefficients should, in principle, be obtainable from suitable experiments. Then we have nN' free coefficients $U_{\alpha\beta}$ which must be chosen in such a way that $\mathrm{Re}\gamma_i \geq 0$ for all Floquet exponents. Of course, this is a not always solvable; linear optimization problem which can be treated by several standard methods are briefly discussed in Chap. 10.

Time-delayed feedback control methods may allow us to stabilize periodic orbits in chaotic systems [65]. It is interesting that the shape of the unstable trajectory remains unchanged also in the presence of a finite control force. Several applications of this control concepts have been discussed in the context of balancing by humans [68], the control of mechanical oscillating metal beams [70], or of CO_2 lasers systems [69].

5.3.2 Synchronization

Synchronization is a widespread phenomenon observed between coupled systems. For example the well-known Belousov–Zhabotinskii chemical reaction can be chaotic, but it is spatially uniform [57]. Hence, all spatial regions are obviously synchronized with each other, even if the basic dynamics is a chaotic motion. However, synchronization is not universal. In other circumstances the uniformity of the Belousov–Zhabotinskii reaction becomes unstable and a pronounced spatiotemporal dynamics occurs.

The first quantitative observation of a synchronization phenomenon is attributed to Huygens in 1673 during his experiments for developing improved pendulum clocks [72]. Two clocks were found to oscillate with the same frequency due to the very weak coupling in terms of the nearly imperceptible oscillations of the trestle on which both clocks were hanging. But we realize again that synchronization is not a universal phenomena, because it was observed by Huygens only if the individual frequencies of the clocks are almost coincided.

The problem of a control of chaotic systems can at least be partially solved by the application of synchronization effects. This is possible if one of the coupled systems serves as a controller that is connected to the system. The goal of the control is to make this system follow a prescribed time evolution, i.e., a tracking protocol. We may interpret this behavior as a synchronization of the system under control with the dynamics of the controller.

Here, we will discuss a very simple control mechanism using synchronization effects. Let us assume we have a system the dynamics of which is described by the evolution equation (2.53) and a well-defined optimal control problem. The solution of this problem are the control equations which are defined by the optimal evolution equation (2.74), the control law (2.75), and the adjoint evolution equation (2.70). As discussed in Chap. 2, the control law is usually a set of algebraic relations between the vectorial control functions u^*, the generalized momenta P^*, and the state vector X^* which may be used for the elimination of the control function $u^* = u^*(X^*, P^*, t)$. Unfortunately, this representation is not unique. Because of the fact that the differential equations (2.74), (2.75), and (2.70) solve the optimal control problem, we obtain the complete solution $(X^*(t), P^*(t), u^*(t))$. From here, we have infinite number of ways to eliminate the time in $u^*(t)$ partially or completely by suitable combinations of X^* and P^*. Hence, there exist infinite number of different relations $u^* = u^*(X^*, P^*, t)$. The main problem of using synchronization effects for driving a system to the optimal dynamics is to find an appropriate equation connecting u^*, X^*, and P^*. This usually requires heuristic experience.

Let us assume that we have such a relation. Hence, we obtain the optimal evolution equations

$$\dot{X}^* = F(X^*, u^*(X^*, P^*, t), t) = \widetilde{F}(X^*, P^*, t) \tag{5.65}$$

and the corresponding adjoint evolution equations which may be written in the form

$$\dot{P}^* = V(X^*, u^*(X^*, P^*, t), t) = \widetilde{V}(X^*, P^*, t) \ . \tag{5.66}$$

Both equations may be implemented in a computer (the controller) which is connected with the real system, described also by the evolution equation (2.53). The control mechanism of the system may now be constructed in such a way that the control function u is given by

$$u = u^*(X, P^*, t) \ , \tag{5.67}$$

where X belongs to the current state of the system while the momenta P^* are the "control signals" coming from the computer[9]. Thus, the system dynamics is given by

$$\dot{X} = F(X, u^*(X, P^*, t), t) = \widetilde{F}(X, P^*, t) \ . \tag{5.68}$$

Formally, we now have two coupled systems the dynamics of which is defined by (5.65) and (5.68). The first, the active part, given by the controller, drives the second, which is called the passive part[10].

If the initial conditions of the real system are equal to that of the optimal trajectory, $X(0) = X^*(0)$, the future evolution of the controlled system is, of course, determined by $X(t) = X^*(t)$. But in the more realistic case that the initial conditions are not equal, synchronization can help drive the system to the nominal trajectory $X^*(t)$.

The evolution of the difference $Y(t) = X(t) - X^*(t)$ between the system evolution and the nominal curve is given by

$$\dot{Y}(t) = \widetilde{F}(X^*(t) + Y(t), P^*(t), t) - \widetilde{F}(X^*(t), P^*(t), t) \ . \tag{5.69}$$

Thus, we conclude that synchronization of the nominal system and the controlled system occurs if (5.69) has a stable fixed point for $Y = 0$. For small $Y(t)$ this can often be proved by using the linear stability analysis which requires that the matrix $\partial \widetilde{F}(X^*, P^*, t) / \partial X^*$ is negative definite along the optimal trajectory.

In order to get an impression about the importance of a appropriate choice of the control law $u^* = u^*(X^*, P^*, t)$, we consider a standard linear quadratic problem with constant coefficients[11]. Here we have

$$F(X, u, t) = AX + Bu \tag{5.70}$$

---

[9] We suppose that the computer is fast enough to communicate with the system in a real time mode.

[10] The physical structure of the coupling suggests that the two coupled systems apparently represent a special version of an open-loop control. But this is not correct, because the control function u and the state X are originally connected via the feedback law (5.67).

[11] That is the optimal regulator problem; see Sect. 3.3.

and the control law $u^* = -R^{-1}B^TGX^*$; see (3.55) or $u^* = R^{-1}B^TP^*$, and (3.46). The matrix R is due to the quadratic performance and G is the solution of the corresponding algebraic Ricatti equation; see Sect. 3.3.1. Now we may combine both equivalent representations to

$$u^* = \alpha R^{-1}B^TP^* + (\alpha - 1)R^{-1}B^TGX^* \tag{5.71}$$

with an arbitrary parameter α. Thus, (5.67) reads

$$u = \alpha R^{-1}B^TP^* + (\alpha - 1)R^{-1}B^TGX \ , \tag{5.72}$$

and we obtain

$$\widetilde{F}(X, P^*, t) = \left[A + (\alpha - 1)R^{-1}B^TG\right] X + \alpha BR^{-1}B^TP^* \ . \tag{5.73}$$

Hence, the stability matrix is given by

$$M = \frac{\partial \widetilde{F}(X^*, P^*, t)}{\partial X^*} = A + (\alpha - 1)R^{-1}B^TG \ . \tag{5.74}$$

For $\alpha = 1$ we get $M = A$ and the stability of the synchronized system is the same as the stability of the free system, which means if the system is unstable, the synchronization can never be successful. For $\alpha = 0$ we always obtain stable behavior of the controlled system and therefore of the synchronization procedure. The only condition which must be satisfied is that the pair (A, B) is reachable. For all other values of the parameter α, the decision if the synchronized system is stable or not depends on the structure of the controlled system.

Finally, we remark that the application of synchronization techniques on control theoretical problems is a well-established topic in nonlinear dynamics [2–4, 8]. Several controlling strategies have been investigated to drive a response system into a synchronized state [5–7].

References

1. A.N. Kolmogorov: Dokl. Akad. Nauk. USSR **98**, 527 (1954)
2. A. Isidori: *Nonlinear Control Systems* (Springer, Berlin Heidelberg New York, 1989)
3. R. Konnur: Phys. Rev. Lett. **77**, 2937 (1996)
4. W.S. Levine: *The Control Handbook* (CRC Press, New York, 1996)
5. T.C. Newell, P.M. Alsing, A. Grvrielides, V. Kovanis: Phys. Rev. E **49**, 313 (1994)
6. Y.-C. Lai, C. Grebogi: Phys. Rev. Lett. **77**, 5047 (1996)
7. T.C. Newell, P.M. Alsing, A. Grvrielides, V. Kovanis: Phys. Rev. E **51**, 2963 (1995)
8. H. Nijmeijer, A.J. van der Schaft: *Nonlinear Dynamical Control Systems* (Springer, Berlin Heidelberg New York, 1990)
9. H.G. Schuster: *Deterministic Chaos: An Introduction*, 2nd edn (VCH Verlagsgesellschaft, Weinheim, 1988)

10. S. Tezuka: *Uniform Random Numbers: Theory and Practice* (Kluwer Academic Publishers, Dordrecht, 1995)
11. A. Quetelet, P.F. Verhulst: Annuaire de l'Académie Royale des Sciences de Belgique **16**, 97 (1850)
12. H.O. Peitgen, P.H. Richter: *The Beauty of Fractals* (Springer, Berlin Heidelberg New York, 1986)
13. S. Grossmann, S. Thomae: Z. Naturforsch. **32A**, 1353 (1977)
14. S.H. Strogatz: *Nonlinear Dynamics and Chaos* (Addison-Wesley, Reading, MA, 1994)
15. M.J. Feigenbaum: J. Stat. Phys. **19**, 25 (1978)
16. P. Coullet, J. Tresser: J. Phys. (Paris) **C5**, 25 (1978)
17. R.M. May: Nature **261**, 459 (1976)
18. J. Robinson: *Economics of Imperfect Competition* (Macmillian, London, 1933)
19. T. Mitra, H.Y. Wan: Rev. Econ. Stud. **52**, 263 (1985)
20. T. Mitra, H.Y. Wan: J. Econ. Theory **40**, 229 (1986)
21. T.N. Srinivasan: Econometrica **32**, 358 (1964)
22. H. Uzawa: Rev. Econ. Stud. **31**, 1 (1964)
23. T. Puu: Chaos, Solitons and Fractals **5**, 35 (1995)
24. U. Frisch, D. d'Humiéres, B. Hasslacher, P. Lallemand, Y. Pomeau, J.-P. Rivet: Complex Syst. **1**, 649 (1987)
25. O. Biham, A.A. Middleton, D. Levine: Phys. Rev. A **46**, R6124 (1992)
26. D. Chowdbury, J. Kertezs, K. Nagel, L. Santen, A. Schadschneider: Phys. Rev. E **61**, 3270 (2000)
27. M. Cremer, A.D. May: *An Extended Traffic Model for Freeway Control.* Technical Report UCB-ITS-RR-85-7 (Institute of Transportation Studies, University of California, Berkeley, CA)
28. K. Nagel, M. Schreckenberg: J. Phys. I **2**, 2221 (1992)
29. P. Bantay, I.M. Janosi: Phys. Rev. Lett. **68**, 2058 (1992)
30. D.J. Amit: *Modeling Brain Function: The World of Attractor Neural Networks* (Cambridge University Press, Cambridge, 1989)
31. E. Basar: *Chaos in Brain Function* (Springer, Berlin Heidelberg New York, 1990)
32. G. Peng, H.J. Heermann: Phys. Rev. E **49**, 1796 (1994)
33. M. Schulz, S. Trimper: J. Phys. A: Math. Gen. **33**, 7289 (2000)
34. S. Chen, H. Chen, D. Martinez, W. Matthaeus: Phys. Rev. Lett. **67**, 3776 (1991)
35. U. Frisch, B. Hasslacher, Y. Pomeau: Phys. Rev. Lett. **56**, 1505 (1986)
36. G.B. Ermentrout, L. Edlestein-Keshet: J. Theoret. Biol. **160**, 97 (1993)
37. Z.H. Olami, J.S. Feder, K. Christensen: Phys. Rev. Lett. **68**, 1244 (1992)
38. M. Markus, B. Hess: Nature (London) **347**, 56 (1990)
39. M.A. Nowak, R.M. May: Nature (London) **359**, 826 (1992)
40. O. Baran, C.C. Wan, R. Harris: comp-gas/9805001
41. C.C. Barton, P.R. La Pointe: *Fractals in the Earth Science* (Plenum Press, New York, 1995)
42. A. Crisanti, M.H. Jensen, A Vulpiani, G. Paladin: Phys. Rev. A **46**, R7363 (1992)
43. D. Sornette, P. Miltenberger, C. Vanneste: Pure Appl. Geophys. **142**, 491 (1994)
44. J. Baez, J. Gilliam: Lett. Math. Phys. **31**, 205 (1994)
45. L. Wagner: Phys. Rev. E **49** 2115 (1994)
46. L. Kadanoff, G. McNamara, G. Zanetti: Phys. Rev. A **40**, 4527 (1989)
47. S. Wolfram: *Theory and Application of Cellular Automata* (World Scientific, Singapore, 1986)

48. S. Wolfram: Cellular *Automata and Complexity* (Addison-Wesley, Reading, MA, 1994)
49. M. Sipper, E. Ruppin: Physica D **99**, 428 (1997)
50. D. Stauffer: J. Phys. A: **24**, 909 (1991)
51. S. Wolfram: Nature **311**, 419 (1984)
52. V.I. Arnold: Uspekhi Mat. Nauk. **18**, 13 (1963)
53. J. Moser: Nachr. Akad. Wiss. Göttingen No. 1 (1962)
54. M. Henon, C. Heiles: Astron. J. **69**, 73 (1964)
55. J. Ford, G. Lunsford: Phys. Rev. A **1**, 59 (1970)
56. A.I. Lotka: J. Am. Chem. Soc. **42**, 1595 (1920)
57. J.-C. Raux, R.H. Simoyi, H.L. Swinney: Physica D **8**, 257 (1983)
58. C.W. Clark: *Mathemathical Bioeconomics* (Wiley, New York, 1976)
59. S. Dasgupta, T. Mitra: On price characterization of optimal plans in a multisector economy. In: *Economic Theory and Policy* ed by B. Dutta, S. Gangopadhyay, D. Mookherjee, D. Ray (Oxford University Press, Bombay, 1990), p. 122
60. F.J. Romeiras, C. Grebogi, E. Ott, W.P. Dayawansa: Physica D **58**, 165 (1992)
61. A. Locatelli: *Optimal Control* (Birkhäuser, Basel, 2001)
62. W.M. Wonham: *Linear Multivariable Control: A Geometrical Approach* (Springer, Berlin Heidelberg New York, 1974)
63. D.G. Luenberger: IEEE Trans. Automat. Control **AC-16**, 596 (1971)
64. K. Ogata: *Modern Control Engineering* (Prentice-Hall, Englewood Cliffs, NJ, 1990)
65. K. Pyragas: Phys. Lett. A **170**, 421 (1992)
66. G. Floquet: Annales École Normale **12**, 47 (1883)
67. J.K. Hale, S.M. Verduyn Lunel: *Introduction to Functional Differential Equations* (Springer, Berlin Heidelberg New York, 1993)
68. F. Schürer: Zur Theorie des Balancierens: Math. Nachr. **1**, 295 (1948)
69. S. Bielawski, D. Derozier, P. Glorieux: Phys. Rev. E **49**, R971 (1994)
70. T. Hikihara, T. Kawagoshi: Phys. Lett. A **211**, 29 (1996)
71. W. Just: Principles of time delayed feedback control. In: *Handbook of Chaos Control*, ed by H.G. Schuster (Wiley-VCH, Weinheim, 1999)
72. C. Huygens: Die Pendeluhr: Horologium oscillatorium (1673). In: *Ostwald's Klassiker der exakten Wissenschaften*, vol. 192 (A. Heckscher und A. von Öttingen, Leipzig, 1913)

6

Nonequilibrium Statistical Physics

6.1 Statistical Approach to Phase Space Dynamics

6.1.1 The Probability Distribution

Although most many-body systems are beyond the technical possibilities of the mathematical calculus of mechanics, we are able to calculate the properties of large systems by applying methods belonging to the statistical physics. In order to do this, we should have a general concept describing complex systems at microscopic scales. This description should fulfill two conditions. On the one hand, the approach should consider the apparent unpredictability of the chaotic motion of many-body systems, and on the other hand, it should be a starting point for establishing relations between various relevant quantities at the macroscopic level.

As already mentioned in Chap. 1, a concrete prediction about the microscopic movement of all particles is impossible if the initial condition cannot be determined exactly. However, we may give the probability for the realization of a certain microscopic state either on the basis of the preparation of the complex system or due to a suitable empirical estimation.

The intuitive notation of the probability is clear in dice games or coin-tossing. The probability is empirically defined by the relative frequency of a given realization repeating the game an infinite number of times. Probability reflects our partial ignorance, as in the outcome of a dice game. This frequency concept of probability introduced at the beginning of the last century is the appropriate basis for standard physical concepts, in particular the ensemble theory, but it is not very suitable for the characterization of complex systems like biological systems, climate systems, or traffic networks and financial markets. In fact, we are neither able to determine frequencies by a successive repetition in the sense of a classical scientific experiment nor we have enough information about possible outcomes. An alternative way to overcome this dilemma is the interpretation of the probability as a degree of belief that an

M. Schulz: *Control Theory in Physics and other Fields of Science*
STMP **215**, 149–191 (2006)

event will occur. This concept was the original point of view of Bayes and it combines a priori judgements and scientific information in a natural way.

Bayesian statistics is very general and can be applied to any thinkable process, independent of the feasibility of repeating the process under the conserved conditions. This definition is in particular suitable for systems with a very high degree of complexity, for which a repetition in the sense of a scientific experiment is almost always impossible.

In order to formulate the probability concept more precisely, let us come back to our general agreement given in Chap. 1. There we have defined the complete phase space $\mathbb{P}$ containing all degrees of freedom. In the sense of a physical description of an arbitrary system on the basis of first principles, each vector $\Gamma \in \mathbb{P}$ characterizes a certain microscopic state of the system.[1] Since we use here a strictly classical approach, Γ consists of all coordinates and momenta of all elementary particles of the system, $\Gamma(t) = \{q_1, \ldots, q_N, p_1, \ldots, p_N\}$. Then we have split $\mathbb{P}$ into the subspace $\mathbb{P}_{\text{rel}}$ of the relevant degrees of freedom X and the complementary subspace $\mathbb{P}/\mathbb{P}_{\text{rel}}$ of the irrelevant degrees of freedom $\Gamma_{\text{irr}} \cdot \mathbb{P}/\mathbb{P}_{\text{rel}}$. In all previous chapters we had assumed tacitly that either the state of the system was in fact a microscopic state or the variables of the state vector was almost decoupled from the irrelevant variables. The latter case is typically for the most applications of a deterministic control theory. This decoupling was more or less an empirically founded assumption. Nevertheless, there are many situations in which the relevant variables are connected by deterministic relations, despute the obvious presence of a large set of irrelevant degrees of freedom. Mechanical constructions, chemical reactions, and hydrodynamic systems are classical examples of apparently deterministic systems, although these systems consist of an immensely large set of internal, microscopic atomic and subatomic degrees of freedom.

However, there are also situations where the decoupling procedure does not work. In addition, we would establish the empirical assumption of a possible decoupling of relevant and irrelevant degrees of freedom by well founded physical arguments. To this aim we need as a first step an appropriate representation combining the uncertainty in determining an initial state with the still deterministic microscopic equations of motion. One possible start point is the application of the probability concepts in terms of the set theory. This method has the advantage that the axiomatic basis of the probability theory, firstly given by Kolmogorov [20], can be connected directly with physical first principles.

---

[1] In the context of classical physics, this representation is not a completely satisfactory concept, since all quantum mechanical effects are neglected. However, it is always possible to obtain a similarly consistent formulation of the following considerations on the basis of quantum physics [3, 4].

The elements of such a theory are denoted with respect to the probabilistic properties as events. In particular, each microscopic state Γ of the phase space is denoted as a certain event of the underlying microscopic system.

Events form various sets including the set of all events Ω and the set of no events. For instance, an arbitrary region of the phase space corresponds to such a set of events and the whole phase space has to be interpreted as the set Ω. All possible sets of events form a closed system under the operations of union and intersection.

From this abstract point of view, the probability is now defined as a measure $P(A)$ of the expected frequency or of our degree of belief for the appearance of an arbitrary event contained in the set A. The measure $P(A)$ is always nonnegative, $P(A) \geq 0$, and normalized, $P(\Omega) = 1$. Furthermore, if A and B are two non-overlapping sets, $A \cap B = \emptyset$, the probability that an event is contained in $A \cup B$ is the probability that the event is an element of either A or B. This leads to the main axiom of probability theory:

$$P(A \cup B) = P(A) + P(B) \quad \text{for} \quad A \cap B = \emptyset \,. \tag{6.1}$$

We generalize this relation to a countable collection of non-overlapping sets A_i $(i = 1, 2, \ldots)$ such that $A_i \cap A_j = \emptyset$ for all $i \neq j$ and obtain

$$P\left(\bigcup_i A_i\right) = \sum_i P(A_i) \,. \tag{6.2}$$

After this short excursion into the set theory we can now continue our original problem. We consider the set of events $A(\Gamma)$ that the system is within an infinitesimal volume element $\mathrm{d}\Gamma = \prod_{i=1}^{N} [\mathrm{d}q_i \mathrm{d}p_i]$ of the phase space $\mathbb{P}$ centered on the microscopic state Γ.

The main problem is to assign the probability measure $\mathrm{d}P(A(\Gamma), t)$ that at time t a microscopic state is an element of $A(\Gamma)$. This is in case of Bayesian statistics an 'a priori' probability, which is simply assumed with respect to our experience and in the case of the frequency concept an 'a posteriori' probability which is available after a sufficiently large number of observations or experiments. It is convenient to write:

$$\mathrm{d}p = \rho(q_1, \ldots, q_s, p_1, \ldots, p_s, t)\mathrm{d}q_1 \ldots \mathrm{d}q_s \mathrm{d}p_1 \ldots \mathrm{d}p_s = \rho(\Gamma, t)\mathrm{d}\Gamma \,, \tag{6.3}$$

where the function $\rho(\Gamma, t)$ is denoted as probability density for the outcome of the state Γ at time t. By definition, the density is a nonnegative quantity. Each finite region R of the phase space is a union of infinitesimal volume elements. Due to (6.2), the probability of finding a microscopic state in this region at time t is:

$$P(R, t) = \int_R \mathrm{d}\Gamma \rho(\Gamma, t) \,. \tag{6.4}$$

If we expand the region R over the whole phase space, we receive the normalization condition:

$$\int \rho(\Gamma, t) \mathrm{d}\Gamma = 1 \,. \tag{6.5}$$

This equation corresponds to $P(\Omega) = 1$ in the language of set theory, reflecting our knowledge that the certainty for finding the system somewhere in the phase space is always true.

Finally, it should be remarked that the probabilistic description based on Bayesian statistics is the only concept when we do not have enough information about the distribution of the states of a realistic system in the phase space; and even if we have this information, it is still a convenient representation for complex systems which approaches the frequency concept with increasing degree of information. From a pure mathematical point of view, both concepts are completely equivalent and only the interpretation of that what probability means differs essentially. A physical theory would always prefer the frequency picture, because it corresponds to the experimental root of physics.

The concrete mathematical structure of a microscopically derived probability distribution may be too complicated for a further treatment. But the probabilistic concept itself permits the theoretical solution to the general problem that initial conditions are measurable only with a finite accuracy.

6.2 The Liouville Equation

We want to address ourselves to the task whether we can determine the evolution of the probability distribution for a given microscopic system. To this aim we assume that an initial state Γ_0 was realized with a certain probability $\rho(\Gamma_0, 0)\mathrm{d}\Gamma$.

In course of the deterministic microscopic motion, the initial state is shifted into another microscopic state along the trajectory $\Gamma(t) = \{q_i(t), p_i(t)\}$ with the boundary condition $\Gamma(0) = \Gamma_0$. In other words, the probability density ρ is conserved along each trajectory of the complex system. This circumstance requires

$$\frac{\mathrm{d}\rho}{\mathrm{d}t} = \frac{\partial \rho}{\partial t} + \sum_{i=1}^{N} \left[\frac{\partial \rho}{\partial q_i} \dot{q}_i + \frac{\partial \rho}{\partial p_i} \dot{p}_i \right] = 0 \,. \tag{6.6}$$

After replacing the velocities $\dot{q}_i$ and forces $\dot{p}_i$ by (1.1) we arrive at

$$\frac{\partial \rho}{\partial t} + \hat{L}\rho = 0 \,, \tag{6.7}$$

where we have introduced the Liouvillian of the system

$$\hat{L} = \sum_{i=1}^{N} \left[\frac{\partial H}{\partial p_i} \frac{\partial}{\partial q_i} - \frac{\partial H}{\partial q_i} \frac{\partial}{\partial p_i} \right] \,. \tag{6.8}$$

Relation (6.7) is called the Liouville equation and is the most important equation of statistical physics, just as the Schrödinger equation is the main equation of quantum mechanics. The Liouvillian plays the same role that the Hamiltonian plays in Newtonian mechanics. The Hamiltonian fixes the rules of evolution of any microscopic state of the underlying system. In statistical physics, the Liouvillian again defines the equations of motion[2], which is now represented by the distribution function ρ.

For all microscopic systems the mechanical and the statistical representations of the evolution are equivalent. The difference between both descriptions lies in the definition of what we call the objects of evolution: points in phase space are the objects of classical mechanics while distribution functions are the objects of statistical physics.

The meaning of the Liouville equation for the evolution of complex systems lies in the combination of the probabilistic and deterministic features of the evolution process. For an illustration, let us assume that a certain initial situation at time $t = 0$ can be described at the microscopic level by a probability distribution $\rho(\Gamma, 0)$ as initial degree of belief in the sense of Bayesian statistics. Then the Liouville equation represents a deterministic map of $\rho(\Gamma, 0)$ onto the probability distribution $\rho(\Gamma, t)$ at a later time $t > 0$. In other words, the Liouville equation conserves our degree of belief.

6.3 Generalized Rate Equations

6.3.1 Probability Distribution of Relevant Quantities

In the microscopic probability distribution $\rho(\Gamma, t)$, all degrees of freedom are contained equally. Such a function, even if we would be able to determine it, is of course impractical and therefore unusable for the analysis and control of complex systems because of the large number of the contained degrees of freedom.

In general, we are interested in the description of complex systems only on the basis of the relatively small number of relevant degrees of freedom. Such an approach may be denoted as a kind of reductionism. Unfortunately, we are not able to give an unambiguous definition about what degree of freedom is relevant for the description of a complex system and what degree of freedom is irrelevant. As we have mentioned in Chap. 1, the relevant quantities are introduced empirically in accordance with the underlying problem.

To proceed, we take into account the formal splitting (1.3). This allows us to write the microscopic probability density as $\rho(\Gamma, t) = \rho(X, \Gamma_{\text{irr}}, t)$. In order to eliminate the irrelevant degrees of freedom, we integrate the probability density over Γ_{irr}

$$p(X, t) = \int d\Gamma_{\text{irr}} \rho(X, \Gamma_{\text{irr}}, t) \quad . \tag{6.9}$$

[2] The equations of motion are the characteristic equations of the Liuoville equation.

The remaining probability density $p(X,t)$ is more suitable for describing complex systems. The elimination of more or less all microscopic, irrelevant degrees of freedom corresponds to the transition from the microscopic level to a macroscopic representation. By definition, the probability density $p(X,t)$ is also normalized

$$\int \mathrm{d}X p(X,t) = \int \mathrm{d}X \mathrm{d}\Gamma_{\mathrm{irr}} \rho(X, \Gamma_{\mathrm{irr}}, t) = \int \mathrm{d}\Gamma \rho(\Gamma, t) = 1 \; . \tag{6.10}$$

The integration over all irrelevant degrees of freedom means that we suppose a maximal measure of ignorance over these quantities.

We may think about this description in geometrical terms. The system of relevant degrees of freedom can be represented by a point in the corresponding N_{rel}-dimensional subspace of the phase space. Obviously, an observer records apparently unpredictable behavior of the evolution of the relevant quantities at the macroscopic scale, if he disposes only about the relevant data. That is because of the fact that the dynamical evolution of the relevant quantities is governed by the hidden irrelevant degrees of freedom on microscopic scales. Thus, different microscopic trajectories in the phase space can lead to the same macroscopic results in the subspace, and vice versa, identical macroscopic initial configurations may develop in different directions (Fig. 6.1).

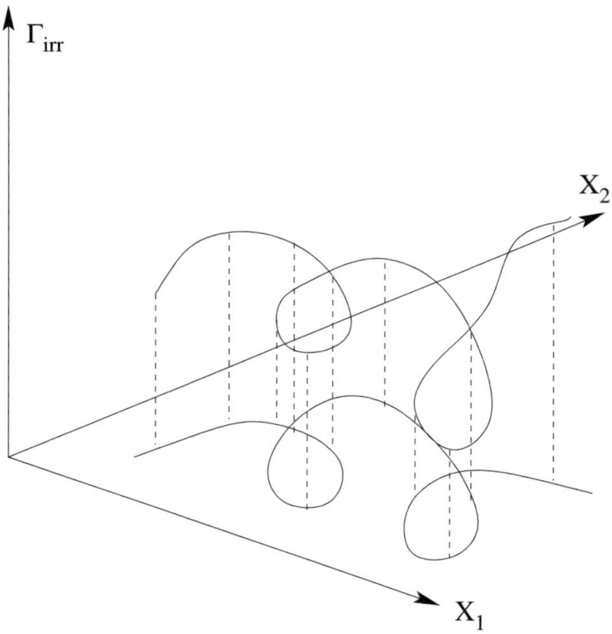

Fig. 6.1. Schematic projection of a trajectory from the full phase space onto the subspace of relevant degrees of freedom

We are not able to predict completely the later evolution even if we would know precisely the initial conditions. In other words, the restriction onto the subspace of relevant quantities leads to a permanent loss of the degree of information.

The average of an arbitrary function $f(\Gamma)$ is obtained by adding all values of $f(\Gamma)$ considering the statistical weight $\rho(\Gamma, t)\, \mathrm{d}\Gamma$. Hence

$$\overline{f}(t) = \int \mathrm{d}\Gamma \rho(\Gamma, t)\, f(\Gamma) \ . \tag{6.11}$$

The mean value may be a time-dependent quantity due to the time dependence of the probability density. If the function f depends only on the relevant degrees of freedom, i.e., $f = f(YX)$, then we get

$$\overline{f}(t) = \int \mathrm{d}\Gamma \rho(\Gamma, t)\, f(X) = \int \mathrm{d}X p(X, t)\, f(X) \ . \tag{6.12}$$

In this expression, the dynamics of the irrelevant degrees of freedom is again hidden in the distribution function $p(X, t)$. Obviously, the relevant probability density satisfies all necessary conditions for a sufficient description of a system on the level of the selected set of relevant degrees of freedom.

6.3.2 The Formal Solution of the Liouville Equation

We can formally integrate the Liouville equation (6.7) to obtain the solution

$$\rho(\Gamma, t) = \exp\left\{-\hat{L}t\right\} \rho(\Gamma, 0) \ . \tag{6.13}$$

This expression considers the microscopic equations of motion due to the concrete structure of the Liouvillian $\hat{L}$. The operator $\exp\{-\hat{L}t\}$ is referred to as the time propagator associated with the dynamical variables of the system. For a better understanding of the meaning of the time propagator, let us expand the exponential function in powers of t

$$\rho(\Gamma, t) = \left[1 - \hat{L}t + \frac{1}{2!}\left(\hat{L}t\right)^2 - \frac{1}{3!}\left(\hat{L}t\right)^3 + \cdots\right] \rho(\Gamma, 0) \ . \tag{6.14}$$

The right-hand side may be interpreted as a perturbative solution obtained from a successive integration of the Liouville equation. To demonstrate this, we write the Liouville equation as an integral equation. Then we are able to construct the map

$$\rho^{(n+1)}(\Gamma, t) = \rho(\Gamma, 0) - \int_0^t \hat{L}\rho^{(n)}(\Gamma, \tau)\, \mathrm{d}\tau \tag{6.15}$$

with the initial function $\rho^{(0)}(\Gamma, t) = \rho(\Gamma, 0)$. The series $\rho^{(0)}, \rho^{(1)}, \ldots, \rho^{(n)}, \ldots$ eventually converges against the solution $\rho(\Gamma, t)$ of the Liouville equation. In fact we receive

$$
\begin{aligned}
\rho^{(1)} &= \rho^{(0)} - \hat{L}\rho^{(0)}t \\
\rho^{(2)} &= \rho^{(0)} - t\hat{L}\rho^{(0)} + \frac{t^2}{2}\hat{L}^2\rho^{(0)} \\
\rho^{(3)} &= \rho^{(0)} - t\hat{L}\rho^{(0)} + \frac{t^2}{2}\hat{L}^2\rho^{(0)} - \frac{t^3}{6}\hat{L}^3\rho^{(0)} . \\
&\vdots
\end{aligned}
\tag{6.16}
$$

As expected, the solutions $\rho^{(0)}, \rho^{(1)}, \rho^{(2)}, \ldots$ of the hierarchical system (6.15) are identical with the first terms of expansion (6.14).

Unfortunately, for complex systems the formal solution (6.13) is in general too complicated to be useful in practice. In order to describe the dynamical behavior of such systems, we must look for alternative ways.

6.3.3 The Nakajima–Zwanzig Equation

Obviously, the knowledge of the relevant probability density $p(X,t)$ is a sufficient presupposition for the study of complex systems on the level of the chosen relevant degrees of freedoms. Our previous knowledge allows us to derive this function from the complete microscopic probability distribution function $\rho(\Gamma,t)$. For this purpose we would have to solve the Liouville equation with all the microscopic degrees of freedom at first. Then, in the subsequent step, we would be able to remove the irrelevant degrees of freedom from the microscopic distribution function by integration.

To avoid this unrealistic procedure we want to answer the question whether one can find an equation that describes the evolution of $p(X,t)$ and that contains exclusively relevant degrees of freedom.

Of course we can also remove the relevant degrees of freedom from every given microscopic probability distribution so that we arrive at the distribution function of the irrelevant degree of freedom

$$
\rho_{\mathrm{irr}}(\Gamma_{\mathrm{irr}}, t) = \int \mathrm{d}X \rho(X, \Gamma_{\mathrm{irr}}, t) \ , \tag{6.17}
$$

where we have to consider the normalization condition

$$
\int \mathrm{d}\Gamma_{\mathrm{irr}} \rho_{\mathrm{irr}}(\Gamma_{\mathrm{irr}}, t) = 1 \ . \tag{6.18}
$$

The product of the probability distributions (6.17) at the initial time $t = t_0$ and (6.9) at the time t is again a probability density

$$
\widetilde{\rho}(X, \Gamma_{\mathrm{irr}}, t, t_0) = \rho_{\mathrm{irr}}(\Gamma_{\mathrm{irr}}, t_0)\, p(X,t) \ . \tag{6.19}
$$

Of course, this surrogate probability distribution is no longer identical to the microscopic probability density $\rho(\Gamma,t)$. But the average values of any functions of relevant degrees of freedom calculated by an application of the density $\widetilde{\rho}$ remain unchanged in comparison with the use of ρ. Indeed, we get

$$p(X,t) = \int d\Gamma_{\text{irr}}\rho(X,\Gamma_{\text{irr}},t) = \int d\Gamma_{\text{irr}}\widetilde{\rho}(X,\Gamma_{\text{irr}},t,t_0) . \tag{6.20}$$

The generation of the surrogate probability distribution $\widetilde{\rho}$ is usually denoted as projection formalism. This procedure may be symbolically expressed by an application of a projection operator onto the probability distribution function

$$\widetilde{\rho}(X,\Gamma_{\text{irr}},t,t_0) = \hat{P}\rho(\Gamma,t) , \tag{6.21}$$

where we have introduced the special projection operator

$$\hat{P} \ldots = \rho_{\text{irr}}(\Gamma_{\text{irr}},t_0)\int d\Gamma_{\text{irr}} \ldots . \tag{6.22}$$

Apart from $\hat{P}$ we still need the complementary operator $\hat{Q} = 1 - \hat{P}$. Using (6.22) it is simple to demonstrate that these operators have the following 'idempotent' properties:

$$\hat{P}^2 = \hat{P} \quad \hat{Q}^2 = \hat{Q} \quad \text{and} \quad \hat{P}\hat{Q} = \hat{Q}\hat{P} = 0 , \tag{6.23}$$

typically for all projection operators. The first equation is a direct consequence of (6.22), while the last two follow from

$$\hat{Q}^2 = 1 - 2\hat{P} + \hat{P}^2 = 1 - 2\hat{P} + \hat{P} = 1 - \hat{P} = \hat{Q} \tag{6.24}$$

and

$$\hat{Q}\hat{P} = \hat{P} - \hat{P}^2 = \hat{P} - \hat{P} = 0 . \tag{6.25}$$

We now return to the question, how to describe the time-dependent evolution of the relevant probability density. To proceed, we need some information about the initial distribution at time t_0. While we can provide a meaningful initial distribution for simple physical systems due to the realization of an arbitrary number of repeatable experiments, we must fall back on more or less accurate estimations depending on the respective level of experience if we want to describe more complex phenomena of, for example, biological or meteorological character. The distribution of the relevant degrees of freedom can be fixed relatively simply: we assume that the values X_0 of all relevant degrees of freedom are well known at the initial time t_0. Therefore, we can write

$$p(X,t_0) = p(X,t_0 \mid X_0,t_0) = \delta(X - X_0) . \tag{6.26}$$

In this context, $p(X,t \mid X_0,t_0)$ means the probability density of the relevant degrees of freedom at time t, while the initial state was X_0. In principle, the following procedure also works for all other possible initial distributions. We will see later that all these cases can be mapped onto (6.26). On the other hand, we have no essential information about the irrelevant degrees of freedom.

However, we may assume that relevant and irrelevant degrees of freedom are uncorrelated at least for the initial state. Here the idea of Bayesan statistics comes again into its own. The statistical independence of relevant and irrelevant degrees can be neither verified nor rejected. It is an 'a priori' assumption reflecting the degree of our belief. Considering these assumptions the initial microscopic probability distribution can be written as

$$\rho(\Gamma, t_0) = \rho_{\mathrm{irr}}(\Gamma_{\mathrm{irr}}, t_0)\,\delta(X - Y_0) = \widetilde{\rho}(X, \Gamma_{\mathrm{irr}}, t_0, t_0) \tag{6.27}$$

with the property

$$\hat{P}\rho(\Gamma, t_0) = \rho(\Gamma, t_0) \ . \tag{6.28}$$

Now we apply the projection operator $\hat{P}$ to the Liouville equation (6.7) and obtain

$$\begin{aligned}\frac{\partial \hat{P}\rho(\Gamma, t)}{\partial t} &= -\hat{P}\hat{L}\left(\hat{P} + \hat{Q}\right)\rho(\Gamma, t) \\ &= -\hat{P}\hat{L}\hat{P}\rho(\Gamma, t) - \hat{P}\hat{L}\hat{Q}\rho(\Gamma, t) \ .\end{aligned} \tag{6.29}$$

We replace $\rho(\Gamma, t)$ in the second term of the right-hand side by the formal solution (6.13) of the Liouville equation where the initial time t_0 is taken into account. Then we arrive at

$$\hat{P}\hat{L}\hat{Q}\rho(\Gamma, t) = \hat{P}\hat{L}\hat{Q}\mathrm{e}^{-\hat{L}(t-t_0)}\rho(\Gamma, t_0) \ . \tag{6.30}$$

For further treatment of this expression we need the identity

$$\mathrm{e}^{-\hat{L}(t-t_0)} = \mathrm{e}^{-\hat{L}_1(t-t_0)} - \int_{t_0}^{t} \mathrm{d}t' \mathrm{e}^{-\hat{L}_1(t-t')}\hat{L}_2\mathrm{e}^{-\hat{L}(t'-t_0)} \ , \tag{6.31}$$

where we have split the Liouvillian into two arbitrary parts $\hat{L}_1$ and $\hat{L}_2$ via $\hat{L} = \hat{L}_1 + \hat{L}_2$. This identity may be checked by the derivative with respect to the time

$$\begin{aligned}-\hat{L}\mathrm{e}^{-\hat{L}(t-t_0)} &= -\hat{L}_1\mathrm{e}^{-\hat{L}_1(t-t_0)} - \hat{L}_2\mathrm{e}^{-\hat{L}(t-t_0)} \\ &\quad + \hat{L}_1 \int_{t_0}^{t} \mathrm{d}t' \mathrm{e}^{-\hat{L}_1(t-t')}\hat{L}_2\mathrm{e}^{-\hat{L}(t'-t_0)} \ .\end{aligned}$$

Then, substituting the integral kernel using (6.31), we obtain

$$\begin{aligned}-\hat{L}\mathrm{e}^{-\hat{L}\Delta t} &= -\hat{L}_1\mathrm{e}^{-\hat{L}_1\Delta t} - \hat{L}_2\mathrm{e}^{-\hat{L}\Delta t} + \hat{L}_1\left[\mathrm{e}^{-\hat{L}_1\Delta t} - \mathrm{e}^{-\hat{L}\Delta t}\right] \\ &= -\hat{L}_2\mathrm{e}^{-\hat{L}\Delta t} - \hat{L}_1\mathrm{e}^{-\hat{L}\Delta t} \\ &= -\hat{L}\mathrm{e}^{-\hat{L}\Delta t}\end{aligned} \tag{6.32}$$

with $\Delta t = t - t_0$. Thus identity (6.31) is proven. In particular, if we replace $\hat{L}_1$ by $\hat{L}\hat{Q}$ and $\hat{L}_2$ by $\hat{L}\hat{P}$, we get

$$\mathrm{e}^{-\hat{L}(t-t_0)} = \mathrm{e}^{-\hat{L}\hat{Q}(t-t_0)} - \int_{t_0}^{t} \mathrm{d}t' \mathrm{e}^{-\hat{L}\hat{Q}(t-t')}\hat{L}\hat{P}\mathrm{e}^{-\hat{L}(t'-t_0)} \ . \tag{6.33}$$

We substitute (6.33) in (6.30) to obtain

$$\hat{P}\hat{L}\hat{Q}\rho(\Gamma,t) = \hat{P}\hat{L}\hat{Q}\mathrm{e}^{-\hat{L}\hat{Q}(t-t_0)}\rho(\Gamma,t_0) - \int_{t_0}^{t} \mathrm{d}t'\hat{P}\hat{L}\hat{Q}\mathrm{e}^{-\hat{L}\hat{Q}(t-t')}\hat{L}\hat{P}\mathrm{e}^{-\hat{L}(t'-t_0)}\rho(\Gamma,t_0) \; . \tag{6.34}$$

The first term on the right-hand side disappears. This property follows from a Taylor expansion of the exponential function. The expansion is apparently an infinite series, but by (6.28) we know that all coefficients must vanish identically as a result of (6.23). To go further, we write the integral kernel in a more symmetric form. Considering $\hat{Q} = \hat{Q}^2$, we conclude

$$\begin{aligned}\hat{Q}\mathrm{e}^{-\hat{L}\hat{Q}\tau} &= \hat{Q}\left[1 - \tau\hat{L}\hat{Q} + \frac{\tau^2}{2}\hat{L}\hat{Q}\hat{L}\hat{Q} + \cdots\right] \\ &= \hat{Q}\left[1 - \tau\hat{Q}\hat{L}\hat{Q} + \frac{\tau^2}{2}\hat{Q}\hat{L}\hat{Q}^2\hat{L}\hat{Q} + \ldots\right]\hat{Q} = \hat{Q}\mathrm{e}^{-\hat{Q}\hat{L}\hat{Q}\tau}\hat{Q} \; .\end{aligned} \tag{6.35}$$

From (6.13) we see that

$$\hat{P}\hat{L}\hat{Q}\rho(\Gamma,t) = -\int_{t_0}^{t} \mathrm{d}t'\hat{P}\hat{L}\hat{Q}\mathrm{e}^{-\hat{Q}\hat{L}\hat{Q}(t-t')}\hat{Q}\hat{L}\hat{P}\rho(\Gamma,t') \tag{6.36}$$

and coming back to (6.29), we obtain

$$\frac{\partial\hat{P}\rho(\Gamma,t)}{\partial t} = -\hat{P}\hat{L}\hat{P}\rho(\Gamma,t) + \int_{t_0}^{t} \mathrm{d}t'\hat{P}\hat{L}\hat{Q}\mathrm{e}^{-\hat{Q}\hat{L}\hat{Q}(t-t')}\hat{Q}\hat{L}\hat{P}\rho(\Gamma,t') \; . \tag{6.37}$$

When this relationship is integrated over all irrelevant degrees of freedom, we obtain a closed linear integro-differential equation for the probability distribution function of the relevant degrees of freedom. Considering (6.22), we get

$$\begin{aligned}\frac{\partial p(X,t)}{\partial t} &= -\int \mathrm{d}\Gamma_{\mathrm{irr}}\left[\hat{L}\rho_{\mathrm{irr}}(\Gamma_{\mathrm{irr}},t_0)\right]p(X,t) \\ &\quad + \int_{t_0}^{t} \mathrm{d}t' \int \mathrm{d}\Gamma_{\mathrm{irr}}\left[\hat{L}\hat{Q}\mathrm{e}^{-\hat{Q}\hat{L}\hat{Q}(t-t')}\hat{Q}\hat{L}\rho_{\mathrm{irr}}(\Gamma_{\mathrm{irr}},t_0)\right]p(X,t')\end{aligned} \tag{6.38}$$

or, more formally,

$$\begin{aligned}\frac{\partial p(X,t \mid X_0,t_0)}{\partial t} &= -\hat{M}(t_0)\,p(X,t \mid X_0,t_0) \\ &\quad + \int_{t_0}^{t} \mathrm{d}t'\hat{K}(t_0,t-t')\,p(X,t' \mid X_0,t_0) \; ,\end{aligned} \tag{6.39}$$

where we have introduced the frequency operator

$$\hat{M}(t_0) = \int \mathrm{d}\Gamma_{\text{irr}} \left[\hat{L}\rho_{\text{irr}}(\Gamma_{\text{irr}}, t_0) \right] \tag{6.40}$$

and the memory operator

$$\hat{K}(t_0, t - t') = \int \mathrm{d}\Gamma_{\text{irr}} \left[\hat{L}\hat{Q} \exp\left\{ -\hat{Q}\hat{L}\hat{Q}(t - t') \right\} \hat{Q}\hat{L}\rho_{\text{irr}}(\Gamma_{\text{irr}}, t_0) \right] . \tag{6.41}$$

This equation is called the Nakajima–Zwanzig equation [3, 5] or the generalized rate equation. The Nakajima–Zwanzig equation is still a proper relation, although it describes apparently only the evolution of the relevant probability distribution function. However, the complete dynamics of the irrelevant degrees of freedom including their interaction with the relevant degrees of freedom is in particular hidden in the memory operator.

The dependence of the operators $\hat{M}$ and $\hat{K}$ on the initial time t_0 is a remarkable property, which reflects the fact that a complex system does not necessarily have to be in a stationary state. Therefore, completely different developments of the probability density $p(X, t \mid Y_0, t_0)$ may be observed for the same system and for the same initial conditions but for different initial times.

The Nakajima–Zwanzig equation allows the prediction of the further evolution of the relevant probability distribution function, presupposed we are able to determine the exact mathematical structure of the frequency and memory operators. In principle, we are also able to derive more general evolution equations than the Nakajima–Zwanzig equation, e.g., by use of time-dependent projectors or of projection operators which depend even on the relevant probability distribution function. But then the useful convolution property is lost, which characterizes the memory term in (6.39). Additionally all evolution equations obtained by projection formalisms are physically equivalent and mathematically accurate, so that from this point of view also none of the possible evolution equations possesses a possible preference.

The main problem is however the determination of the operators $\hat{M}$ and $\hat{K}$. The complete determination of these quantities equals the solution of the Liouville equation. Consequently, this method is unsuitable for systems with a sufficiently high degree of complexity. But we can try to approach these operators of the Nakajima–Zwanzig equation in a heuristic manner using empirical experiences and mathematical considerations. Physical intuition plays an important role at several stages of this approach [6–8]. In this way one can combine certain model conceptions and real observations and arrive at a comparatively reasonable approximation of the accurate evolution equation. Here it then becomes also important that what projection formalism one uses. However, for the majority of the considered problems, (6.39) is quite a suitable equation which gives us the opportunity for a further progress.

6.4 Notation of Probability Theory

6.4.1 Measures of Central Tendency

In the future we will almost always use the space of the relevant degrees of freedom. Therefore we will abandon an extra designation of all quantities which are related to this space. We now speak, again as in the case of deterministic systems, of an N-dimensional state X and of the corresponding phase space instead of relevant degrees of freedom and of their corresponding subspace of dimension N_{rel}. Only if the possibility of a mistake exists will we use the old notation.

Suppose that we consider a probability distribution function $p(X,t)$ with only one (relevant) degree of freedom. Generally, each multivariable probability density may be reduced to such a single variable function by integration over all degrees of freedom except one.

Let us now answer the following question. What is the typical value of the outcome of a given problem concerning a sufficiently complex system if we know the probability distribution function $p(X,t)$? Unfortunately, there is no unambiguous answer. The quantity used most frequently for the characterization of the central tendency is the mean or average

$$\overline{X(t)} = \int \mathrm{d}X p(X,t)\, X \; . \tag{6.42}$$

There are other two major measures of central tendency. The probability $P_<(x,t)$ gives the time-dependent fraction of events with values less then x,

$$P_<(x,t) = \int\limits_{-\infty}^{x} \mathrm{d}X p(X,t) \; . \tag{6.43}$$

The function $P_<(x,t)$ increases monotonically with x from 0 to 1. Using (6.43), the central tendency may be characterized by the median $X_{1/2}(t)$. The median is the halfway point in a graded array of values,

$$P_<(X_{1/2},t) = \frac{1}{2} \; . \tag{6.44}$$

Finally, the most probable value $X_{\max}(t)$ is another quantity describing the mean behavior. This quantity maximizes the density function

$$\left.\frac{\partial p(X,t)}{\partial X}\right|_{X=X_{\max}(t)} = 0 \; . \tag{6.45}$$

If this equation has several solutions, the most probable value $X_{\max}(t)$ is the one with the largest p. Apart from unimodal symmetric probability distribution functions, the three quantities differ. These differences are important for the interpretation of empirical averages obtained from a finite number of observations.

For a few trials, the most probable value will be sampled first and the average made on a few such measures will not be far from $x_{\max}(t)$. In contrast, the empirical average determined from a large but finite number of observations approaches progressively the true average $\overline{X(t)}$.

6.4.2 Measure of Fluctuations around the Central Tendency

We consider again only one degree of freedom. When repeating an observation of this variable several times, one expects them to be within an interval anchored at the central tendency. The width of this interval is a measure of the deviations from the central tendency. A possible measure of this width is the average of the absolute value of the spread defined by

$$D_{\mathrm{sp}}(t) = \int_{-\infty}^{\infty} \mathrm{d}X \left| X - X_{1/2}(t) \right| p(X,t) \ . \tag{6.46}$$

The absolute value of the spread does not exist for probability distribution functions decaying as or slower than X^{-2} for a large X. Another measure is the standard deviation σ. This quantity is the square root of the variance σ^2

$$\sigma^2 = \int_{-\infty}^{\infty} \mathrm{d}X \left[X - \overline{X(t)} \right]^2 p(X,t) \ . \tag{6.47}$$

The standard deviation such as for probability densities $p(X,t)$ with tails decaying as or slower than X^{-3} does not always exist.

6.4.3 Moments and Characteristic Functions

Now we come back to the general case of a multivariable probability distribution function $p(X,t)$ with $X = \{X_1, X_2, \ldots, X_N\}$. The moments of order n are defined by the average

$$m^{(n)}_{\alpha_1 \alpha_2 \ldots \alpha_n}(t) = \int \mathrm{d}X \left[\prod_{k=1}^{n} X_{\alpha_k} \right] p(X,t) \ . \tag{6.48}$$

The first moment $m^{(1)}_{\alpha}(t)$ is the mean $\overline{X_\alpha(t)}$ of component α. Therefore, the formal vector $\left\{ m^{(1)}_1(t), m^{(1)}_2(t), \ldots \right\}$ defines in generalization of (6.42) the central tendency of the underlying dynamics. The second moment $m^{(2)}_{\alpha\beta}(t)$ corresponds to the average

$$m^{(2)}_{\alpha\beta}(t) = \int \mathrm{d}X p(X,t) X_\alpha X_\beta \ . \tag{6.49}$$

These quantities are also denoted as components of the correlation matrix. For definition (6.48) to be meaningful, the integral on the right-hand side must be convergent, which means that a necessary condition for the existence of a moment of order n is that the probability density function decays faster than $|X|^{-n-N}$ for $|X| \to \infty$. This is trivially obeyed for probability distribution functions which vanish outside a finite region of the phase space.

Statistical problems are often discussed in terms of moments because they avoid the difficult problem of determining the full functional behavior of the probability density. In principle, the knowledge of all the moments is in many realistic cases equivalent to that of the probability distribution function. However, the strict equivalence between the knowledge of all the moments and the probability density requires further constraints.

The moments are closely related to the characteristic function which is defined as the Fourier transform of the probability distribution

$$\hat{p}(k,t) = \int \mathrm{d}X \exp\{\mathrm{i}kX\} p(X,t) \tag{6.50}$$

with the N-dimensional vector $k = \{k_1, k_2, \ldots, k_N\}$. From here we obtain the inverse relation

$$p(X,t) = \frac{1}{(2\pi)^N} \int \mathrm{d}k \exp\{-\mathrm{i}kX\} \hat{p}(k,t) \ . \tag{6.51}$$

Thus, the normalization condition (6.10) is equivalent to $\hat{p}(0,t) = 1$ and the moments of the probability density can be obtained from derivatives of the characteristic function at $k = 0$

$$m^{(n)}_{\alpha_1\alpha_2\ldots\alpha_n}(t) = (-i)^n \prod_{l=1}^{n} \left(\frac{\partial}{\partial k_{\alpha_l}} \right) \hat{p}(k,t) \bigg|_{k=0} \ . \tag{6.52}$$

If all moments exist, the characteristic function may be also presented as the series expansion

$$\hat{p}(k,t) = \sum_{n=0}^{\infty} \sum_{\{\alpha_1,\alpha_2,\ldots\alpha_n\}} \frac{i^n}{n!} m^{(n)}_{\alpha_1\alpha_2\ldots\alpha_n}(t) \left(\prod_{l=1}^{n} k_{\alpha_l} \right) \ . \tag{6.53}$$

The inversion formulas show that different characteristic functions arise from different probability distribution functions, i.e., the characteristic function $\hat{p}(k,t)$ is truly characteristic. Additionally, the straightforward derivation of the moments by (6.52) makes any determination of the characteristic function directly relevant to measurable quantities.

6.4.4 Cumulants

Another important function is the cumulant generating function which is defined as the logarithm of the characteristic function

$$\Phi(k,t) = \ln \hat{p}(k,t) \ . \tag{6.54}$$

This leads to the introduction of the cumulants $c_{\alpha_1\alpha_2\ldots\alpha_n}(t)$ as derivatives of the cumulant generating function at $k = 0$

$$c^{(n)}_{\alpha_1\alpha_2\ldots\alpha_n}(t) = (-\mathrm{i})^n \prod_{l=1}^{n} \left(\frac{\partial}{\partial k_{\alpha_l}}\right) \Phi(k,t)\bigg|_{k=0} . \tag{6.55}$$

Each cumulant of order n is a combination of moments of orders $l \leq n$, as can be seen by substitution of (6.53) and (6.54) into (6.55). We get for the first cumulants

$$\begin{aligned} c^{(1)}_{\alpha} &= m^{(1)}_{\alpha} \\ c^{(2)}_{\alpha\beta} &= m^{(2)}_{\alpha\beta} - m^{(1)}_{\alpha} m^{(1)}_{\beta} \\ c^{(3)}_{\alpha\beta\gamma} &= m^{(3)}_{\alpha\beta\gamma} - m^{(2)}_{\alpha\beta} m^{(1)}_{\gamma} - m^{(2)}_{\beta\gamma} m^{(1)}_{\alpha} - m^{(2)}_{\gamma\alpha} m^{(1)}_{\beta} + 2 m^{(1)}_{\alpha} m^{(1)}_{\beta} m^{(1)}_{\gamma} . \\ &\vdots \end{aligned} \tag{6.56}$$

The first-order cumulants are the averages of the single components X_α. The second-order cumulants define the covariance matrix with the elements

$$c^{(2)}_{\alpha\beta} = m^{(2)}_{\alpha\beta} - m^{(1)}_{\alpha} m^{(1)}_{\beta} . \tag{6.57}$$

The covariance is a generalized measure of the degree to which the values X deviate from the central tendencies. In particular, for a single variable X, the second-order cumulant is equivalent to the variance σ^2. Higher-order cumulants contain information of decreasing significance. Especially, if all higher-order cumulants vanishes, we can easily deduce by using (6.54) and (6.51) that the corresponding probability density $p(X,t)$ is a Gaussian probability distribution

$$p(X,t) = \frac{1}{\sqrt{\det 2\pi\widetilde{\sigma}}} \exp\left\{-\frac{1}{2}\left(X - m^{(1)}\right)\widetilde{\sigma}^{-1}\left(X - m^{(1)}\right)\right\} . \tag{6.58}$$

Note that the theorem of Marcienkiewicz [10] shows that either all but the first two cumulants vanish or there is an infinite number of nonvanishing cumulants. In other words, the cumulant generating function cannot be a polynomial of degree greater than 2.

Obviously, higher-order cumulants characterize the natural deviation from the Gaussian behavior. In the case of a single variable X the normalized third-order cumulant $\lambda_3 = c^{(3)}/\sigma^3$ is called the skewness while $\lambda_4 = c^{(4)}/\sigma^4$ is denoted as excess kurtosis. The skewness is a measure for the asymmetry of the probability distribution function. For symmetric distributions, the excess kurtosis quantifies the first correction to the Gaussian behavior.

6.5 Combined Probabilities

6.5.1 Conditional Probability

As discussed in the previous paragraph, $p(X,t \mid X_0,t_0)$ is the probability density that the system in the state X_0 at time t_0 will be in the state X at time $t > t_0$. Hence,

$$P(R,t \mid X_0,t_0) = \int_R \mathrm{d}X p(X,t \mid X_0,t_0) \tag{6.59}$$

is the probability that the system occupies an arbitrary state of the region R at time t if the system was in the state X_0 at time t_0. This is a special kind of conditional probability which is directly related to the time development of a complex system.

More general, the conditional probability may be defined in the language of set theory. Here $P(A \mid B)$ is the probability that an event contained in the set A appears under the condition that we know it was also contained in the set B.

In particular, we can interpret A as the set of all trajectories of the system, which touch at time t the region R, while B is the set of trajectories, which go at time t_0 through the point X_0. In this sense each trajectory is an event. Both A and B are subsets of the set Ω of all trajectories. Then $P(A \mid B) = P(R,t \mid X_0,t_0)$ may be understood as the probability that any trajectory of B also belongs to A. In particular we receive therefore the normalization condition $P(\Omega \mid B) = 1$ which allows us to conclude

$$\int \mathrm{d}X p(X,t \mid X_0,t_0) = 1 \,. \tag{6.60}$$

Statistical independence means $P(A \mid B) = P(A)$, i.e., the knowledge that one event occurs in B does not change the probability that it occurs in A. If $P(A \mid B) > P(A)$, we say that A and B are positively correlated while $P(A \mid B) < P(A)$ corresponds to a negative correlation between A and B.

6.5.2 Joint Probability

Let us now consider an event which is an element of the set A as well as of the set B. Then the event is also contained in $A \cap B$. The probability $P(A \cap B)$ is called the joint probability that the event is contained in both classes. Conditional probabilities, the joint probabilities, and the usual probabilities or unconditional probabilities become connected on a very natural type

$$P(A \cap B) = P(A \mid B)P(B) = P(B \mid A)P(A) \,. \tag{6.61}$$

This representation allows a natural definition of statistically independent events. Obviously, statistical independence requires simply $P(A \cap B) = P(A)P(B)$ and therefore $P(A \mid B) = P(A)$ and $P(B \mid A) = P(B)$. For example, the probability that a system stays in the infinitesimal small volume $\mathrm{d}X$ of its phase space at time t and it was in the volume $\mathrm{d}X_0$ at the initial time t_0 is a typical problem to consider. The corresponding (infinitesimal) joint probability may be written as $\mathrm{d}P(X,t;X_0,t_0) = p(X,t;X_0,t_0)\,\mathrm{d}X\mathrm{d}X_0$ with

$$p(X,t;X_0,t_0) = p(X,t \mid X_0,t_0)\,p(X_0,t_0) \,. \tag{6.62}$$

Suppose we know all sets B_i which could condition the appearance of an event in the set A. The B_i should be mutually exclusive, $B_i \cap B_j = \emptyset$ for all $i \neq j$, and exhaustive, $\bigcup_i B_i = \Omega$. Thus we obtain

$$P(A) = P(A \cap \Omega) = P\left(A \cap \bigcup_i B_i\right) = P\left(\bigcup_i (A \cap B_i)\right) . \tag{6.63}$$

If we take into account $(A \cap B_i) \cap (A \cap B_j) = \emptyset$ we obtain due to (6.2) and (6.61)

$$P(A) = \sum_i P(A \cap B_i) = \sum_i P(A \mid B_i) P(B_i) . \tag{6.64}$$

This general relation specifies in the case of the probability density immediately to

$$p(X,t) = \int p(X,t \mid X_0,t_0) p(X_0,t_0) \, \mathrm{d}X_0 . \tag{6.65}$$

Because of the symmetry $P(A \cap B) = P(B \cap A)$ we also get

$$p(X_0,t_0) = \int p(X,t \mid X_0,t_0) p(X_0,t_0) \, \mathrm{d}X . \tag{6.66}$$

Due to (6.60), the last equation is a simple identity. Equation (6.62) permits in particular the extension of the initial condition (6.26) on any probability distributions.

This constitutes a warning that it is always preferable to represent each joint probability distribution function $p(X,t;Z,\tau)$ in the form (6.62). If we want to generally determine this joint probability for $t > \tau > t_0$, then we must calculate the integral

$$p(X,t;Z,\tau) = \int \mathrm{d}X_0 p(X,t \mid Z,\tau;X_0,t_0) \times p(Z,\tau \mid X_0,t_0) p(X_0,t_0) \tag{6.67}$$

in which the conditional probability $p(X,t \mid Z,\tau;X_0,t_0)$ occurs. The reason for the more complicated structure consists in the fact that the deterministic character of the microscopic dynamics is possibly partially conserved on the level of the relevant degrees of freedom. Hence, a certain memory of the initial information remains, which is, for instance, expressed by the appearance of the memory kernel $\hat{K}(t-t')$ in the Nakajima–Zwanzig equation. This effect indicates a possible feedback between the relevant degrees of freedom via the hidden irrelevant degrees of freedom. Only if this feedback disappears, $p(X,t \mid Z,\tau;X_0,t_0) = p(X,t \mid Z,\tau)$ applies and the simpler relationship $p(X,t;Z,\tau) = p(X,t \mid Z,\tau)\,\rho(Z,\tau)$ becomes valid for arbitrary points in time (Fig. 6.2). On the other hand, (6.62) is always valid. The correctness of this relation is justified by the fact that relevant and irrelevant degrees of freedom are assumed to be initially uncorrelated even if former information is unknown. We point out again that this assumption has to be understood in the sense of the Bayesian definition of statistics.

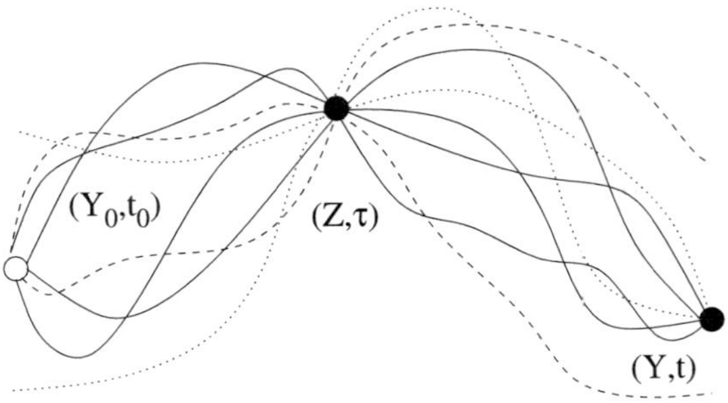

Fig. 6.2. Possible contributions to the joint probability density $p(Y,t;Z,\tau;Y_0,t_0)$. The integration over all positions Y_0 leads to $p(Y,t;Z,\tau)$. Only the full trajectories contribute to the conditional probability density $p(Z,\tau \mid Y_0,t_0)$ as well as to the conditional probability density $p(Y,t \mid Z,\tau;Y_0,t_0)$. These events form together with the probability density $p(Y_0,t_0)$ the joint probability $p(Y,t;Z,\tau;Y_0,t_0)$. The *dashed curves* are also contained in $p(Z,\tau \mid Y_0,t_0)$ but not in $p(Y,t;Z,\tau)$. Roughly speaking, they are filtered out due to conditional probability $p(Y,t;\mid Z,\tau;Y_0,t_0)$ in the expression $p(Y,t;Z,\tau;Y_0,t_0) = p(Y,t;\mid Z,\tau;Y_0,t_0)p(Z,\tau \mid Y_0,t_0)p(Y_0,t_0)$. On the other hand, the product $p(Y,t;\mid Z,\tau)p(Z,\tau \mid Y_0,t_0)p(Y_0,t_0)$ also considers the *dotted lines* which contribute particularly to $p(Y,t \mid Z,\tau)$. Thus, we have to expect $p(Y,t;\mid Z,\tau;Y_0,t_0) \neq p(Y,t;\mid Z,\tau)$. The equivalence between both quantities holds only if no memory effect appears

6.6 Markov Approximation

We once again return to the problem of the selection of relevant degrees of freedom. If we define the relevant degrees of the complex system in such a way that all these variables change relatively slow compared to the irrelevant degrees of freedom, then the memory kernel (6.41) of the Nakajima–Zwanzig equation may approach

$$\hat{K}(t_0,t-t') = \hat{K}(t_0)\,\delta(t-t') \ . \tag{6.68}$$

This representation is called the Markov approximation [4,12]. The assumption of such a separation between slow, relevant time scales and fast, irrelevant time scales is at least an appropriate approximation for many complex systems. But it should be remarked that there is really no such thing as a system with the Markov character. If we observe the system on a very fine time scale, the immediate history will almost certainly be required to predict the probabilistic development. In other words, there is a certain characteristic time during which the previous history is important. However, systems whose memory time is so small may be, on the time scale on which we carry out observations, assumed to be a Markov-like system. We substitute (6.68) in the Nakajima–Zwanzig equation (6.39) to get

$$\frac{\partial p(X,t \mid X_0,t_0)}{\partial t} = -\hat{L}_{\text{markov}}\,p(X,t \mid X_0,t_0) \tag{6.69}$$

with the Markovian $\hat{L}_{\text{markov}} = \hat{M}(t_0) - \hat{K}(t_0)$. However, we also know about situations where a part of the irrelevant degrees of freedom is considerably slower than the relevant degrees of freedom and only the remaining part of the irrelevant degrees of freedom contributes to the fast dynamics. In these cases, it seems to be more favorable to derive the evolution equation for the probability density $p(X, t \mid X_0, t_0)$ by the use of time-dependent projectors capturing the effects of the slow irrelevant dynamics.

Such generalization basically changes nothing of the general procedure of the separation of the time scales, except for the occurrence of an explicit time dependence of the operator $\hat{L}_{\text{markov}}(t)$. Therefore, we can also use the Markov approximation for these problems. However, the concept of the separation of time scales fails or becomes uncontrolled if a suitable set of irrelevant degrees of freedom offers characteristic time scales similar to those of the relevant degrees of freedom. By assuming an infinitesimal time interval $\mathrm{d}t$ we obtain from (6.69)

$$p(X, t + \mathrm{d}t \mid X_0, t_0) = \left[1 - \hat{L}_{\text{markov}}(t)\,\mathrm{d}t\right] p(X, t \mid X_0, t_0) \ . \tag{6.70}$$

In general, we may express the operator $1 - \hat{L}_{\text{markov}}(t)\,\mathrm{d}t$ by an integral representation using the transition function $U_{\text{markov}}(X, t + dt \mid Z, t)$

$$p(X, t + \mathrm{d}t \mid X_0, t_0) = \int \mathrm{d}Z U_{\text{markov}}(X, t + \mathrm{d}t \mid Z, t)\, p(Z, t \mid X_0, t_0) \ . \tag{6.71}$$

We multiply (6.71) with the initial distribution function $p(X_0, t_0)$ and integrate over all configurations X_0. Considering (6.65) we get

$$p(X, t + \mathrm{d}t) = \int \mathrm{d}Z U_{\text{markov}}(X, t + \mathrm{d}t \mid Z, t)\, p(Z, t) \ . \tag{6.72}$$

Thus, the integral kernel $U_{\text{markov}}(X, t + \mathrm{d}t \mid Z, t)$ can interpreted as the conditional probability density $p(X, t + \mathrm{d}t \mid Z, t)$ for a transition from the state Z at time t to the state X at time $t + \mathrm{d}t$. This necessitates a further explanation. We remember that (6.67) requires the more general relation

$$\begin{aligned} p(X, t + \mathrm{d}t) = \int \mathrm{d}X_0 \int \mathrm{d}Z p(X, t + \mathrm{d}t \mid Z, t; X_0, t_0) \\ \times\, p(Z, t \mid X_0, t_0)\, p(X_0, t_0) \ . \end{aligned} \tag{6.73}$$

A simple comparison between (6.72) and (6.73) leads to the necessary condition $p(X, t + \mathrm{d}t \mid Z, t; X_0, t_0) = p(X, t + \mathrm{d}t \mid Z, t)$. This is simply another formulation of the Markov property. It is, even by itself, extremely powerful. In particular, this property means that we can define higher conditional and joint probabilities in terms of the simple conditional probability. To obtain a general relation between the conditional probabilities at different times, we shift the time $t \to t + \mathrm{d}t$ in (6.71) and obtain

$$\begin{aligned} p(X, t + 2dt \mid Y_0, t_0) = \int dY p(X, t + 2dt \mid Y, t + dt) \\ \times\, p(Y, t + dt \mid X_0, t_0) \ . \end{aligned} \tag{6.74}$$

On the other hand, the transformation $\mathrm{d}t \to 2\mathrm{d}t$ leads to

$$p(X, t + 2dt \mid Y_0, t_0) = \int dZ p(X, t + 2dt \mid Z, t) p(Z, t \mid Y_0, t_0) \ . \tag{6.75}$$

so that we obtain from (6.71), (6.74), and (6.75)

$$p(X, t + 2\mathrm{d}t \mid Z, t) = \int \mathrm{d}Y p(X, t + 2\mathrm{d}t \mid Y, t + \mathrm{d}t) \times p(Y, t + \mathrm{d}t \mid Z, t) \ . \tag{6.76}$$

When repeating this procedure infinite times, one obtains a relation for finite time differences

$$p(X, t \mid Z, t'') = \int \mathrm{d}Y p(X, t \mid Y, t') p(Y, t' \mid Z, t'') \ , \tag{6.77}$$

which is the Chapman–Kolmogorov equation. This equation is a rather complex nonlinear functional equation relating all conditional probabilities obtained from a given Markovian to each other.

This is a remarkable result: the conditional probability density obtained from an arbitrary Markovian must satisfy the Chapman–Kolmogorov equation. In addition, the Chapman–Kolmogorov equation is an important criterion for the presence of the Markov property. Whenever empirically determined conditional probabilities satisfy (6.77) we are able to introduce the Markov property.

6.7 Generalized Fokker–Planck Equation

6.7.1 Differential Chapman–Kolmogorov Equation

The determination of the Markovian for a given process on the basis of a microscopic theory is probably excluded. Therefore, we are always dependent on empirical considerations and observations. It would be reasonable if we would know some rules from which the Markovian $\hat{L}_{\mathrm{markov}}$ can be constructed.

For all evolution processes with Markov properties the parameter-free Chapman–Kolmogorov equation (6.77) is a universal relation. However, the Chapman–Kolmogorov equation has many solutions. In particular, for a given dimension N of the state X, every solution of (6.69) must also be a solution of (6.77), independent of the special mathematical structure of the operator $\hat{L}_{\mathrm{markov}}$. Therefore, we can possibly use this equation to obtain information about the general mathematical structure of $\hat{L}_{\mathrm{markov}}$. To do so we follow Gardiner [11] and define the subsequent quantities for all $\varepsilon > 0$:

$$F_\alpha(Z, t) = \lim_{\delta t \to 0} \frac{1}{\delta t} \int\limits_{|Y - Z| < \varepsilon} \mathrm{d}Y p(Y, t + \delta t \mid Z, t) \Delta Y_\alpha + o(\varepsilon) \ , \tag{6.78}$$

and

$$D_{\alpha\beta}(Z,t) = \lim_{\delta t\to 0}\frac{1}{\delta t}\int\limits_{|Y-Z|<\varepsilon} \mathrm{d}Y p(Y,t+\delta t \mid Z,t)\,\Delta Y_\alpha \Delta Y_\beta + o(\varepsilon)\ , \quad (6.79)$$

where we have used the notation $\Delta Y_\alpha = Y_\alpha - Z_\alpha$. Furthermore we introduce

$$W(Y \mid Z;t) = \lim_{\delta t\to 0}\frac{1}{\delta t} p(Y,t+\delta t \mid Z,t) \quad (6.80)$$

for $|Y - Z| > \varepsilon$. We will see later that these quantities were chosen in a very natural way. They can be obtained directly from observations or they are defined by suitable model assumptions.

If we are able to build the Markovian $\hat{L}_{\mathrm{markov}}$ by the exclusive use of these quantities, we have reached our goal. Note that possible higher-order coefficients must vanish for $\varepsilon \to 0$. For instance, the third-order quantity defined by

$$C_{\alpha\beta\gamma}(Z,t) = \lim_{\delta t\to 0}\frac{1}{\delta t}\int\limits_{|Y-Z|<\varepsilon} \mathrm{d}Y p(Y,t+\delta t \mid Z,t)\,\Delta Y_\alpha \Delta Y_\beta \Delta Y_\gamma \quad (6.81)$$

may be approximated by $|C_{\alpha\beta\gamma}| \simeq |D_{\alpha\beta}|\,\varepsilon = o(\varepsilon)$. Thus, if $D_{\alpha\beta}$ exists, the coefficient $C_{\alpha\beta\gamma}$ is of an order of magnitude $o(\varepsilon)$ and disappears for $\varepsilon \to 0$. To proceed, we consider the time evolution of the average of an arbitrary function f which is twice continuously differentiable

$$\begin{aligned}
&\frac{\partial}{\partial t}\int \mathrm{d}Z f(Z)\, p(Z,t \mid Y,t') \\
&= \lim_{\delta t\to 0}\frac{1}{\delta t}\int \mathrm{d}Z f(Z)\left[p(Z,t+\delta t \mid Y,t') - p(Z,t \mid Y,t')\right]\mathrm{d}Z \\
&= \lim_{\delta t\to 0}\frac{1}{\delta t}\iint \mathrm{d}Z\mathrm{d}X f(Z)\, p(Z,t+\delta t \mid X,t)\, P(X,t \mid Y,t') \\
&- \lim_{\delta t\to 0}\frac{1}{\delta t}\iint \mathrm{d}Z\mathrm{d}X f(X)\, p(Z,t+\delta t \mid X,t)\, P(X,t \mid Y,t')\ ,
\end{aligned} \quad (6.82)$$

where we have used the Chapman–Kolmogorov equation (6.77) in the first term and the normalization condition (6.60) to produce the corresponding terms in (6.82). Since $f(Z)$ is twice continuously differentiable, we may write

$$\begin{aligned}
f(Z) &= f(X) + \sum_\alpha \frac{\partial f(X)}{\partial X_\alpha}\left[Z_\alpha - X_\alpha\right] \\
&+ \frac{1}{2}\sum_{\alpha\beta}\frac{\partial^2 f(X)}{\partial X_\alpha \partial X_\beta}\left[Z_\alpha - X_\alpha\right]\left[Z_\beta - X_\beta\right] + R(Z,X)\ ,
\end{aligned} \quad (6.83)$$

where the reminder function $R(Z,X)$ vanishes for $|X - Z| = \varepsilon \to 0$ as $o(\varepsilon^2)$. We now divide the integrals in (6.82) into two regions $|X - Z| \le \varepsilon$ and $|X - Z| > \varepsilon$ and substitute (6.83) into (6.82):

$$\begin{aligned}
&\frac{\partial}{\partial t}\int \mathrm{d}Z f(Z)\, p(Z,t \mid Y,t') \\
&= \lim_{\delta t\to 0}\frac{1}{\delta t}\iint_{|Z-X|<\varepsilon} \mathrm{d}Z\mathrm{d}X p(Z,t+\delta t \mid X,t)\, p(X,t \mid Y,t') \\
&\quad\times\left\{\sum_{\alpha}\frac{\partial f(X)}{\partial X_\alpha}[Z_\alpha - X_\alpha] + \frac{1}{2}\sum_{\alpha\beta}\frac{\partial^2 f(X)}{\partial X_\alpha \partial X_\beta}[Z_\alpha - X_\alpha][Z_\beta - X_\beta]\right\} \\
&\quad+ \lim_{\delta t\to 0}\frac{1}{\delta t}\iint_{|Z-X|<\varepsilon} \mathrm{d}Z\mathrm{d}X R(Z,X)\, p(Z,t+\delta t \mid X,t)\, p(X,t \mid Y,t') \\
&\quad+ \lim_{\delta t\to 0}\frac{1}{\delta t}\iint_{|Z-X|<\varepsilon} \mathrm{d}Z\mathrm{d}X f(X)\, p(Z,t+\delta t \mid X,t)\, p(X,t \mid Y,t') \\
&\quad+ \lim_{\delta t\to 0}\frac{1}{\delta t}\iint_{|Z-X|\geq\varepsilon} \mathrm{d}Z\mathrm{d}X f(Z)\, p(Z,t+\delta t \mid X,t)\, p(X,t \mid Y,t') \\
&\quad- \lim_{\delta t\to 0}\frac{1}{\delta t}\iint \mathrm{d}Z\mathrm{d}X f(X)\, p(Z,t+\delta t \mid X,t)\, p(X,t \mid Y,t')\,.
\end{aligned} \tag{6.84}$$

Let us now compute the limit $\varepsilon \to 0$ line by line. The first term of this expression can be transformed in the following way. We take the limit $\delta t \to 0$ inside the integral to obtain with the help of (6.78) and (6.79)

$$\begin{aligned}
(1) &= \int \mathrm{d}X p(X,t \mid Y,t') \\
&\quad\times\left\{\sum_{\alpha} F_\alpha(X,t)\frac{\partial f(X)}{\partial X_\alpha} + \frac{1}{2}\sum_{\alpha\beta} D_{\alpha\beta}(X,t)\frac{\partial^2 f(X)}{\partial X_\alpha \partial X_\beta}\right\}\,.
\end{aligned} \tag{6.85}$$

The second term of (6.84) disappears for $\varepsilon \to 0$ due to $R(Z,X) \sim o(|Z-X|^2)$. The third term and the fifth term can be collected in one expression. Thus, we get

$$\begin{aligned}
(3)+(5) = -\lim_{\delta t\to 0}\frac{1}{\delta t}\iint_{|Y-Z|\geq\varepsilon} & \mathrm{d}Z\mathrm{d}X f(X)\, p(Z,t+\delta t \mid X,t) \\
& \times p(X,t \mid Y,t')\,,
\end{aligned} \tag{6.86}$$

or with the use of (6.80) and considering $\varepsilon \to 0$

$$(3)+(5) = -\mathcal{H}\iint \mathrm{d}Z\mathrm{d}X f(X)\, W(Z \mid X;t)\, p(X,t \mid Y,t')\ . \tag{6.87}$$

Notice that we use the symbol $\mathcal{H}$ to indicate the principal value integral. We finally get for the fourth term

$$(4) = \mathcal{H}\iint \mathrm{d}X\mathrm{d}Z f(Z)\, W(Z \mid X;t)\, p(X,t \mid Y,t')\ . \tag{6.88}$$

We put these all results together to obtain

$$\int \mathrm{d}X \frac{\partial}{\partial t} p(X,t \mid Y,t') f(X) = \int \mathrm{d}X p(X,t \mid Y,t')$$
$$\times \left\{ \sum_\alpha F_\alpha(X,t) \frac{\partial f(X)}{\partial X_\alpha} + \frac{1}{2} \sum_{\alpha\beta} D_{\alpha\beta}(X,t) \frac{\partial^2 f(X)}{\partial X_\alpha \partial X_\beta} \right\}$$
$$+ \mathcal{H} \iint \mathrm{d}Z\mathrm{d}X \left[f(X) - f(Z) \right] W(X \mid Z; t) p(Z,t \mid Y,t') \tag{6.89}$$

and after integrating by parts we get

$$\frac{\partial}{\partial t} \int dX f(X) p(X,t \mid Y,t') = \int dX f(X)$$
$$\times \left\{ -\sum_\alpha \frac{\partial}{\partial X_\alpha} F_\alpha(X,t) + \frac{1}{2} \sum_{\alpha\beta} \frac{\partial^2}{\partial X_\alpha \partial X_\beta} D_{\alpha\beta}(X,t) \right\} p(X,t \mid Y,t')$$
$$+ \mathcal{H} \iint dZdX \left[f(X) - f(Z) \right] W(X \mid Z; t) p(Z,t \mid Y,t') \; . \tag{6.90}$$

Finally we consider that we have chosen the function f to be arbitrary. We can then deduce that the conditional probability fulfils the relation

$$\frac{\partial}{\partial t} p(X,t \mid Y,t') = -\sum_\alpha \frac{\partial}{\partial X_\alpha} F_\alpha(X,t) p(X,t \mid Y,t')$$
$$+ \frac{1}{2} \sum_{\alpha\beta} \frac{\partial^2}{\partial X_\alpha \partial X_\beta} D_{\alpha\beta}(X,t) p(X,t \mid Y,t')$$
$$+ \mathcal{H} \int \mathrm{d}Z \left[W(X \mid Z; t) p(Z,t \mid Y,t') - W(Z \mid X; t) p(X,t \mid Y,t') \right] \; . \tag{6.91}$$

This equation is the differential form of the Chapman–Kolmogorov equation which is also denoted in the literature as the forward differential Chapman–Kolmogorov equation. The right-hand side of this equation defines the general structure of the Markovian $\hat{L}_{\mathrm{markov}}$.

If we want to specify the differential Chapman–Kolmogorov equation we must consider that by definition the components $D_{\alpha\beta}(X,t)$ must form a positive definite matrix and that $W(X \mid Z; t)$ must be a nonnegative function. Then it can be shown under certain conditions that a nonnegative solution to the differential Chapman–Kolmogorov equation exists and that this solution also satisfies the Chapman–Kolmogorov equation. The conditions to be satisfied are the initial condition

$$p(X,t \mid Y,t) = \delta(X - Y) \; , \tag{6.92}$$

which follows directly from (6.65), and any appropriate boundary conditions.

We may also derive the backward differential Chapman–Kolmogorov equations which give the time evolution with respect to the initial variables of $p(X,t \mid Y,t')$. To this aim we consider

$$\begin{aligned}\frac{\partial}{\partial t'} p(X,t \mid Y,t') &= \lim_{\delta t \to 0} \frac{1}{\delta t'} \left[p(X,t \mid Y,t') - p(X,t \mid Y,t' - \delta t') \right] \\ &= \lim_{\delta t \to 0} \frac{1}{\delta t'} \int \mathrm{d}Z p(Z,t' \mid Y,t' - \delta t') p(X,t \mid Y,t') \\ &- \lim_{\delta t \to 0} \frac{1}{\delta t'} \int \mathrm{d}Z p(X,t \mid Z,t') p(Z,t' \mid Y,t' - \delta t') \end{aligned} \quad (6.93)$$

by use of the normalization condition (6.60) in the first term and the Chapman–Kolmogorov equation (6.77) in the second term. It is easy to show that we can carry out the infinitesimal shift $p(Z,t' \mid Y,t'-\delta t') \to p(Z,t'+\delta t' \mid Y,t')$ without a noticeable change of (6.93). Hence we get

$$\begin{aligned}\frac{\partial}{\partial t'} p(X,t \mid Y,t') = \lim_{\delta t \to 0} \int \mathrm{d}Z \, & \frac{p(Z,t'+\delta t' \mid Y,t')}{\delta t'} \\ \times & \left[p(X,t \mid Y,t') - p(X,t \mid Z,t') \right] \end{aligned} \quad (6.94)$$

and, therefore, using similar techniques to those used for the derivation of (6.91)

$$\begin{aligned}\frac{\partial}{\partial t'} p(X,t \mid Y,t') &= -\sum_{\alpha} F_{\alpha}(Y,t') \frac{\partial}{\partial Y_{\alpha}} p(X,t \mid Y,t') \\ &- \frac{1}{2} \sum_{\alpha\beta} D_{\alpha\beta}(Y,t') \frac{\partial^2}{\partial Y_{\alpha} \partial Y_{\beta}} p(X,t \mid Y,t') \\ &+ \mathcal{H} \int \mathrm{d}Z W(Z \mid Y;t') \left[p(X,t \mid Y,t') - p(X,t \mid Z,t') \right] . \end{aligned} \quad (6.95)$$

This equation is called the backward differential Chapman–Kolmogorov equation. We remark that the forward and the backward equations are equivalent to each other. The main difference is which set of variables is held fixed. For the forward equation, solutions exist for $t \geq t'$ and (6.92) is the initial condition with respect to the free variables (X,t). In the case of the backward equation, we hold (X,t) fixed so that since the backward equation expresses development in $t' \leq t$, (6.92) is the final condition of (6.95).

6.7.2 Deterministic Processes

As we have stressed several times in the previous chapters, there are also phenomena in complex systems, which may be described by a completely deterministic motion. If the corresponding processes also possess a Markov character, then they can be described by a differential Chapman–Kolmogorov equation with $B_{\alpha\beta} = 0$ and $W(Y \mid Z;t) = 0$. It remains the equation

$$\frac{\partial}{\partial t} P(X,t \mid Y,t') = -\sum_{\alpha} \frac{\partial}{\partial X_{\alpha}} F_{\alpha}(X,t) P(X,t \mid Y,t') \ . \quad (6.96)$$

The solution to this equation with the initial condition (6.92) is

$$P(X,t \mid Z,t') = \delta\left(X - \widetilde{X}(t)\right) . \tag{6.97}$$

That means the system moves along a certain trajectory $\widetilde{X}(t)$ without any uncertainty. The curve $\widetilde{X}(t)$ is a solution of the ordinary differential equations

$$\frac{\mathrm{d}\widetilde{X}_\alpha(t)}{\mathrm{d}t} = F_\alpha(\widetilde{X}(t),t) \tag{6.98}$$

with the initial conditions

$$\widetilde{X}(t') = Y . \tag{6.99}$$

Equation (6.98) are also denoted as kinetic equations. These differential equations are however generally no equations of motion in the sense of the classical mechanics. In particular the kinetic equations are always irreversible[3], i.e., they are not invariant under reversal of the time direction.

On the other hand, these equations represent the repeatedly used assumption of the decoupling of relevant and irrelevant degrees of freedom. With the knowledge of the above-discussed projection formalism we are able to refine this statement. Decoupling means that the influence of the irrelevant degrees of freedom to the relevant degrees of freedom is reflected in simple parameters of the evolution equations, for example, friction coefficients, reaction rates, or viscosities, so that one gets in fact the impression of a deterministic motion of the relevant degrees of freedom. The condition for the applicability of such deterministic evolution equations are obviously small values of the so-called diffusion coefficients $B_{\alpha\beta}$ and the transition rates $W(Y \mid Z; t)$.

To demonstrate the validity of (6.97) we point out first that for $t = t'$ the initial conditions (6.99) lead to $P(X,t' \mid Z,t') = \delta(X - Z)$. The proof of all other times is best obtained by direct substitution. We see that

$$\begin{aligned}
\frac{dP(X,t \mid Y,t')}{dt} &= -\sum_\alpha \left[\frac{\partial}{\partial X_\alpha}\delta\left(X - \widetilde{X}(t)\right)\right]\frac{d\widetilde{X}_\alpha(t)}{dt} \\
&= -\sum_\alpha \frac{\partial}{\partial X_\alpha}\left[\delta\left(X - \widetilde{X}(t)\right)F_\alpha(\widetilde{X}(t),t)\right] \\
&= -\sum_\alpha \frac{\partial}{\partial X_\alpha}\left[F_\alpha(X,t)P(X,t \mid Y,t')\right]
\end{aligned} \tag{6.100}$$

leads to the expected identity. It should be remarked that the methods of characteristics can be used to obtain (6.98) from (6.96) in a direct way.

6.7.3 Markov Diffusion Processes

If we assume the quantities $W(X \mid Z; t)$ to be zero, the differential Chapman–Kolmogorov equation reduces to the so-called Fokker–Planck equation [2]

[3] For example, an excellent mechanical construction of a pendulum also cannot avoid the occurence of weak friction effects.

$$\frac{\partial}{\partial t} p(X,t \mid Y,t') = -\sum_{\alpha} \frac{\partial}{\partial X_\alpha} F_\alpha(X,t) p(X,t \mid Y,t') + \sum_{\alpha\beta} \frac{1}{2} \frac{\partial^2}{\partial X_\alpha \partial X_\beta} D_{\alpha\beta}(X,t) p(X,t \mid Y,t') \; . \tag{6.101}$$

The functions $F_\alpha(X,t)$ are the components of the drift vector and the $D_{\alpha\beta}(X,t)$ are known as components of the diffusion matrix. As mentioned above, the diffusion matrix is symmetric and positive semi-definite as a result of definition (6.79). All processes which can be described by an equation of type (6.101) are called Markov diffusion processes.

The general solution of this differential equation cannot be given explicitly. However, we can use certain similarities in the mathematical structure of the Fokker–Planck equation and the Schrödinger equation of quantum mechanics in order to transfer well-known solution methods.

But there are some important differences between both equations, which have to do with the operator structure of the right-hand side. Each Schrödinger equation always requires a self-adjoint but not necessarily positive definite operator while the differential operator of a Fokker–Planck equation must be positive semi-definite, but not self-adjoint. In the case of constant components F_α and $B_{\alpha\beta}$, the Fokker–Planck equation (6.101) can be solved exactly, subject to the initial condition (6.92), and we arrive at

$$p(X,t \mid Y,t') = \frac{1}{\sqrt{\det(2\pi D \Delta t)}} \exp\left\{ -\frac{1}{2\Delta t} \sum_{\alpha\beta} \Delta X_\alpha D^{-1}_{\alpha\beta} \Delta X_\beta \right\} \tag{6.102}$$

with $\Delta X_\alpha = X_\alpha - Y_\alpha - F_\alpha \Delta t$, $\Delta t = t - t'$, and N the dimension of the state space. That is nothing but a multivariable Gaussian probability distribution function. The initial condition appears for $\Delta t \to 0$ while the Gaussian spreads over the whole space for $\Delta t \to \infty$. The center of the Gaussian moves with the constant velocity F.

6.7.4 Jump Processes

Finally we consider the case $F_\alpha = D_{\alpha\beta} = 0$. We now have

$$\frac{\partial}{\partial t} p(X,t \mid Y,t') = \mathcal{H} \int \mathrm{d}Z W(X \mid Z;t) \, p(Z,t \mid Y,t') - \mathcal{H} \int \mathrm{d}Z W(Z \mid X;t) \, p(X,t \mid Y,t') \; . \tag{6.103}$$

This is a so-called Master (or rate) equation. The initial condition of this equation is again given by (6.92). In order to discuss the underlying processes described by master equations we solve (6.103) approximately in first order to a small time interval δt. The short-time solution to this equation with the initial condition (6.92) is

$$p(X, t' + \delta t \mid Y, t') = \delta(X - Y)\left[1 - \mathcal{H}\int \mathrm{d}Z W(Z \mid Y; t)\,\delta t\right] + W(X \mid Y; t)\,\delta t\,. \tag{6.104}$$

The first contribution corresponds to a finite probability for the system to stay at the original position Y in the state space. This probability decreases with increasing time. The probability that the system does not remain at Y is given by the second term of (6.104). Hence, a characteristic path of the system through the space of state will consist of a series of discontinuous jumps whose distribution is given by $W(X \mid Y; t)$. For this reason, processes described by master equations are denoted as jump processes.

The master equation (6.103) may be specified to the case where the state space consists of discrete numbers only. Then, the master equation takes the form

$$\frac{\partial}{\partial t} p_{nn'}(t, t') = \sum_m \left[W_{nm}(t)\, p_{mn'}(t, t') - W_{mn}(t)\, p_{nn'}(t, t')\right] \tag{6.105}$$

with $W_{nm}(t) = W(n \mid m; t)$ and $p_{nm}(t, t') = p(n, t \mid m, t')$. In this representation, the concept of jump processes becomes particularly clear. But it should remarked again that pure jump processes can occur even in a continuous state space.[4]

6.8 Correlation and Stationarity

6.8.1 Stationarity

The macroscopic dynamics, i.e., the dynamics on the level of the relevant variables of a certain system is stationary in a strict sense if all joint probabilities are invariant under a time shift Δt

$$p(X_1, t_1; \ldots, X_n, t_n; \ldots) = p(X_1, t_1 + \Delta t; \ldots, X_n, t_n + \Delta t; \ldots)\,. \tag{6.106}$$

From here we conclude $p(X, t) = p(X)$ so that due to (6.61) all conditional probabilities are also invariant under a time shift.

Furthermore, the definition of stationarity implies that the operators of the Nakajima–Zwanzig equation (6.39) depends no longer on the initial time, $\hat{M}(t_0) = \hat{M}$ and $\hat{K}(t_0, t - t') = \hat{K}(t - t')$, and that the coefficients of the differential Chapman–Kolmogorov equation (6.91) are simple functions of the states, i.e., $F_\alpha(X, t) = F_\alpha(X)$, $D_{\alpha\beta}(X, t) = D_{\alpha\beta}(X)$, and $W(X \mid Y; t) = W(X \mid Y)$. Finally, stationarity means that all moments and cumulants have constant values.

[4] Of course, the trajectory of a large but finite complex system has no real jumps. The appearance of jumps is the result of the Markov approximation. If we take into account the exact equation of motion, the jumps correspond to relatively fast but continuous changes of the macroscopic state during a short time period.

There exist several other definitions of stationary processes which are, in fact, less restrictive. For instance, an nth order stationary process arises when (6.106) holds only for joint probability distribution functions of less than $n+1$ points in time, while asymptotically stationary processes are observed only for infinitely large shifts Δt.

6.8.2 Correlation

The knowledge of moments and cumulants does not tell a great deal about the dynamics of a given system. For instance, all moments have constant values in the special case of stationary systems. What would be of interest are measurable quantities obtained from joint probabilities (or alternatively from the conditional probabilities) which simultaneously give information about the state of the system at several points in time. The simplest quantity is the correlation function

$$\overline{f(t)g(t')} = \int \mathrm{d}X\mathrm{d}X' f(X)g(X')p(X,t;X',t') \tag{6.107}$$

between two arbitrary functions f and g of the state of the system. A special case is the autocorrelation function $\overline{f(t)f(t')}$ which considers the same quantity at different time points. Higher correlations can be constructed in a similar way

$$\overline{f_1(t_1)\dots f_n(t_n)} = \int \prod_i^n \left[\mathrm{d}X^{(i)} f_i\left(X^{(i)}\right)\right] p(X^{(1)},t_1;\dots;X^{(n)},t_n)\,. \tag{6.108}$$

A very natural class of correlation functions may be obtained by identifying the functions f_i with the components X_α of the state vector. These correlation functions may be interpreted as generalized moments, $\overline{X_{\alpha_1}(t_1)\dots X_{\alpha_n}(t_n)}$, which approach the standard definition (6.48) for $t_1 = \dots = t_n = t$. Analogously we can design combinations of correlation functions, which may be understood as a generalization of cumulants (6.55). For instance, the generalized covariance functions are defined by

$$C_{\alpha\beta}(t,t') = \overline{X_\alpha(t)X_\beta(t')} - \overline{X_\alpha(t)}\;\overline{X_\beta(t')}\,, \tag{6.109}$$

which are useful to consider for processes with average values different from zero. Because of (6.106) stationarity always requires that the correlation functions are invariant under a time shift. In particular, we get for stationary processes $\overline{X_\alpha(t)X_\beta(t')} = \overline{X_\alpha(t-t')X_\beta(0)}$ and $C_{\alpha\beta}(t,t') = C_{\alpha\beta}(t-t',0)$ which is simply denoted as $C_{\alpha\beta}(t-t')$.

Finally, we introduce the so-called correlation time. For the sake of simplicity we concentrate here on the discussion of the autocovariance function $C(t-t')$ of a stationary process in a one-dimensional state space. The symmetry of this function immediately requires

$$C(t-t') = C(t'-t) = C(\mid t-t' \mid)\,. \tag{6.110}$$

One possibility which provides a measure for the corresponding correlation time τ_c is the integral

$$\tau_c = C^{-1}(0) \int_0^\infty C(t) \mathrm{d}t \,. \tag{6.111}$$

This definition is independent of the precise functional form of the autocovariance function. The correlation time τ_c may have a finite, infinite, or indeterminate value.

The last case is more or less irrelevant for the majority of complex systems. It corresponds, for instance, to periodic autocovariance functions, e.g., $C(t) \sim \cos(\omega t)$.

A divergent time scale $\tau_c \to \infty$ indicates the existence of a dominant correlation. Processes characterized by such an autocorrelation function are said to be long-range correlated. For instance, $C(t) \sim t^{-a}$ with $0 < a < 1$ is a typical autocovariance function of long-ranged correlated processes.

When integral (6.111) is finite, the corresponding process shows short-range correlated behavior. In this case, the values of the observed quantities are practically uncorrelated if sequentially realized measurements are separated by a time scale sufficiently longer then τ_c. An important example is the autocovariance of the Ornstein–Uhlenbeck process, $C \sim \exp\{-t/\tau_c\}$, which is often used to model a realistic noise signal.

6.8.3 Spectra

In order to characterize a stationary process, it is very natural to calculate the Fourier transform for the covariance functions $C_{\alpha\beta}(t)$

$$S_{\alpha\beta}(\omega) = \int_{-\infty}^{+\infty} \mathrm{d}t C_{\alpha\beta}(t) \exp\{\mathrm{i}\omega t\} \,. \tag{6.112}$$

Due to the symmetry property $C_{\alpha\beta}(t) = C_{\beta\alpha}(-t)$, the Fourier transforms are self-adjoint, $S_{\alpha\beta}(\omega) = S^*_{\beta\alpha}(\omega)$. $S_{\alpha\beta}(\omega)$ are called the spectral functions of the underlying processes. In addition to the above-introduced classification of correlation processes, the same properties might be investigated in the frequency domain.

To this end, we consider again a stationary process in a one-dimensional state space. We obtain from (6.112) and (6.111) the important relation $\tau_c = S(0)$. Therefore, we conclude that convergent behavior of $S(\omega)$ in low-frequency regions indicates a short-range correlation, while any kind of divergence is related to long-range correlations.

6.9 Stochastic Equations of Motions

6.9.1 The Mori–Zwanzig Equation

In the beginning of this book we already referred to the fact that the microscopic mechanical (or quantum-mechanical) equations of motion and the Liouville equation are equivalent representations of a given system. Subsequently, we demonstrated that the Liouville equation can be reduced to a Nakajima–Zwanzig equation, which contains only the relevant degrees of freedom representing a suitable description of the system on a more or less macroscopic level. It is reasonable to ask whether one can also reduce the complicated system of microscopic mechanical equations of motion to a macroscopic description. To this end, we introduce a set of linearly independent, differentiable functions $G_\alpha(t)$ $(\alpha = 1, \ldots, M)$, which are assumed to be functions of the microscopic state $\Gamma = \{q_1, \ldots, p_N\}$, $G_\alpha(t) = G_\alpha(\Gamma(t))$. All these functions are denoted as relevant variables representing macroscopically observable or measurable quantities of our interest. The time evolution of each G_α is ruled by the microscopic equations of motion

$$\frac{\mathrm{d}G_\alpha}{\mathrm{d}t} = \sum_i^N \left[\frac{\partial G_\alpha}{\partial q_i} \dot{q}_i + \frac{\partial G_\alpha}{\partial p_i} \dot{p}_i \right] = \sum_i^N \left[\frac{\partial G_\alpha}{\partial q_i} \frac{\partial H}{\partial p_i} - \frac{\partial G_\alpha}{\partial p_i} \frac{\partial H}{\partial q_i} \right] , \tag{6.113}$$

where we have used Hamilton's equations (1.1) describing the evolution of all $2N$ microscopic coordinates q_i and momenta p_i. From (6.113) we get

$$\frac{\mathrm{d}G_\alpha}{\mathrm{d}t} = \hat{L} G_\alpha , \tag{6.114}$$

where the Liouvillian $\hat{L}$ is defined as in (6.8). Equation (6.114) is formally integrated to yield

$$G_\alpha(t) = \exp\{\hat{L}(t - t_0)\} G_\alpha^0 , \tag{6.115}$$

and $G_\alpha^0 = G_\alpha(\Gamma_0)$ are fixed by the microscopic initial state $\Gamma_0 = \Gamma(t_0)$. In other words, the relevant quantities $G_\alpha(t)$ of a given system are unique functions of the initial state, $G_\alpha(t) = G_\alpha(t, \Gamma_0)$. To proceed we should eliminate all irrelevant degrees of freedom contained in the Liouvillian by the use of an appropriate projection formalism.

A convenient starting point for this intention is the introduction of a scalar product (A, B). As is easily verified, the scalar product properties are satisfied by identifying, for instance, the scalar product with the average

$$(A, B) = \int \mathrm{d}\Gamma_0 A(\Gamma_0) B(\Gamma_0) p(\Gamma_0, t_0) \tag{6.116}$$

considering the probability distribution function at the initial time t_0. This representation has the advantage that the scalar products

$$(G_\alpha(t), G_\beta(t')) = \int \mathrm{d}\Gamma_0 G_\alpha(t, \Gamma_0) G_\beta(t', \Gamma_0) p(\Gamma_0, t_0) \tag{6.117}$$

are identical to the correlation functions, $(G_\alpha(t), G_\beta(t')) = \overline{G_\alpha(t)G_\beta(t')}$. In order to interpret (6.117) we must consider that each microscopic initial state Γ_0 defines a unique trajectory of the system through the phase space while the statistical weight of each trajectory is given by $p(\Gamma_0, t_0)\,\mathrm{d}\Gamma_0$.

We remark that many other possibilities defining a suitable scalar product exist. For the following derivation the precise structure of the scalar product is of secondary interest.

The central point is the introduction of an appropriate projection operator $\hat{P}$. We use the definition [4, 17]

$$\hat{P} = \sum_{\alpha\beta} G_\alpha^0 H_{\alpha\beta}(G_\beta^0, \ldots) , \tag{6.118}$$

where $H_{\alpha\beta}$ are the components of the inverse of the $M \times M$ matrix formed by the scalar products (G_α^0, G_β^0). Obviously, we get $\hat{P}G_\alpha^0 = G_\alpha^0$ and therefore $\hat{P}^2 G_\alpha^0 = G_\alpha^0$. Thus, the projection operator and the corresponding complementary operator $\hat{Q} = 1 - \hat{P}$ fulfil relations (6.23). Let us now consider the formal solution (6.115) of the equation of motion and insert the identity operator $\hat{P} + \hat{Q}$ after the propagator $\exp\{\hat{L}(t-t_0)\}$. We obtain

$$\begin{aligned}\frac{\mathrm{d}G_\alpha(t)}{\mathrm{d}t} &= \exp\{\hat{L}(t-t_0)\}\left(\hat{P}+\hat{Q}\right)\hat{L}G_\alpha^0 \\ &= \exp\{\hat{L}(t-t_0)\}\hat{P}\hat{L}G_\alpha^0 + \exp\{\hat{L}(t-t_0)\}\hat{Q}\hat{L}G_\alpha^0 .\end{aligned} \tag{6.119}$$

The first term can be written as

$$\exp\{\hat{L}(t-t_0)\}\hat{P}\hat{L}G_\alpha^0 = \sum_{\beta\gamma} \Omega_{\alpha\gamma} G_\gamma(t) ,$$

where we have introduced the $M \times M$ frequency matrix $\Omega_{\alpha\gamma}$

$$\Omega_{\alpha\gamma} = \sum_\beta H_{\gamma\beta}(G_\beta^0, \hat{L}G_\alpha^0) . \tag{6.120}$$

The second term in equation (6.119) can be rearranged by using the identity

$$\mathrm{e}^{\hat{L}(t-t_0)} = \mathrm{e}^{\hat{L}_1(t-t_0)} + \int_{t_0}^{t} \mathrm{d}t' \mathrm{e}^{\hat{L}(t-t')} \hat{L}_2 \mathrm{e}^{\hat{L}_1(t'-t_0)} \tag{6.121}$$

with $\hat{L} = \hat{L}_1 + \hat{L}_2$. This identity may be checked in a similar way as (6.31) by a derivative with respect to time. If we replace $\hat{L}_1$ by $\hat{Q}\hat{L}$ and $\hat{L}_2$ by $\hat{P}\hat{L}$, we arrive at

$$\begin{aligned}\mathrm{e}^{\hat{L}(t-t_0)}\hat{Q}\hat{L}G_\alpha^0 &= \mathrm{e}^{\hat{Q}\hat{L}(t-t_0)}\hat{Q}\hat{L}G_\alpha^0 \\ &\quad + \int_{t_0}^{t} \mathrm{d}t' \mathrm{e}^{\hat{L}(t-t')}\hat{P}\hat{L}\mathrm{e}^{\hat{Q}\hat{L}(t'-t_0)}\hat{Q}\hat{L}G_\alpha^0 .\end{aligned} \tag{6.122}$$

Because of properties (6.23) the first contribution can be transformed into

$$f_\alpha(t) = \mathrm{e}^{\hat{Q}\hat{L}(t-t_0)}\hat{Q}\hat{L}G_\alpha^0 = \hat{Q}\mathrm{e}^{\hat{Q}\hat{L}(t-t_0)}\hat{Q}\hat{L}G_\alpha^0 \,. \tag{6.123}$$

This quantity is referred to as the fluctuating force or as the residual force. By construction, the time evolution of $f_\alpha(t)$ from its initial value $\hat{Q}\hat{L}G_\alpha^0$ is ruled by the anomalous propagator $\exp\{\hat{Q}\hat{L}(t-t_0)\}$ rather than by the usual one $\exp\{\hat{L}(t-t_0)\}$. The presence of the complementary projection operator $\hat{Q}$ has the important consequence that

$$\left(G_\beta^0, f_\alpha(t)\right) = 0 \,. \tag{6.124}$$

From a geometrical point of view, the fluctuating forces $f_\alpha(t)$ are orthogonal to all initial relevant quantities G_β^0 at all times. In other words, the forces evolve in a subspace intrinsically different from the one spanned by the set G_β^0. The second term on the right-hand side of (6.122) can be transformed into a more convenient form. We write

$$\begin{aligned} &\mathrm{e}^{\hat{L}(t'-t_0)}\hat{P}\hat{L}\mathrm{e}^{\hat{Q}\hat{L}(t-t')}\hat{Q}\hat{L}G_\alpha^0 \\ &= \mathrm{e}^{\hat{L}(t'-t_0)}\sum_{\gamma\beta} G_\gamma^0 H_{\gamma\beta}(G_\beta^0, \hat{L}\hat{Q}\mathrm{e}^{\hat{Q}\hat{L}(t-t')}\hat{Q}\hat{L}G_\alpha^0) \\ &= \sum_{\beta\gamma} G_\gamma(t') H_{\gamma\beta}(G_\beta^0, \hat{L}\hat{Q}\mathrm{e}^{\hat{Q}\hat{L}(t-t')}\hat{Q}\hat{L}G_\alpha^0) \,, \end{aligned} \tag{6.125}$$

where we have used (6.115). As a result, the equation of motion (6.119) can be written as

$$\frac{\mathrm{d}G_\alpha(t)}{\mathrm{d}t} = \sum_\gamma \left[\Omega_{\alpha\gamma} G_\gamma(t) + \int_{t_0}^{t} \mathrm{d}t' K_{\alpha\gamma}(t-t') G_\gamma(t') \right] + f_\alpha(t) \,, \tag{6.126}$$

where we have introduced the quantity

$$\begin{aligned} K_{\alpha\gamma}(t-t') &= \sum_\beta H_{\gamma\beta}(G_\beta^0, \hat{L}\hat{Q}\mathrm{e}^{\hat{Q}\hat{L}(t-t')}\hat{Q}\hat{L}G_\alpha^0) \\ &= \sum_\beta H_{\gamma\beta}(G_\beta^0, \hat{L}f_\alpha(t-t')) \,, \end{aligned} \tag{6.127}$$

which is referred to as the memory matrix. It should be pointed out that both, the frequency matrix and the memory matrix, still depend on the initial time t_0. Equation (6.126) is denoted as the generalized Langevin equation or as the Mori–Zwanzig equation [18, 19]. No approximation has taken into account in the previous derivation, so that (6.126) is still equivalent to the mechanical equations of motion (6.113).

A special situation occurs in the case of stationarity. Then we obtain

$$K_{\alpha\gamma}(t-t') = -\sum_\beta H_{\gamma\beta}(f_\beta(t'), f_\alpha(t)) \,, \tag{6.128}$$

where we have used the relation $(G_\beta^0, \hat{L}f_\alpha(t-t')) = -\,(f_\beta(t'), f_\alpha(t))$ which can be checked straightforwardly. For the sake of simplicity we set $t_0 = 0$. Then we obtain due to the stationarity

$$(f_\beta(t'), f_\alpha(t)) = (f_\beta(0), f_\alpha(t-t')) = (\hat{Q}\hat{L}G^0_\beta, f_\alpha(t-t')) \,. \tag{6.129}$$

It is easily demonstrated that

$$(\hat{Q}\hat{L}G^0_\beta, f_\alpha(t-t')) = (\hat{L}G^0_\beta, \hat{Q}f_\alpha(t-t')) = (\hat{L}G_\beta(0), f_\alpha(t-t'))\,, \tag{6.130}$$

and therefore $(f_\beta(t'), f_\alpha(t)) = (\hat{L}G_\beta(t'), f_\alpha(t))$. Thus we get the desired relation

$$\begin{aligned}(f_\beta(t'), f_\alpha(t)) &= \frac{\mathrm{d}}{\mathrm{d}t'}(G_\beta(t'), f_\alpha(t)) = \frac{\mathrm{d}}{\mathrm{d}t'}(G_\beta(0), f_\alpha(t-t')) \\ &= -\frac{\mathrm{d}}{\mathrm{d}t}(G_\beta(0), f_\alpha(t-t')) = -\frac{\mathrm{d}}{\mathrm{d}t}(G_\beta(t'), f_\alpha(t)) \\ &= -(G_\beta(t'), \hat{L}f_\alpha(t)) = -(G^0_\beta, \hat{L}f_\alpha(t-t'))\,,\end{aligned} \tag{6.131}$$

where we have applied several times the stationarity condition and the formal equation of motion (6.114) which is valid, of course, for all dynamic quantities of the system.

From (6.126) it is straightforward to obtain the corresponding equation for the correlation functions $\overline{G_\alpha(t)G_\beta(t_0)}$. Exploiting the orthogonality of the residual forces and the relevant quantities (6.124), we obtain

$$\begin{aligned}\frac{\mathrm{d}}{\mathrm{d}t}\overline{G_\alpha(t)G_\beta(t_0)} &= \sum_\gamma \Omega_{\alpha\gamma}\overline{G_\gamma(t)G_\beta(t_0)} \\ &\quad + \sum_\gamma \int_{t_0}^{t} \mathrm{d}t' K_{\alpha\gamma}(t-t')\overline{G_\gamma(t')G_\beta(t_0)}\,.\end{aligned} \tag{6.132}$$

This equation is still linear in the correlation functions and may be solved by standard methods. Equation (6.132) can also be derived from the Nakajima–Zwanzig equation (6.39) in a similar form.

The remaining problem is the specification of the frequency and memory matrices. Definitions (6.120) and (6.127) are in general too complicated to be useful in practice. In particular, if we want to describe the control of complex systems the dynamic of which is given by evolution equations of type (6.126) and (6.132), we require alternative methods in order to approximate these quantities. Physical intuition and empirical knowledge from experiments and observations play very important roles at several stages of these approaches [5,9].

In a correct framework, the results of the formalism are particularly rewarding because of their simple mathematical form. Of course, the obtained results cannot be claimed to be the output of a real theory firmly rooted on microscopic intuition and reasoning, but the general structure of (6.126) and (6.132) is motivated by universal principles of physics.

6.9.2 Separation of Time Scales

Suppose that we have included in the set $\{G_\alpha\}$ all the dynamical variables with a time dependence much slower than any microscopic time scale predictable from the Liouvillian. These relevant quantities substantially determine the macroscopic behavior of the system. Since the evenly discussed projection formalism gives no particular hint for the selection of these slow variables, we have to deal with problems which also occurred with the introduction of the Markov approximation of the Nakajima–Zwanzig equation. Especially, the choice of which variables are actually slow is largely guided by the problem in mind.

After we determined the slow quantities as relevant variables, we can assume that the projection formalism collects the fast dynamics more or less in the residual forces due to the very complicated time dependence governed by the anomalous propagator $\exp\{\hat{Q}\hat{L}(t-t')\}$. From a macroscopic point of view, the residual forces behave apparently as random functions. As a consequence, all the elements of the memory matrix (6.128) are likely to be characterized by decay times considerably shorter than those associated with the elements $\overline{G_\alpha(t)G_\beta(t')}$ of the correlation matrix.

Thus we may assume that over the characteristic time scales of $\overline{G_\alpha(t)G_\beta(t')}$ the decay time of the memory matrix is so short that $K_{\alpha\gamma}(t-t')$ may be approximately written as

$$K_{\alpha\gamma}(t-t') = K^0_{\alpha\gamma}\delta(t-t') \,. \tag{6.133}$$

This estimation is again called the Markov approximation or the separation of time scales. The representation (6.133) requires that the residual force correlations are of a δ-type, $\overline{f_\alpha(t)f_\beta(t')} \sim \delta(t-t')$ and furthermore that the condition $\bar{f}_\alpha(t) = 0$ holds, which can always be satisfied after realizing the shifts $f_\alpha \to f_\alpha - \bar{f}_\alpha$ and the corresponding changes of G_α. As a consequence of (6.133), the Mori–Zwanzig equations (6.126) now read

$$\frac{\mathrm{d}G_\alpha(t)}{\mathrm{d}t} = \sum_\gamma \tilde{\Omega}_{\alpha\gamma}G_\gamma(t) + f_\alpha(t) \tag{6.134}$$

with $\tilde{\Omega}_{\alpha\gamma} = \Omega_{\alpha\gamma} + K^0_{\alpha\gamma}$. It is seen that the separation of time scales yields a complete loss of memory effects in the Mori–Zwanzig equations. The system of ordinary linear differential equations (6.134) is a linearized version of a set of the so-called Langevin equation. In particular, the Markov approximation reduces (6.132) to

$$\frac{\mathrm{d}}{\mathrm{d}t}\overline{G_\alpha(t)G_\beta(t_0)} = \sum_\gamma \tilde{\Omega}_{\alpha\gamma}\overline{G_\gamma(t)G_\beta(t_0)} \,, \tag{6.135}$$

representing a simple homogeneous system of linear differential equations with constant coefficients.

6.9.3 Wiener Process

For our further discussion, let us now study the properties of the trajectories of the so-called normalized Wiener process $W(t)$. This process satisfies a Fokker–Planck equation (6.101) in which there is only one variable W, the drift coefficient is zero, and the diffusion coefficient is 1,

$$\frac{\partial}{\partial t} p\left(W, t \mid W', t'\right) = \frac{1}{2} \frac{\partial^2}{\partial W^2} p\left(W, t \mid W', t'\right) . \tag{6.136}$$

All trajectories connecting the state W' at time t' with the state W at time t contribute to the conditional probability density $p\left(W, t \mid W, t'\right)$. In order to be able to discuss the properties of these trajectories we have to solve (6.136) under the initial condition $p\left(W, t' \mid W', t'\right) = \delta\left(W - W'\right)$. This is a standard procedure leading to the well-known Gaussian

$$p\left(W, t \mid W', t'\right) = \frac{1}{\sqrt{2\pi(t - t')}} \exp\left\{-\frac{(W - W')^2}{2(t - t')}\right\} . \tag{6.137}$$

Thus, if the process has arrived the state W' at time t', the averaged state at time $t > t'$ is given by

$$\overline{W} = \int W p\left(W, t \mid W', t'\right) \mathrm{d}W = W' \tag{6.138}$$

while the variance (6.47) becomes

$$\sigma^2 = \int\limits_{-\infty}^{\infty} \mathrm{d}W \left[W - \overline{W}\right]^2 p\left(W, t \mid W', t'\right) \mathrm{d}W = t - t' . \tag{6.139}$$

It is easy to see that for $\delta t = t - t' \to 0$ equation (6.139) yield $|W - W'| \sim \sqrt{\delta t} \to 0$ and $|\mathrm{d}W(t)/\mathrm{d}t| \sim |W - W'| / \delta t \to \infty$. Therefore, each trajectory of a Wiener process is a continuous but not differentiable path.

Let us now determine the autocorrelation function of the Wiener process on the condition that the initial value of the process is $W_0 = W(t_0)$. The corresponding joint probability density is given by

$$p\left(W, t; W', t' | W_0 t_0\right) = p\left(W, t \mid W', t'\right) p\left(W', t' \mid W_0, t_0\right) \tag{6.140}$$

so that

$$\begin{aligned} \overline{W(t)W(t')} &= \int\int W W' p\left(W, t \mid W', t'\right) p\left(W', t' \mid W_0, t_0\right) \mathrm{d}W \mathrm{d}W' \\ &= \min\left(t - t_0, t' - t_0\right) + W_0^2 . \end{aligned} \tag{6.141}$$

We conclude that the autocovariance function of the Wiener process is given by $C(t, t') = \min\left(t - t_0, t' - t_0\right)$. As a final point we should note that infinitesimal changes $\mathrm{d}W(t) = W(t + \mathrm{d}t) - W(t)$ satisfy

$$\overline{\mathrm{d}W(t)} = 0 \quad \text{and} \quad \overline{\mathrm{d}W(t)^2} = \mathrm{d}t \tag{6.142}$$

due to (6.138) and (6.139) while (6.141) leads to

$$\overline{\mathrm{d}W(t)\mathrm{d}W(t')} = 0 \quad \text{for} \quad t \neq t' . \tag{6.143}$$

Higher orders vanish as $\overline{\mathrm{d}W(t)^f} \sim \mathrm{d}t^{f/2} = o(\mathrm{d}t)$ for $f > 2$. The simplest way of characterizing these results is to say that $\mathrm{d}W(t)$ is an infinitesimal element of order 1/2, i.e., $\mathrm{d}W(t) \sim \sqrt{\mathrm{d}t}$, and that in calculating differentials, infinitesimal elements of order higher than 1 are discarded so that $\mathrm{d}W(t)^{2+n} \sim \mathrm{d}t^{1+n/2} \to 0$ for all $n > 0$. From here we obtain the important result that the stochastic fluctuations of $\mathrm{d}W(t)$ causes $\overline{\mathrm{d}W(t)/\mathrm{d}t} = 0$ while $\overline{|\mathrm{d}W(t)/\mathrm{d}t|} \sim \mathrm{d}t^{-1/2}$ diverges [14,15].

6.9.4 Stochastic Differential Equations

The linearity of the Langevin equation (6.134) is a consequence of the projection formalism introduced in the previous sections. In many practical cases, we have to deal with nonlinear Langevin equations. These equations may be derived in a more or less intuitive manner, but they are rarely based on a real theoretical framework only. However, in the case of Markov processes the Langevin equations can be obtained severely from the corresponding Fokker–Planck equations.

To proceed, we now consider a system of stochastic differential equations that generalizes the linear system (6.134),

$$\dot{X}_\alpha(t) = F_\alpha(X(t)) + \sum_{k=1}^{R} d_{\alpha,k}(X(t))\eta_k(t) \,. \tag{6.144}$$

Here, $F_\alpha(X)$ and $d_{\alpha,k}(X)$ are differentiable functions of the N-dimensional state vector X while $\eta_k(t)$ $(k = 1, \ldots, R)$ are linearly independent stochastic functions.

Equations of such a type are also denoted as Langevin equations. In principle, these equations can be derived formally from (6.134) in a heuristic way. To this aim we take into account a set of N relevant quantities G_α. These relevant quantities may be specified as functions of the state vector X, $G_\alpha(t) = G_\alpha(X(t))$. We substitute (6.144) into $\dot{G}_\alpha = \sum_\beta(\partial G_\alpha/\partial X_\beta)\dot{X}_\beta$ and compare the result with (6.134). This allows us to identify

$$\sum_\beta \frac{\partial G_\alpha}{\partial X_\beta} F_\beta = \sum_\beta \tilde{\Omega}_{\alpha\beta} G_\beta \quad \text{and} \quad f_\alpha(t) = \sum_{\beta,k} \frac{\partial G_\alpha}{\partial X_\beta} d_{\beta,k}\eta_k(t) \,. \tag{6.145}$$

The first equation defines the functions $F_\alpha(X)$ while the second one requires a further explanation. The fluctuation forces $f_\alpha(t)$ are assumed in the context of the Markov approximation as fast varying quantities with a more or less stochastic character, but they can nevertheless be structured still from the relatively slow relevant quantities and the fast irrelevant variables. Even the macroscopically uncontrollable dynamics of the irrelevant degrees of freedom is the reason for the apparently stochastic behavior of the fluctuation forces $f_\alpha(t)$. Therefore the separation

$$f_\alpha(t) = \sum_{k=1}^{R} \tilde{B}_{\alpha,k}(X(t))\eta_k(t) \tag{6.146}$$

of the fluctuation forces into bilinear combinations of independent stochastic functions $\eta_k(t)$ which are assumed to be exclusively controlled by the dynamics of the irrelevant degrees of freedom and systematic terms $\tilde{B}_{\alpha,k} = \sum_\beta (\partial G_\alpha / \partial X_\beta) d_{\beta,k}$ controlled by the relevant quantities, and the state vector, respectively, is a natural ansatz, which is not in disagreement with the requirements of the projection formalism.

It should be noticed that (6.146) is only an obvious assumption, which is perhaps supported by empirical experience. Equation (6.146) cannot be generally derived so far in the framework of a closed theory.

As mentioned above, the Markov approximation (6.133) requires

$$\overline{f_\alpha(t) f_\beta(t')} \sim \delta(t - t') \quad \text{and} \quad \bar{f}_\alpha(t) = 0 \,. \tag{6.147}$$

This δ-function character is also transferred to the stochastic functions $\eta_k(t)$. In general, we are able to specify the stochastic functions to

$$\overline{\eta_k(t)\eta_{k'}(t')} = \delta_{kk'}\delta(t - t') \quad \text{and} \quad \bar{\eta}_k(t) = 0 \tag{6.148}$$

by a suitable choice of the functions $\tilde{B}_{\alpha,k}(X)$ in (6.146).

Let us return to the discussion of the stochastic differential equation (6.144). Unfortunately, this equation as it stands has no meaning, and we do not know how to deal with it. The reason is that the δ-character of the correlation functions (6.148) causes jumps in the state vector $X(t)$ such that the value of $X(t)$ at time t is not well defined. Basically, the problem arises from the fact that the stochastic functions $\eta_k(t)$ change substantially during an infinitesimally small time interval so that the equation does not specify what value of $d_{\alpha,k}(X)$ should be used in the product $d_{\alpha,k}(X(t))\eta_k(t)$.

For a better understanding of the problem, we divide the time axis into infinitesimally small subintervals of length $\mathrm{d}t$ by means of partitioning points t_i with $t_{i+1} = t_i + \mathrm{d}t$ and define intermediate points τ_i such that $t_i < \tau_i < t_{i+1}$. Then we obtain from (6.144)

$$X_\alpha(t_{i+1}) = X_\alpha(t_i) + F_\alpha(X(t_i))\mathrm{d}t + \sum_{k=1}^{R} \int_{t_i}^{t_{i+1}} d_{\alpha,k}(X(t'))\eta_k(t')\mathrm{d}t' \tag{6.149}$$

or due to the mean value theorem

$$X_\alpha(t_{i+1}) = X_\alpha(t_i) + F_\alpha(X(t_i))\mathrm{d}t + \sum_{k=1}^{R} d_{\alpha,k}(X(\tau_i)) \int_{t_i}^{t_{i+1}} \eta_k(t')\mathrm{d}t' \,. \tag{6.150}$$

The main problem comes from the integral

$$\mathrm{d}W_k(t_i) = \int_{t_i}^{t_{i+1}} \eta_k(t')\mathrm{d}t' \,. \tag{6.151}$$

It is easy to see that the application of (6.148) leads to

$$\overline{\mathrm{d}W_k(t_i)\mathrm{d}W_{k'}(t_j)} = \delta_{kk'}\delta_{ij}\mathrm{d}t \quad \text{and} \quad \overline{\mathrm{d}W_k(t_i)} = 0 \,. \tag{6.152}$$

This is nothing but the mean and the correlation of infinitesimal changes of independent Wiener processes $W_k(t)$ $(k = 1, \ldots, R)$. Now we can rewrite (6.150) to obtain

$$X_\alpha(t_{i+1}) = X_\alpha(t_i) + F_\alpha(X(t_i))\mathrm{d}t + \sum_{k=1}^{R} d_{\alpha,k}(X(\tau_i))\mathrm{d}W_k(t_i) \,. \tag{6.153}$$

However, we can evaluate $d_{\alpha,k}(X(\tau_i))$ at an arbitrary intermediate time τ_i. It is clear that the choice of this intermediate time has important consequences for numerical simulations of stochastic processes. Two general concepts were established.

In the Ito interpretation [1, 16], the value of $X(\tau_i)$ in taken before the jump. That means explicitly

$$X_\alpha(t_{i+1}) = X_\alpha(t_i) + F_\alpha(X(t_i))\mathrm{d}t + \sum_{k=1}^{R} d_{\alpha,k}(X(t_i))\mathrm{d}W_k(t_i) \tag{6.154}$$

or in the time-continuous notation

$$\mathrm{d}X_\alpha(t) = F_\alpha(X(t))\mathrm{d}t + \sum_{k=1}^{R} d_{\alpha,k}(X(t))\mathrm{d}W_k(t) \,. \tag{6.155}$$

On the other hand, in the Stratonovich interpretation [13, 16], we take the mean of $X(t)$ before and after the jump so that $X(\tau_i) = (X(t_{i+1}) + X(t_i))/2$, i.e.,

$$X_\alpha(t_{i+1}) = X_\alpha(t_i) + F_\alpha(X(t_i))\mathrm{d}t + \sum_{k=1}^{R} d_{\alpha,k}\left(\frac{X(t_i) + X(t_{i+1})}{2}\right)\mathrm{d}W_k(t_i) \,. \tag{6.156}$$

The time-continuous representation of the Stratonovich stochastic differential equation now reads

$$\mathrm{d}X_\alpha(t) = F_\alpha(X(t))\mathrm{d}t + \sum_{k=1}^{R} d_{\alpha,k}\left(X(t) + \frac{1}{2}\mathrm{d}X(t)\right)\mathrm{d}W_k(t) \,. \tag{6.157}$$

Neglecting $\mathrm{d}X(t)$ on the right-hand side of (6.157) one apparently obtains the same time-continuos notation as (6.155). This constitutes a warning that both representations are identical. That is always not the case. A solution of the Ito stochastic differential equation (6.155) with given F_α and $d_{\alpha,k}$ is not a solution of the same equation with the same coefficients, but interpreted to be of the Stratonovich type. However, both solutions become identical if the coefficients of an Ito stochastic differential equation

$$\mathrm{d}X_\alpha = F_\alpha^{\mathrm{Ito}}(X)\mathrm{d}t + \sum_{k=1}^{R} d_{\beta,k}^{\mathrm{Ito}}(X)\mathrm{d}W_k \tag{6.158}$$

are related to the coefficients of a Stratonovich stochastic differential equation

$$\mathrm{d}X_\alpha = F_\alpha^{\mathrm{St}}(X)\mathrm{d}t + \sum_{k=1}^{R} d_{\alpha,k}^{\mathrm{St}}\left(X + \frac{1}{2}\mathrm{d}X\right)\mathrm{d}W_k \tag{6.159}$$

by a certain transformation. To find this relation, we expand the Stratonovich representation up to the first order in $\mathrm{d}X(t)$ and consider that $X(t)$ is also a solution of (6.158):

$$\begin{aligned}
\mathrm{d}X_\alpha &= F_\alpha^{\mathrm{St}}(X)\mathrm{d}t + \sum_{k} d_{\alpha,k}^{\mathrm{St}}\left(X + \frac{1}{2}\mathrm{d}X\right)\mathrm{d}W_k \\
&= F_\alpha^{\mathrm{St}}(X)\mathrm{d}t + \sum_{k} d_{\alpha,k}^{\mathrm{St}}(X)\mathrm{d}W_k + \frac{1}{2}\sum_{k,\beta}\frac{\partial d_{\alpha,k}^{\mathrm{St}}(X)}{\partial X_\beta}\mathrm{d}X_\beta\,\mathrm{d}W_k \\
&= F_\alpha^{\mathrm{St}}(X)\mathrm{d}t + \sum_{k} d_{\alpha,k}^{\mathrm{St}}(X)\mathrm{d}W_k \\
&\quad + \frac{1}{2}\sum_{k,\beta}\frac{\partial d_{\alpha,k}^{\mathrm{St}}(X)}{\partial X_\beta} a_\alpha^{\mathrm{Ito}}(X)\mathrm{d}t\mathrm{d}W_k \\
&\quad + \frac{1}{2}\sum_{k,j,\beta}\frac{\partial d_{\alpha,k}^{\mathrm{St}}(X)}{\partial X_\beta} d_{\beta,j}^{\mathrm{Ito}}(X)\mathrm{d}W_j\,\mathrm{d}W_k\ .
\end{aligned} \tag{6.160}$$

Considering $\mathrm{d}W \sim \mathrm{d}t^{1/2}$, the third term is of order $\mathrm{d}t^{3/2}$ and can be neglected. But the last term contains $\mathrm{d}W_j\ \mathrm{d}W_k \sim \mathrm{d}t$ and, therefore, it must be taken into accout. We use the splitting

$$\mathrm{d}W_j\ \mathrm{d}W_k = \overline{\mathrm{d}W_j(t)\mathrm{d}W_k(t)} + \mathrm{d}\varphi_{kj}(t)\ . \tag{6.161}$$

Since $\mathrm{d}W_j\ \mathrm{d}W_k$ is of order $\mathrm{d}t$, but $\mathrm{d}W_k$ is of order $\mathrm{d}t^{1/2}$, the fluctuations of $\mathrm{d}W_j\ \mathrm{d}W_k$ can be neglected compared with the fluctuations of $\mathrm{d}W_k$. Thus it remains the systematic part and we obtain with (6.152)

$$\mathrm{d}X_\alpha = \left[F_\alpha^{\mathrm{St}}(X) + \frac{1}{2}\sum_{k,\beta}\frac{\partial d_{\alpha,k}^{\mathrm{St}}(X)}{\partial X_\beta} d_{\beta,k}^{\mathrm{Ito}}(X)\right]\mathrm{d}t + \sum_{k} d_{\alpha,k}^{\mathrm{St}}(X)\mathrm{d}W_k\ . \tag{6.162}$$

A comparison of (6.162) with (6.158) leads to the desired transformations between the coefficients of Ito stochastic differential equations and Stratonovich stochastic differential equations describing the same process. We obtain conditions

$$d_{\alpha,k}^{\mathrm{St}} = d_{\alpha,k}^{\mathrm{Ito}} \tag{6.163}$$

and

$$F_\alpha^{\mathrm{St}} = F_\alpha^{\mathrm{Ito}} - \frac{1}{2}\sum_{\beta,k} d_{\beta,k}^{\mathrm{Ito}}\frac{\partial d_{\alpha,k}^{\mathrm{Ito}}}{\partial X_\beta} \tag{6.164}$$

Finally we remark that independently of the interpretation of the stochastic differential equation (6.144), the coefficients $F_\alpha(X)$ and $d_{\alpha,k}(X)$ can be extended to explicitly time-dependent functions $F_\alpha(X,t)$ and $d_{\alpha,k}(X,t)$. Such an extension is motivated above all by the fact that possibly a part of the irrelevant variables possesses relatively slow time scales in the order of the magnitude of the characteristic time of the relevant quantities.

6.9.5 Ito's Formula and Fokker–Planck Equation

Let us consider an arbitrary differentiable function $f(X)$ where $X = X(t)$ is the solution of the Ito stochastic differential equation (6.154). For the sake of simplicity we consider only one relevant degree of freedom and only one Wiener process. Then the Ito differential equation reads

$$\mathrm{d}X = F(X,t)\mathrm{d}t + d(X,t)\mathrm{d}W(t) \,. \tag{6.165}$$

The differential $\mathrm{d}f$ is defined as

$$\mathrm{d}f = f(X(t+\mathrm{d}t)) - f(X(t)) = f(X+\mathrm{d}X) - f(X) \,. \tag{6.166}$$

We expand (6.166) to second-order in $\mathrm{d}X$

$$\mathrm{d}f = \frac{\partial f(X)}{\partial X}\mathrm{d}X + \frac{1}{2}\frac{\partial^2 f(X)}{\partial X^2}\mathrm{d}X^2 + o\left(\mathrm{d}X^2\right) \tag{6.167}$$

and replace $\mathrm{d}Y$ by (6.165). Considering again $\mathrm{d}W \sim \mathrm{d}t^{1/2}$, we expand (6.167) up to first order in $\mathrm{d}t$

$$\mathrm{d}f = \frac{\partial f(X)}{\partial X}\left[F(X,t)\mathrm{d}t + d(X,t)\mathrm{d}W\right] + \frac{1}{2}\frac{\partial^2 f(X)}{\partial X^2}\left[b(X,t)\mathrm{d}W\right]^2 \,, \tag{6.168}$$

where all other terms have been discarded since they are of higher order. This relation is known as Ito's formula.

Now we perform the average with respect to all realizations of the Wiener process $W(t)$ and obtain

$$\overline{\frac{\mathrm{d}f}{\mathrm{d}t}} = \overline{\frac{\partial f(X)}{\partial X}F(X,t)} + \frac{1}{2}\overline{\frac{\partial^2 f(X)}{\partial X^2}d^2(X,t)} \,. \tag{6.169}$$

For the determination of the averages we used the Ito calculus, especially (6.152) and the property that the value of $X(t)$ is taken before the jump $\mathrm{d}W(t)$. The last remark means that the actual value of $X(t)$ and the value of the subsequent jump $\mathrm{d}W(t)$ are statistically independent.

On the other hand, $X(t)$ has the conditional probability density $p\left(X,t\,|\,X',t'\right)$. If the evolution starts from the initial state $X(t') = X'$, then the averages are given by

$$\overline{\frac{\mathrm{d}f}{\mathrm{d}t}} = \int \frac{\partial}{\partial t}p\left(X,t \mid X',t'\right) f(X)\mathrm{d}X \tag{6.170}$$

and

$$\overline{\frac{\partial f(X)}{\partial X}F(X,t)} = \int \frac{\partial f(X)}{\partial X}F(X,t)p\left(X,t \mid X',t'\right)\mathrm{d}X$$
$$= -\int f(X)\frac{\partial}{\partial X}\left[F(X,t)p\left(X,t \mid X',t'\right)\right]\mathrm{d}X \tag{6.171}$$

and

$$\overline{\frac{\partial^2 f(X)}{\partial X^2}d^2(X,t)} = \int \frac{\partial^2 f(X)}{\partial X^2}d^2(X,t)P\left(X,t \mid X',t'\right)\mathrm{d}X$$
$$= \int f(X)\frac{\partial^2}{\partial X^2}\left[d^2(X,t)P\left(X,t \mid X',t'\right)\right]\mathrm{d}X\,. \tag{6.172}$$

Putting all these results together and integrating by parts, we arrive at

$$\int \mathrm{d}X f(X)\frac{\partial}{\partial t}p\left(X,t \mid X',t'\right)$$
$$= \int \mathrm{d}X f(X)\frac{1}{2}\frac{\partial^2}{\partial X^2}\left[d^2(X,t)p\left(X,t \mid X',t'\right)\right]$$
$$- \int \mathrm{d}X f(X)\frac{\partial}{\partial X}\left[F(X,t)p\left(X,t \mid X',t'\right)\right]\,. \tag{6.173}$$

Now we consider that we have chosen the function $f(X)$ to be arbitrary. Hence, we conclude that the conditional probability satisfies the equation

$$\frac{\partial}{\partial t}p\left(X,t \mid X',t'\right) = \frac{1}{2}\frac{\partial^2}{\partial X^2}\left[d^2(X,t)p\left(X,t \mid X',t'\right)\right]$$
$$- \frac{\partial}{\partial X}\left[F(X,t)p\left(X,t \mid X',t'\right)\right]\,. \tag{6.174}$$

Obviously, we get a complete equivalence between the stochastic differential equation (6.165) and the Fokker–Planck equation (6.174). This result can be generalized for the case of a system of N stochastic differential equations with R Wiener processes. The set of differential equations may be given by (6.144). The corresponding Fokker–Planck equation then reads

$$\frac{\partial}{\partial t}p\left(X,t \mid X',t'\right) = \frac{1}{2}\sum_{\alpha,\beta}\frac{\partial^2}{\partial X_\alpha \partial X_\beta}\left[D_{\alpha\beta}(X,t)p\left(X,t \mid X',t'\right)\right] \tag{6.175}$$
$$- \sum_{\alpha}\frac{\partial}{\partial X_\alpha}\left[F_\alpha(X,t)p\left(X,t \mid X',t'\right)\right] \tag{6.176}$$

with

$$B_{\alpha\beta}(X,t) = \sum_{k=1}^{\mu} d_{\alpha,k}(X,t)d_{\beta,k}(X,t)\,. \tag{6.177}$$

A similar connection between the Stratonovich stochastic differential equations and the corresponding Fokker–Planck equation can be obtained by application of the converting rules (6.164) and (6.163). Thus, the system of stochastic differential equations (6.144) in the Stratonovich interpretation is related to the Fokker–Planck equation

$$
\begin{aligned}
&\frac{\partial}{\partial t} p\left(X, t \mid X^{\prime}, t^{\prime}\right) \\
&=\frac{1}{2} \sum_{\alpha, \beta, k} \frac{\partial}{\partial X_{\alpha}}\left[d_{\alpha, k}(X, t) \frac{\partial}{\partial X_{\beta}}\left(d_{\beta, k}(X, t) p\left(X, t \mid X^{\prime}, t^{\prime}\right)\right)\right] \\
&-\sum_{\alpha} \frac{\partial}{\partial X_{\alpha}}\left[F_{\alpha}(X, t) p\left(X, t \mid X^{\prime}, t^{\prime}\right)\right] . \qquad (6.178)
\end{aligned}
$$

Finally it should be noticed that the connection between the stochastic differential equations and Fokker–Planck equations allows us to create representative trajectories for a given Fokker–Planck equation by numeric simulations.

References

1. L. Arnold: *Stochastic Differential Equations* (Wiley-Interscience, New York, 1974)
2. H. Risken: *The Fokker–Planck Equation* (Springer, Berlin Heidelberg New York, 1984)
3. E. Fick, G. Sauermann: *Quantenstatistik DYNAMISCHER Prozesse I* (Akademische Verlagsgesellschaft, Leipzig, 1983)
4. E. Fick, G. Sauermann: *Quantenstatistik Dynamischer Prozesse IIa* (Akademische Verlagsgesellschaft, Leipzig, 1983)
5. M. Schulz: *Statistical Physics and Economics* (Springer, Berlin Heidelberg New York, 2003)
6. W. Götze, L. Sjögren; J. Non-Cryst. Solids **131–133**, 161 (1991)
7. W. Götze, L. Sjögren, Rep. Prog. Phys. **55**, 241 (1992)
8. K. Kawasaki: Ann. Phys. **61**, 1 (1970)
9. W. Götze, A. Zippelius: Phys. Rev. A **14**, 1842 (1976)
10. J. Marcienkiewicz: Math. Z. **44**, 612 (1939)
11. C.W. Gardiner: *Handbook of Stochastic Methods* (Springer, Berlin Heidelberg New York, 1997)
12. U. Balucanu, M. Zoppi: *Dynamics of the Liquid State* (Clarendon Press, Oxford, 1994)
13. R.L. Stratonovich: *Introduction to the Theory of Random Noise* (Gordon and Breach, New York, 1963)
14. W. Feller: *An Introduction to Probability Theory and Its Applications*, vol. 1, 3rd edn (Wiley, New York, 1968)
15. W. Feller: *An Introduction to Probability Theory and Its Applications*, vol. 2, 2nd edn (Wiley, New York, 1971)
16. N.G. van Kampen: *Stochstic Processes in Physics and Chemistry* (North-Holland, Amsterdam, 1981)
17. H. Mori: Prog. Theor. Phys. **33**, 423 (1965)
18. H. Mori: Prog. Theor. Phys. **34**, 339 (1965)
19. R. Zwanzig: Ann. Rev. Phys. Chem. **16**, 67 (1961)
20. A.N. Kolmogorov: *Foundations of the Theory of Probability* (Chelsa, New York, 1950)

7

Optimal Control of Stochastic Processes

7.1 Markov Diffusion Processes under Control

7.1.1 Information Level and Control Mechanisms

Many control problems appearing for complex systems are subject to imperfectly known disturbances. As we have learned in the previous chapter, these disturbances originate by the hidden irrelevant degrees of freedom. Although we may formally separate the dynamics of relevant and irrelevant variables into different contributions to the equations of motion, see Sect. 1, the time-dependence of the terms concerning the irrelevant degrees of freedom is still open. Unfortunately, this problem cannot solved generally, although the Mori–Zwanzig equations define a formal expression for the residual forces (6.123). To proceed, it is necessary to make some approximations about these existing, but usually 'a priori' unknown forces. The simplest, but most common approach is the Markov approximation discussed in Sects. 6.6 and 6.9.2. This case allows the introduction and application of the very powerful Ito calculus or alternatively the use of Fokker–Planck equations in order to describe the dynamics of the relevant variables driven by external random sources simulating the effects of the irrelevant degrees of freedom. A large number of physical processes in complex systems is sufficiently described by the Markov approach. As a standard problem we refer to the Brownian motion, originally observed by Ingenhousz and Brown [1] and first described by Einstein [2]. Therefore, we focus in the following mainly on Markov processes.

Let us now assume that we know from experimental investigations or appropriate assumptions the mathematical structure[1] of the equations of

[1] The detailed determination of the coefficients a and b of (6.155) is by no means a simple problem. Usually, one uses an appropriate theoretical model containing several free parameters. Then, one determines theoretically several moments and cumulants and compares this with experimentally obtained results. However, this problem is often hard to handle, especially if the a and b are complicated functions of the state X.

M. Schulz: *Control Theory in Physics and other Fields of Science*
STMP **215**, 193–212 (2006)

motion (6.155) describing a stochastically driven process . Then, a control of this process can be realized under three essentially different conditions:

- The controller has no information about the stochastic driving terms during the whole control period. In this case he chooses as control a function of time determined before the control starts. Obviously, this is an open loop control of a stochastic process.
- The controller has information about the current state and the history, but not about the future evolution of the stochastic terms. In this case he reacts onto a current situation. This is a typical closed loop or feedback control.
- The controller is able to calculate some information about the future evolution of the driving terms from the current state and the history of the driving terms. In other words, the controller has a partial knowledge about the intrinsic dynamics of the hidden degrees of freedom of the complex system. It is clear that under this circumstance, the driving terms are no longer stochastic in the sense of the Wiener Process. This is a closed loop foresighted control.

Only the first two types of control are related to real Markov processes. Any predictions which are necessary for the last type of control require the presence of a memory as discussed in Sects. 6.3.3 and 6.9.1.

7.1.2 Path Integrals

We had already discussed in general that each complex system on the macroscopic level can be described by a probabilistic theory. In this sense the probability distribution function $p(X,t)$ is the central quantity of a quantitative description. Additionally, as a generalization of $p(X,t)$, we may introduce the joint probability density $p(X^{(M)},t_M;X^{(M-1)},t_{M-1};\ldots;X^{(0)},t_0)$ with $M+1$ sequenced points in time. The knowledge of this function allows formally the determination of the probabilistic weight for a trajectory passing the points $X^{(0)} = X(t_0)$ at t_0, $X^{(1)} = X(t_1)$ at t_1, and so on. In this way, we may interpret the ordered set $[X]_M = \{X(t_0), X(t_1), \ldots, X(t_{M-1}), X(t_M)\}$ as a discrete path which is realized with the probability

$$p(X^{(M)},t_M;X^{(M-1)},t_{M-1};\ldots;X^{(0)},t_0)\,. \tag{7.1}$$

Obviously, in the joint probability representation each path is now one possible event.

Using conditional probabilities, the joint probability for a discrete path of $M+1$ points can be written as follows:

$$\begin{aligned}
p([X]_M) &= p\left(X^{(M)},t_M;\ldots;X^{(0)},t_0\right) \\
&= p\left(X^{(M)},t_M \mid X^{(M-1)},t_{M-1};\ldots;X^{(0)},t_0\right) \\
&\quad \times p\left(X^{(M-1)},t_{M-1} \mid X^{(M-2)},t_{M-2};\ldots\right)
\end{aligned}$$

$$\vdots$$
$$\times\, p\left(X^{(1)}, t_1 \mid X^{(0)}, t_0\right) p\left(X^{(0)}, t_0\right) . \tag{7.2}$$

Thus the Markov condition, see Sect. 6.5, immediately requires

$$p([X]) = \prod_{j=0}^{M-1} p\left(X^{(j+1)}, t_{j+1} \mid X^{(j)}, t_j\right) . \tag{7.3}$$

Setting now $t_0 = 0$, $t_M = T$, $t_j = jT/M$, and taking the limit $M \to \infty$ one obtains the joint probability of a certain trajectory[2] $X(t)$ of the system moving from $X(0)$ to $X(T)$

$$p([X]) = \lim_{M\to\infty} \prod_{j=0}^{M-1} p\left(X(t_{j+1}), t_{j+1} \mid X(t_j), t_j\right) . \tag{7.4}$$

In principle, (7.3) is a generalization of the Chapman–Kolmogorov equation (6.77). The conditional probability $p\left(X^{(j+1)}, t_{j+1} \mid X^{(j)}, t_j\right)$ follows from the solution of the Fokker–Planck equation (6.101) or in general, from the solution of the differential Chapman–Kolmogorov equation (6.91). An alternative way to obtain an explicit expression for this quantity is to write

$$\begin{aligned} p\left(X, t \mid X', t'\right) &= \int \mathrm{d}X'' \delta\left(X - X''\right) p\left(X'', t \mid X', t'\right) \\ &= \left.\overline{\delta\left(X - X(t)\right)}\right|_{X', t'} . \end{aligned} \tag{7.5}$$

The last expression is the conditional average considering the initial condition $X(t') = X'$ of the N-dimensional δ-function[3]. For infinitesimally small time periods, $\delta t = t_{j+1} - t_j = T/M$, we obtain from (6.155)

$$\begin{aligned} &p\left(X, t \mid X', t'\right) \\ &= \overline{\delta\left(X - \left(X' + F(X', t')\delta t + \sum_{k=1}^{R} d_k(X', t')\mathrm{d}W_k(t')\right)\right)} , \end{aligned} \tag{7.6}$$

where the average is taken with respect to the R independent realizations W_k of the Wiener process. Note that F and d_k are in this representation N-component vectors, $F = \{F_1, F_2, \ldots, F_N\}$, and $d_k = \{d_{1,k}, d_{2,k}, \ldots, d_{N,k}\}$. The standard Fourier representation of the δ-function yields

---

[2] Obviously, the trajectory $X(t)$ corresponds to the sequence $[X] = [X]_\infty$.

[3] This means that the argument of this function is an N-component vector and the components contribute to the N-dimensional δ-function as follows

$$\delta(X - Y) = \prod_{\alpha=1}^{N} \delta(X_\alpha - Y_\alpha) ,$$

where the δ-function at the right-hand side is the standard Dirac delta function.

$$
\begin{aligned}
p\left(X,t \mid X',t'\right) &= \int \frac{\mathrm{d}^N Q}{(2\pi)^N} \\
&\overline{\times \exp\left\{\mathrm{i}Q\left(X - \left(X' + F(X',t')\delta t + \sum_k d_k(X',t')\mathrm{d}W_k(t')\right)\right)\right\}} \\
&= \int \frac{\mathrm{d}^N Q}{(2\pi)^N} \exp\left\{\mathrm{i}Q\left(\frac{\delta X}{\delta t} - F(X',t')\right)\delta t\right\} \\
&\times \prod_{k=1}^{R} \overline{\exp\left\{-\mathrm{i}Q d_k(X',t')\mathrm{d}W_k(t')\right\}}
\end{aligned}
\tag{7.7}
$$

with $\delta X = X - X'$. Considering (6.137), we obtain

$$
\begin{aligned}
\overline{\exp\left\{-\mathrm{i}Q d_k(X',t')\mathrm{d}W_k(t')\right\}} &= \int \mathrm{d}W \exp\left\{-\mathrm{i}Q d_k(X',t')\left(W - W'\right)\right\} \\
&\times p\left(W,t \mid W',t'\right) \\
&= \int \mathrm{d}\xi \exp\left\{-\mathrm{i}Q d_k(X',t')\xi\right\} \\
&\times \frac{1}{\sqrt{2\pi\delta t}} \exp\left\{-\frac{\xi^2}{2\delta t}\right\} \\
&= \exp\left\{-\frac{1}{2} Q d_k^T(X',t') d_k(X',t') Q \delta t\right\} ,
\end{aligned}
\tag{7.8}
$$

and therefore with (6.177)

$$
\begin{aligned}
p\left(X,t \mid X',t'\right) = \int \frac{\mathrm{d}^N Q}{(2\pi)^N} &\exp\left\{\mathrm{i}Q\left(\frac{\delta X}{\delta t} - F(X',t')\right)\delta t\right\} \\
&\exp\left\{-\frac{1}{2} Q D(X',t') Q \delta t\right\} .
\end{aligned}
\tag{7.9}
$$

Inserting this relation into (7.3), one obtains the joint probability $p\left([X]\right)$ for the realization of a certain path starting in $X(0)$ at $t = 0$ and ending in $X(T)$ at $t = T$:

$$
\begin{aligned}
p\left([X]\right) &= \lim_{M\to\infty} \int \prod_{j=1}^{M} \frac{\mathrm{d}^N Q(t_j)}{(2\pi)^N} \\
&\times \exp\left\{\mathrm{i}\sum_{j=1}^{M}\left[Q(t_j)\left(X(t_j) - X(t_{j-1})\right) - Q(t_j)F(X(t_{j-1}),t_{j-1})\delta t\right]\right\} \\
&\times \exp\left\{-\frac{1}{2}\sum_{j=1}^{M} Q(t_j) D(X(t_{j-1}),t_{j-1}) Q(t_j)\delta t\right\}
\end{aligned}
\tag{7.10}
$$

or symbolically

$$p([X]) = \int DQ \exp\left\{ \mathrm{i} \int_0^T \mathrm{d}t Q(t) \left[\dot{X}(t) - F(X(t), t) \right] \right\} \times \exp\left\{ -\frac{1}{2} \int_0^T \mathrm{d}t Q(t) D(X(t), t) Q(t) \right\} \tag{7.11}$$

with the integral measure

$$DQ = \lim_{M \to \infty} \prod_{j=1}^{M} \frac{\mathrm{d}^N Q(t_j)}{(2\pi)^N} . \tag{7.12}$$

From a mathematical point of view, (7.11) is only a symbolic representation of (7.10). A concrete determination of the quantity $p([X])$ always requires the consideration of the quasi-discrete formulas (7.10). On the other hand, (7.11) is very helpful for the study of general properties of a stochastic processes, the introduction of graph theoretical formulations of a suitable perturbation theory, and renormalization group approaches [4, 6, 10].

7.1.3 Performance

We assume, as in Sect. 2.2.1, that the aim of a stochastic control is also defined by a functional which should be minimized to obtain the optimum control. Since the three different types of control aim, namely, integral functionals, endpoint functionals, and mixed forms can be transformed into each other, we consider only integral functionals as the control aim.

This functional is, similar as (2.52), given by the trajectory $X(t)$, the control $u(t)$, and the control horizon T. In the case of a deterministic control, the constraints, i.e., the equations of motion, the boundary conditions ($X(0) = X_0$ and $X(T) = X_\mathrm{e}$), and the control law $u(t)$ define completely the trajectory of the system. Therefore, it would be possible to compute a unique solution of the equations of motion, $X(t, u, X_0, X_\mathrm{e})$, for an arbitrary control, but fixed boundary conditions and insert this solution in the performance integral (2.52). We obtain the performance along an admissible trajectory

$$R[X, u, T] \to \widetilde{R}[X_0, X_\mathrm{e}, u, T] = \int_0^T \mathrm{d}t \phi(t, X(t, u, X_0, X_\mathrm{e}), u(t)) . \tag{7.13}$$

The minimization of $\widetilde{R}[X_0, X_\mathrm{e}, u, T]$ with respect to all admissible control functions $u(t)$ then yields the wanted optimal control $u^*(t)$. In the case of a stochastic process, one must be more careful. The solution of the equations of motion (6.155) also depends on the concrete realizations of the Wiener process while the end condition is mostly chosen to be free. Thus, we now get

$$\widetilde{R}[X_0, X_\mathrm{e}, u, T] \to \widetilde{R}[u, X_0, W, T] = \int_0^T \mathrm{d}t \phi(t, X(t, u, W, X_0), u(t)) , \tag{7.14}$$

and the performance integral $\widetilde{R}[u, X_0, W, T]$ is now a stochastic quantity. Therefore, it is necessary to average over all realizations of the Wiener process. The simplest requirement for an optimal stochastic control is

$$J(X_0, u, T) = \left.\overline{\widetilde{R}[u, X_0, W, T]}\right|_W \to \inf , \tag{7.15}$$

where the average is taken over all Wiener processes considering the applied control mechanism[4]. However, the average procedure includes another important uncertainty. The minimization of the deterministic performance integral (7.13) with respect to the control function can be transformed into the minimization and maximization, respectively, of

$$\Phi_+ \left(\widetilde{R}[X_0, X_e, u, T]\right) \to \inf \tag{7.16}$$

and

$$\Phi_- \left(\widetilde{R}[X_0, X_e, u, T]\right) \to \sup \tag{7.17}$$

with an arbitrary monotonously increasing function Φ_+ and an arbitrary monotonously decreasing function Φ_-, respectively, without a change of the optimal control. This is not the case in the stochastic optimization. The optimum problems

$$\left.\overline{\Phi_+ \left(\widetilde{R}[u, X_0, W, T]\right)}\right|_W \to \inf \tag{7.18}$$

and

$$\left.\overline{\Phi_- \left(\widetilde{R}[u, X_0, W, T]\right)}\right|_W \to \sup \tag{7.19}$$

now depend on the chosen functions Φ_+ and Φ_-. This may be illustrated by a simple example. Let us consider a certain one-component control leading to the stochastic performance

$$\widetilde{R}[u, X_0, W, T] = \int_0^T u^2(t)\mathrm{d}t + [W(T) - W(0)] \left[\int_0^T (1 + u(t))^2 \mathrm{d}t\right]^{1/2} . \tag{7.20}$$

Then we get for the average over the whole control period

$$\left.\overline{\widetilde{R}[u, X_0, W, T]}\right|_W = \int_0^T u^2(t)\mathrm{d}t \tag{7.21}$$

and consequently the optimal control $u^*(t) = 0$, while the minimization of

[4] This means that in the case of an open loop control, the complete realizations of the Wiener process over the period $[0, T]$ are taken into account while a feedback control requires at a certain time $t > 0$ only the average over the future period $[t, T]$.

$$\overline{\exp\left(\widetilde{R}[u,X_0,W,T]\right)}\bigg|_W = \exp\left(\int\limits_0^T u^2(t)\mathrm{d}t + \frac{T}{2}\int\limits_0^T (1+u(t))^2\mathrm{d}t\right) \tag{7.22}$$

and the optimal control is now given by $u^*(t) = -T/(2+T)$.

7.2 Optimal Open Loop Control

7.2.1 Mean Performance

The open loop control requires that the controller has no information about the stochastic evolution of the driving terms during the whole control period, which means that the optimal control must be a deterministic function of time obtainable from the knowledge of the functional structure about the drift terms and about the diffusion coefficients of the underlying Markov diffusion process. In other words, we now consider the Ito stochastic differential equation

$$\mathrm{d}X(t) = F(X(t),u(t),t)\mathrm{d}t + \sum_{k=1}^{R} d_k(X(t),u(t),t)\mathrm{d}W_k(t) \tag{7.23}$$

with the N-dimensional state vector X, the n-component control function $u(t)$, and R independent realizations of the Wiener process. This is a typical case for physical experiments. An accurate preparation of the initial state $X(0)$, the complete knowledge of the deterministic drift terms F, but incomplete information about the stochastic parts[5], is a standard situation for several physical experiments on mesoscopic scales. Typical examples are tracer diffusion experiments in liquids and amorphous solids where the diffusive particles are not observable during a certain period. An open loop control of such a system is, for example, the induced localization of the particles by external fields close to their injection points under a sufficiently small disturbance of the liquid or solid environment.

The stochastic Ito differential equation (7.23) under control is equivalent to the Fokker–Planck equation

$$\begin{aligned}\frac{\partial}{\partial t}p\left(X,t \mid X',t'\right) &= \frac{1}{2}\sum_{\alpha,\beta}\frac{\partial^2}{\partial X_\alpha \partial X_\beta}\left[B_{\alpha\beta}(X,u(t),t)p\left(X,t \mid X',t'\right)\right] \\ &\quad - \sum_{\alpha}\frac{\partial}{\partial X_\alpha}\left[F_\alpha(X,u(t),t)p\left(X,t \mid X',t'\right)\right] .\end{aligned} \tag{7.24}$$

Both equations, (7.23) and (7.24), correspond to the formal path integral

[5] Only the coupling functions b_k between the noise terms $\mathrm{d}W_k$ and the system are well known.

$$p([X]) = \int DQ \exp\left\{ \mathrm{i} \int_0^T \mathrm{d}t Q(t) \left[\dot{X}(t) - F(X(t), u(t), t) \right] \right\}$$
$$\exp\left\{ -\frac{1}{2} \int_0^T \mathrm{d}t Q(t) D(X(t), u(t), t) Q(t) \right\} \tag{7.25}$$

representing the weight of a certain stochastic trajectory of the system through the phase space. In order to complete the problem, we introduce the initial condition $X(0) = X_0$, while the final position should be open. Finally we have to choose the performance. Here, we use the representation

$$J[X_0, u, T] = \left. \overline{\exp\left\{ - \int_0^T \mathrm{d}t \phi(t, X(t), u(t)) \right\}} \right|_{X(0)=X_0} . \tag{7.26}$$

The average is taken with respect to the external noise. The control aim requires that the averaged performance should become a maximum. The main problem is now to calculate this average. To this aim we consider that $p([X])$ is the statistical weight of a certain admissible trajectory. Thus,

$$J[X_0, u, T] = \int DX \exp\left\{ - \int_0^T \mathrm{d}t \phi(X(t), u(t), t) \right\} p[X] \tag{7.27}$$

with

$$DX = \lim_{M\to\infty} \prod_{j=1}^{M} \mathrm{d}^N X(t_j) . \tag{7.28}$$

We remark that this integral measure must be understood with respect to the time discrete representation used in (7.10). Inserting (7.25) in (7.27), one get the mean performance

$$J[X_0, u, T] = \int DX DQ \exp\left\{ -S\left[X, Q, u, T\right] \right\} \tag{7.29}$$

with the action

$$S\left[X, Q, u, T\right] = \int_0^T \mathrm{d}t L\left(X(t), Q(t), u(t), t\right) \tag{7.30}$$

and the Lagrangian

$$\begin{aligned} L\left(X(t), Q(t), u(t), t\right) = \phi(X(t), u(t), t) + \frac{1}{2} Q(t) D(X(t), u(t), t) Q(t) \\ - \mathrm{i} Q(t) \left[\dot{X}(t) - F(X(t), u(t), t) \right] \end{aligned} \tag{7.31}$$

corresponding to the language used in Chap. 2. In this sense, the 'ghost' variables

$$P(t) = \frac{\partial L}{\partial \dot{X}(t)} = -\mathrm{i}Q(t) \tag{7.32}$$

are called the generalized momenta. Furthermore, the integral (7.29) can be interpreted as taken over the whole generalized phase space $\mathbb{P} \times \overline{\mathbb{P}}$ formed by the phase space[6] $\mathbb{P}$ and the adjoint phase space $\overline{\mathbb{P}}$ corresponding to the set of all admissible momenta $P(t)$.

The remaining problem is the calculation of the optimum control law. The standard way is the evaluation of (7.29) by the Laplace method [3]. This method is sometimes referred to as 'steepest descent'. But this terminology is inadequate for the present case. The idea is very simple. We first determine a provisional optimal control, given by $\widetilde{u}^*$, $\widetilde{X}^*$, and $\widetilde{Q}^*$ (or $\widetilde{P}^*$), which minimizes the action $S[X, Q, u, T]$. Then we expand the action around this so-called tree approximation

$$S[X, Q, u, T] = S\left[\widetilde{X}^*, \widetilde{Q}^*, \widetilde{u}^*, T\right] + \Delta S\left[\xi, \eta, \omega, \widetilde{X}^*, \widetilde{Q}^*, \widetilde{u}^*, T\right] \tag{7.33}$$

with $\xi = X - \widetilde{X}^*$, $\eta = Q - \widetilde{Q}^*$, and $\omega = u - \widetilde{u}^*$. The integral (7.29) now becomes

$$\begin{aligned} J[X_0, u, T] &= \exp\left\{-S\left[\widetilde{X}^*, \widetilde{Q}^*, \widetilde{u}^*, T\right]\right\} \\ &\times \int D\xi D\eta \exp\left\{-\Delta S\left[\xi, \eta, \omega, \widetilde{X}^*, \widetilde{Q}^*, \widetilde{u}^*, T\right]\right\} . \end{aligned} \tag{7.34}$$

The leading contributions to the performance are often determined by the first term in (7.34) while the integral yields only some corrections. If these corrections cannot be neglected, the last step is the determination of the hopefully small contributions ω^* which maximize the performance $\overline{J[u, X, T]}$. Then, the desired optimal control is given by $u^* = \widetilde{u}^* + \omega^*$. The determination of ω^* requires the explicit computation of the second term of (7.34). This can be done by various special techniques which one can find in the widespread literature [4–6]. The presentation of these, partially very powerful, methods goes beyond the scope of this book.

We remark that, however, the calculation of the tree approximation is often a sufficient approach to estimate the optimal control.

7.2.2 Tree Approximation

The above-mentioned provisional optimal control follows directly from the application of the variational principle, discussed in Chap. 2. Hence, we can follow the concept presented there. The Lagrangian (7.31) yields the following Euler–Lagrange equations. The first group corresponds to the evolution equations of the optimum trajectory

$$\dot{\widetilde{X}}^* = F(\widetilde{X}^*, \widetilde{u}^*, t) + D(\widetilde{X}^*, \widetilde{u}^*, t)\widetilde{P}^* , \tag{7.35}$$

[6] For the sake of simplicity we assume $\mathbb{P} = \mathbb{R}^N$.

where we have introduced the N-component momentum state $\widetilde{P}^*(t) = -\mathrm{i}\widetilde{Q}^*(t)$, see (7.32). The second group of equations is given by

$$\dot{\widetilde{P}}^* = \frac{\partial}{\partial \widetilde{X}^*}\left[\phi(\widetilde{X}^*,\widetilde{u}^*,t) - \frac{1}{2}\widetilde{P}^* D(\widetilde{X}^*,\widetilde{u}^*,t)\widetilde{P}^* - F(\widetilde{X}^*,\widetilde{u}^*,t)\widetilde{P}^*\right] \qquad (7.36)$$

while the last group reads

$$\frac{\partial}{\partial \widetilde{u}^*}\phi(\widetilde{X}^*,\widetilde{u}^*,t) = \frac{1}{2}\frac{\partial}{\partial \widetilde{u}^*}\left[\widetilde{P}^* D(\widetilde{X}^*,\widetilde{u}^*,t)\widetilde{P}^* + 2\widetilde{P}^* F(\widetilde{X}^*,\widetilde{u}^*,t)\right] . \qquad (7.37)$$

These equations are completed by the initial conditions

$$\widetilde{X}^*(0) = X_0 \quad \text{and} \quad \widetilde{Q}^*(T) = 0 , \qquad (7.38)$$

which can be obtained following the same considerations as in Sect. 2.4.1. The equations (7.35), (7.36), and (7.37) are the tree approach to the above-introduced open loop stochastic control problem. Although these equations are similar to the optimal control equations (2.70), (2.74), and (2.75) for the deterministic control problem, there are some essential differences:

- The main difference is that the control law, $\widetilde{u}^*(t)$, is only an approximation for the optimal control of the stochastic problem which considers the fluctuations due to the stochastic sources only on a mean field level. We remark, that in case of $D \to 0$, i.e. in case of vanishing noise terms, the evolution equations (7.35), (7.36), and (7.37) converge to (2.74), (2.70), and (2.75) and, furthermore, the provisional optimal control law $\widetilde{u}^*(t)$ approaches the optimal control law $u^*(t)$.
- The second difference belongs to the meaning of the optimal trajectory $\widetilde{X}^*(t)$. While $X^*(t)$ of the deterministic problem corresponds to the trajectory of the system through the phase space in the case of an optimal control, the trajectory $\widetilde{X}^*(t)$ is at best an estimation of the averaged trajectory in the case of an optimal control of the stochastic problem.
- The third problem comes from the fact that the Lagrangian is now a complex quantity. But this problem is only an apparent one.

In order to demonstrate that the complex character of the Lagrangian is not very dangerous, we consider again (7.29) together with (7.30) and (7.31) and execute the integration over Q. Then we obtain

$$J[X_0,u,T] = \int DX \exp\left\{-\overline{S}\left[X,u,T\right]\right\} \mathrm{Det}^{-1/2} D \qquad (7.39)$$

with the reduced action

$$\overline{S}\left[X,u,T\right] = \int_0^T \mathrm{d}t \overline{L}\left(X(t),u(t),t\right) \qquad (7.40)$$

and the corresponding real-valued Lagrangian

$$\overline{L}(X,u,t) = \phi(X,u,t) + \frac{1}{2}\left[\dot{X} - F(X,u,t)\right] D^{-1}(X,u,t)\left[\dot{X} - F(X,u,t)\right] . \tag{7.41}$$

Furthermore, we have used the abbreviation

$$\text{Det}D = \lim_{M\to\infty} \prod_{j=0}^{M-1} \left[(2\pi\delta t)^N \det D(X(t_j),u(t_j),t_j)\right] . \tag{7.42}$$

Note that this path determinant is not considered in the tree approximation but in the subsequent harmonic approach and the perturbation theory. Thus, we get the Euler–Lagrange equations in the tree approximation, namely

$$\frac{\mathrm{d}}{\mathrm{d}t}\sum_{\beta=1}^{N}\left(D^{-1}_{\alpha\beta}\left[\dot{\widetilde{X}}^*_\beta - F_\beta\right]\right) = \frac{\partial\phi}{\partial\widetilde{X}^*_\alpha} - \sum_{\beta,\gamma=1}^{N}\frac{\partial F_\beta}{\partial\widetilde{X}^*_\alpha}D^{-1}_{\beta\gamma}\left[\dot{\widetilde{X}}^*_\gamma - F_\gamma\right] + \frac{1}{2}\sum_{\beta,\gamma=1}^{N}\left[\dot{\widetilde{X}}^*_\beta - F_\beta\right]\frac{\partial D^{-1}_{\beta\gamma}}{\partial\widetilde{X}^*_\alpha}\left[\dot{\widetilde{X}}^*_\gamma - F_\gamma\right] \tag{7.43}$$

and

$$\frac{\partial\phi}{\partial\widetilde{u}^*} + \frac{1}{2}\frac{\partial}{\partial\widetilde{u}^*}\left[\dot{X} - F\right] D^{-1}\left[\dot{X} - F\right] = 0 . \tag{7.44}$$

In order to avoid confusion we have used the component representation in (7.43). Readers may check themselves that both equations are also obtainable from (7.35), (7.36), and (7.37) by elimination of the momenta. Another important property follows for vanishing diffusion coefficients, $D \to 0$. In this case, the integration of (7.29) with respect to Q considering (7.30) and (7.31) leads to

$$\overline{R[u,X,T]} = \int DX \exp\left\{-\int_0^T \mathrm{d}t\phi(X(t),u(t),t)\right\} \times \widehat{\delta}\left(\dot{X}(t) - F(X(t),u(t),t)\right) \tag{7.45}$$

with the formal representation

$$\widehat{\delta}\left(\dot{X}(t) - F(X(t),u(t),t)\right) = \lim_{M\to\infty}\prod_{j=0}^{M-1}\delta\left(X(t_{j+1}) - X(t_j) - F(X(t_j),t_j)\delta t\right) . \tag{7.46}$$

Thus, in case of $B \to 0$, relation (7.45) is equivalent to the performance

$$J[X_0,u,T] = \exp\left\{-\int_0^T \mathrm{d}t\phi(X_{\mathrm{S}}(t),u(t),t)\right\} \tag{7.47}$$

along the solution $X_S(t)$ of the evolution equation $\dot{X} = F(X, u, t)$ but without solving this equation explicitly. Obviously, the exponent of (7.47) is nothing but the deterministic performance integral (7.13) with free end conditions, which means that the open loop stochastic control converges as expected for vanishing diffusion coefficients to the deterministic open loop control. In this special case, the tree approximation becomes a rigorous result. But this statement also means that the tree approach becomes a reasonable approximation for small diffusion coefficients, i.e., for small stochastic perturbations of the system.

Finally, we remark that the Hamilton representation, discussed in Sect. 2.4.2, and the Pontryagin's maximum principle, see Sect. 2.4.3, can also be extended to the tree approximation of the stochastic control problem.

There also exist other approaches to the open loop control problem. A popular alternative method is the expansion of the optimal performance and the control functions in powers of small noise terms [7, 8] or the method of stochastic integration [9].

7.3 Feedback Control

7.3.1 The Control Equation

Now we consider the situation that the controller has information about the current state and the history, but not about the future evolution of the stochastic terms. This is a typical feature of a feedback control. In the literature, two general types of feedback control are discussed [11–13]. We speak about a complete observation if the controller has the full information of the current state $X(t)$ and its history. Otherwise, if the controller have only information about the history and the current values of a certain 'observable' part of the state, we denote this situation as the case of partial observation. In the following we focus mainly on the complete observation case, while partial observations will be considered in the subsequent chapters.

We suppose that the current time is τ. As a criterion to be minimized we now use the expected value of the future performance with respect to the current initial state $X(\tau) = Y$. Thus (7.15) has now the concrete form

$$\begin{aligned} J[Y, \tau, u, T] &= \int_{\tau}^{T} \mathrm{d}t \, \overline{\phi(t, X(t), u(t))}\Big|_{X(\tau)=Y} \\ &= \int_{\tau}^{T} \mathrm{d}t \int \mathrm{d}X \phi(t, X, u(t)) p_u(X, t \mid Y, \tau) \, , \end{aligned} \tag{7.48}$$

where we have supposed that each feedback control u corresponds to a transition probability $p_u(X, t \mid Y, \tau)$ from the state Y at time τ to the state X at time t considering the presence of the control law u. The characteristic structure of the feedback control applied at time t is given by

$$u(t) = u(t, X(t)) \; . \tag{7.49}$$

The aim of an optimal feedback control is now to minimize the conditionally averaged performance $J[Y, \tau, u, T]$. Let us assume that

$$V(Y, \tau, T) = \min_u J[Y, \tau, u, T] \; . \tag{7.50}$$

Thus, an optimal feedback control law u^* has the property that

$$V(Y, \tau, T) = J[Y, \tau, u^*, T] \; . \tag{7.51}$$

In order to obtain an equation for $V(Y, \tau, T)$ we use the following identity:

$$\begin{aligned}
&\int_\tau^t dt' \int dX \frac{\partial}{\partial t'} \left[V(X, t', T) p_u(X, t' \mid Y, \tau) \right] \\
&= \int dX \int_\tau^t dt' \frac{\partial}{\partial t'} \left[V(X, t', T) p_u(X, t' \mid Y, \tau) \right] \\
&= \int dX \; V(X, t, T) p_u(X, t \mid Y, \tau) - \int dX \; V(X, \tau, T) p_u(X, \tau \mid Y, \tau) \\
&= \overline{V(X, t, T)}\Big|_{X(\tau)=Y,u} - V(Y, \tau, T) \; ,
\end{aligned} \tag{7.52}$$

where we have used $P(X, \tau \mid Y, \tau) = \delta(X - Y)$; see (6.92). On the other hand, we obtain from the left-hand side

$$\begin{aligned}
&\int_\tau^t dt' \int dX \frac{\partial}{\partial t'} \left[V(X, t', T) p_u(X, t' \mid Y, \tau) \right] \\
&= \int_\tau^t dt' \; \frac{\partial}{\partial t'} \overline{V(X, t', T)}\Big|_{X(\tau)=Y,u} \\
&+ \int_\tau^t dt' \int dX V(X, t', T) \frac{\partial}{\partial t'} p_u(X, t' \mid Y, \tau) \; .
\end{aligned} \tag{7.53}$$

The second term can be rewritten using the Fokker–Planck equation[7] (6.101)

$$\begin{aligned}
\frac{\partial}{\partial t'} p_u \left(X, t' \mid Y, \tau \right) = &- \sum_\alpha \frac{\partial}{\partial X_\alpha} F_\alpha(X, u, t') p_u \left(X, t' \mid Y, \tau \right) \\
&+ \sum_{\alpha\beta} \frac{1}{2} \frac{\partial^2}{\partial X_\alpha \partial X_\beta} D_{\alpha\beta}(X, u, t') p_u \left(X, t' \mid Y, \tau \right)
\end{aligned} \tag{7.54}$$

[7] We remark that the same procedure can also be carried out for the general case of a differential Chapman–Kolmogorov equation. In this case, the optimal stochastic control equation considers not only the diffusive Wiener processes but also also jump processes.

Thus, we obtain for the second term of (7.53)

$$
\begin{aligned}
(2) = &- \sum_{\alpha} \int_{\tau}^{t} \mathrm{d}t' \int \mathrm{d}X V(X,t',T) \frac{\partial}{\partial X_\alpha} \left[F_\alpha(X,u,t') p_u\left(X,t' \mid Y,\tau\right)\right] \\
&+ \frac{1}{2} \sum_{\alpha\beta} \int_{\tau}^{t} \mathrm{d}t' \int \mathrm{d}X V(X,t',T) \\
&\times \frac{\partial^2}{\partial X_\alpha \partial X_\beta} \left[D_{\alpha\beta}(X,u,t') p_u\left(X,t' \mid Y,\tau\right)\right] \\
= &\sum_{\alpha} \int_{\tau}^{t} \mathrm{d}t' \int \mathrm{d}X \left[\frac{\partial}{\partial X_\alpha} V(X,t',T)\right] F_\alpha(X,u,t') p_u\left(X,t' \mid Y,\tau\right) \\
&+ \frac{1}{2} \sum_{\alpha\beta} \int_{\tau}^{t} \mathrm{d}t' \int \mathrm{d}X \left[\frac{\partial^2}{\partial X_\alpha \partial X_\beta} V(X,t',T)\right] \\
&\times D_{\alpha\beta}(X,u,t') p_u\left(X,t' \mid Y,\tau\right) ,
\end{aligned}
\tag{7.55}
$$

where we have obtained the last expression by integration by parts. Using the so-called backward Focker–Planck operator

$$
\widehat{F}(X,u,t) = \sum_{\alpha} F_\alpha(X,u,t) \frac{\partial}{\partial X_\alpha} + \frac{1}{2} \sum_{\alpha\beta} D_{\alpha\beta}(X,u,t) \frac{\partial^2}{\partial X_\alpha \partial X_\beta} , \tag{7.56}
$$

we obtain

$$
\begin{aligned}
(2) &= \int_{\tau}^{t} dt' \int dX \left[\widehat{F}(X,u,t') V(X,t',T)\right] p_u\left(X,t' \mid Y,\tau\right) \\
&= \int_{\tau}^{t} dt' \, \overline{\widehat{F}(X,u,t') V(X,t',T)}\Big|_{X(\tau)=Y,u} .
\end{aligned}
\tag{7.57}
$$

Inserting this expression in (7.53) we arrive at

$$
\begin{aligned}
&\int_{\tau}^{t} dt' \int dX \frac{\partial}{\partial t'} \left[V(X,t',T) p_u(X,t' \mid Y,\tau)\right] \\
&= \int_{\tau}^{t} dt' \frac{\partial}{\partial t'} \overline{V(X,t',T)}\Big|_{X(\tau)=Y,u} \\
&+ \int_{\tau}^{t} dt' \, \overline{\widehat{F}(X,u,t) V(X,t',T)}\Big|_{X(\tau)=Y,u}
\end{aligned}
\tag{7.58}
$$

and (7.52) can be rewritten as

$$V(Y,\tau,T) = \overline{V(X,t,T)}\Big|_{X(\tau)=Y,u} - \int_{\tau}^{t} dt' \, \frac{\partial}{\partial t'} \overline{V(X,t',T)}\Big|_{X(\tau)=Y,u} - \int_{\tau}^{t} dt' \, \overline{\widehat{F}(X,u,t')V(X,t',T)}\Big|_{X(\tau)=Y,u} . \tag{7.59}$$

On the other hand, we may consider a control

$$\widetilde{u}(t',X) = \begin{cases} u(t',X) & \text{for} \quad \tau \leq t' \leq t \\ u^*(t',X) & \text{for} \quad t \leq t' \leq T . \end{cases} \tag{7.60}$$

Thus, because of (7.48), we now obtain

$$\begin{aligned} J[Y,\tau,\widetilde{u},T] &= \int_{\tau}^{t} dt' \int dX \phi(t',X,u) p_u(X,t' \mid Y,\tau) \\ &\quad + \int_{t}^{T} dt' \int dX' \phi(t',X',u^*) \\ &\quad \times \int dX p_{u^*}(X',t' \mid X,t) p_u(X,t \mid Y,\tau) \\ &= \int_{\tau}^{t} dt' \, \overline{\phi(t',X,u)}\Big|_{X(\tau)=Y,u} \\ &\quad + \int_{t}^{T} dt' \, \overline{\left[\overline{\phi(t',X',u^*)}\Big|_{X'(t)=X,u^*} \right]}\Bigg|_{X(\tau)=Y,u} \\ &= \int_{\tau}^{t} dt' \, \overline{\phi(t',X,u)}\Big|_{X(\tau)=Y,u} + \overline{J[X,t,u^*,T]}\Big|_{X(\tau)=Y,u} \\ &= \int_{\tau}^{t} dt' \, \overline{\phi(t',X,u)}\Big|_{X(\tau)=Y,u} + \overline{V(X,t,T)}\Big|_{X(\tau)=Y,u} , \end{aligned} \tag{7.61}$$

where we have used in the last step (7.51) and in the previous step (7.48). Because of the fact that $\widetilde{u}$ is not the optimal control, we obtain the inequality $V(Y,\tau,T) \leq J[Y,\tau,\widetilde{u},T]$ or

$$V(Y,\tau,T) \leq \int_{\tau}^{t} dt' \, \overline{\phi(t',X,u)}\Big|_{X(\tau)=Y,u} + \overline{V(X,t,T)}\Big|_{X(\tau)=Y,u} . \tag{7.62}$$

Considering (7.59) we get

$$0 \leq \int_{\tau}^{t} dt' \left. \overline{\phi(t', X, u)} \right|_{X(\tau)=Y,u} + \int_{\tau}^{t} dt' \left. \frac{\partial}{\partial t'} \overline{V(X, t', T)} \right|_{X(\tau)=Y,u}$$
$$+ \int_{\tau}^{t} dt' \left. \overline{\widehat{F}(X, u, t') V(X, t', T)} \right|_{X(\tau)=Y,u} . \tag{7.63}$$

The equality holds if $u = u^*$. Finally, we write $t = \tau + \varepsilon$, divide by ε and take the limit $\varepsilon \to 0$. Thus, we obtain

$$0 \leq \frac{\partial}{\partial \tau} V(Y, \tau, T) + \phi(\tau, Y, u) + \widehat{F}(Y, u, \tau) V(Y, \tau, T) . \tag{7.64}$$

Since the first term does not depend on u, the equality is reached if the sum of the second and third terms reaches its minimum. This is, of course, equivalent to the requirement $u = u^*$, i.e., we obtain the equation of the feedback stochastic optimal control

$$\frac{\partial}{\partial t} V(X, t, T) + \min_{u} \left[\phi(t, X, u) + \widehat{F}(X, u, t) V(X, t, T) \right] = 0 , \tag{7.65}$$

where we have replaced $Y \to X$ and $\tau \to t$. On the other hand, the optimal control follows from the relation

$$u^* = \arg \min_{u} \left[\phi(t, X, u) + \widehat{F}(X, u, t) V(X, t, T) \right] . \tag{7.66}$$

We remark that (7.65) is the extension of the Hamilton–Jacobi equation (2.139)) to stochastic processes. In fact, if we identify V with the action S and consider that the derivatives now concern the initial state and the initial time instead of the final state and the final time, we obtain the mapping

$$V \to S \quad \frac{\partial V}{\partial t} \to -\frac{\partial S}{\partial t} \quad \frac{\partial V}{\partial X} \to -\frac{\partial S}{\partial X} \tag{7.67}$$

and therefore with (7.56) considering $D = 0$

$$\frac{\partial S}{\partial t} + \max_{u} \left[F \frac{\partial S}{\partial X} - \phi \right] = \frac{\partial S}{\partial t} + \max_{u} H(X, \frac{\partial S}{\partial X}, u, t) = 0 \tag{7.68}$$

with the Hamiltonian H given by (2.94).

The above-derived control equation (7.65) must be completed by the corresponding boundary conditions. Because of definition (7.48) we immediately get

$$V\left(Y, T, T\right) = 0 . \tag{7.69}$$

Further boundary conditions follow from the Fokker–Planck equation. For example, absorbing boundary conditions for $X = X_0$ requires $P(X, t \mid X_0, \tau) = 0$ and, therefore, $V\left(X_0, \tau, T\right) = 0$.

Let us illustrate the treatment of the control equation by a simple example. We consider the bounded diffusion of a particle under the drift force

$F = -(u_1 - 1)X - u_2$ with $u_2 \geq 0$ and $u_1 \in \mathbb{R}$, while the noise is defined by $u_1 \sigma X \mathrm{d}W$. Thus, we have the Ito stochastic differential equation

$$\mathrm{d}X = [(1 - u_1) X - u_2] \, \mathrm{d}t + u_1 \sigma X \mathrm{d}W \tag{7.70}$$

with the two-component control (u_1, u_2). Thus we obtain the diffusion coefficient $D = (u_1 \sigma X)^2$. In principle, the model corresponds to an Ornstein-Uhlenbeck process in a space-dependent temperature field[8]. Furthermore, we require adsorbing boundary conditions at $X = 0$. The performance to be minimized is the functional

$$J[Y, t, u, T] = \int_t^T \mathrm{d}t' \int \mathrm{d}X u_2^\gamma(t') p_u(X, t' \mid Y, t) \tag{7.71}$$

with $\gamma > 1$, i.e., we are interested in a maximum escape rate over the border $X = 0$ at a minimum external force u_2. Thus, we have the boundary conditions

$$V(0, t, T) = 0 \quad \text{and} \quad V(Y, T, T) = 0 \, . \tag{7.72}$$

The control equation is now given by

$$\frac{\partial V}{\partial t} + \min_u \left[u_2^\gamma + [(1 - u_1) X - u_2] \frac{\partial V}{\partial X} + \frac{1}{2} (u_1 \sigma X)^2 \frac{\partial^2 V}{\partial X^2} \right] \, . \tag{7.73}$$

From here, we find the pre-optimized control functions

$$u_2^{\gamma - 1} = \gamma^{-1} \frac{\partial V}{\partial X} \quad \text{and} \quad u_1 = \frac{1}{\sigma^2 X} \frac{\partial V}{\partial X} \left(\frac{\partial^2 V}{\partial X^2} \right)^{-1} \, . \tag{7.74}$$

We use the ansatz $V(X, t, T) = g(t) X^\gamma$ which satisfies the first boundary condition and which yields $u_2 = g(t)^{1/(\gamma - 1)} X$ and $u_1 = \sigma^{-2} / (\gamma - 1)$, and therefore

$$g' + (1 - \gamma) g^{\gamma/(\gamma - 1)} + \frac{\gamma g}{(\gamma - 1) \sigma^2} \left[(\gamma - 1) \sigma^2 - \frac{1}{2} \right] = 0 \tag{7.75}$$

with the solution

$$g(t) = \left[C \left(1 - \exp\left\{ \gamma \left[2 (\gamma - 1) \sigma^2 - 1 \right] \frac{t - T}{2 (\gamma - 1)^2 \sigma^2} \right\} \right) \right]^{1 - \gamma} \tag{7.76}$$

with

$$C = \frac{2 (\gamma - 1)^2 \sigma^2}{\gamma \left[2 (\gamma - 1) \sigma^2 - 1 \right]} \tag{7.77}$$

also satisfying the second boundary condition. Thus, we have a constant control law for u_1 while the second control law is given by

[8] Note that the Nerst–Einstein relation requires that the diffusion coefficient and the temperature are proportional.

$$u_2 = \left[C \left(1 - \exp\left\{ \gamma \left[2(\gamma - 1)\sigma^2 - 1 \right] \frac{t - T}{2(\gamma - 1)^2 \sigma^2} \right\} \right) \right]^{-1} X \tag{7.78}$$

with singular behavior, $u_2 \sim X(T-t)^{-1}$ for $t \to T$.

Finally, it should be remarked that functional (7.48), which represents the Lagrange formulation of a given control problem, can be also extended to the Bolza formulation with

$$J[Y, \tau, u, T] = \int_{\tau}^{T} \mathrm{d}t\, \overline{\phi(t, X(t), u(t))}\Big|_{X(\tau)=Y} + \overline{\Psi[X(T)]}\Big|_{X(\tau)=Y} \tag{7.79}$$

or to the Meier formulation

$$J[Y, \tau, u, T] = \overline{\Psi[X(T)]}\Big|_{X(\tau)=Y} . \tag{7.80}$$

In both cases we have the boundary conditions

$$V(Y, T, T) = \Psi[Y] . \tag{7.81}$$

While in the case of (7.79) the stochastic control equation (7.65) is still valid, the Meier case (7.80) requires the substitution of $\phi = 0$ in the control equation.

7.3.2 Linear Quadratic Problems

Let us now consider a linear problem defined by the stochastic Ito equation

$$\mathrm{d}X(t) = [A(t)X(t) + B(t)u(t)]\, \mathrm{d}t + \sum_{k=1}^{R} d_k(t) \mathrm{d}W_k(t) \tag{7.82}$$

and the expected system performance

$$\begin{aligned} &J[Y, \tau, u, T] \\ &= \frac{1}{2} \int_{\tau}^{T} \mathrm{d}t \left[\overline{X(t)Q(t)X(t)}\Big|_{X(\tau)=Y} + \overline{u(t)R(t)u(t)}\Big|_{X(\tau)=Y} \right] \\ &+ \frac{1}{2} \overline{X(t)\Omega X(t)}\Big|_{X(\tau)=Y} \end{aligned} \tag{7.83}$$

with the symmetric (and usually positive definite) matrices $Q(t)$ (type $N \times N$) $R(t)$ (type $n \times n$) and Ω (type $N \times N$). Then, the optimal control equation becomes

$$\begin{aligned} 0 &= \frac{\partial}{\partial t} V(X, t, T) \\ &+ \min_u \left[\frac{1}{2} X Q(t) X + \frac{1}{2} u R(t) u + \widehat{F}(X, u, t) V(X, t, T) \right] \end{aligned} \tag{7.84}$$

with

$$\widehat{F}(X,u,t) = \left[XA^T(t) + uB^T(t)\right]\frac{\partial}{\partial X} + \frac{1}{2}\frac{\partial}{\partial X}D(t)\frac{\partial}{\partial X} \,. \tag{7.85}$$

Thus, the pre-optimized control is given by

$$u^{(*)}(t) = -R(t)^{-1}B^T(t)\frac{\partial V(X,t,T)}{\partial X} \,, \tag{7.86}$$

and the control equation now reads

$$\begin{aligned} 0 &= \frac{\partial V}{\partial t} + \frac{1}{2}XQ(t)X - \frac{1}{2}\frac{\partial V}{\partial X}B(t)R(t)^{-1}B^T(t)\frac{\partial V}{\partial X} \\ &\quad + XA^T(t)\frac{\partial V}{\partial X} + \frac{1}{2}\left(\frac{\partial}{\partial X}D(t)\frac{\partial}{\partial X}\right)V \,. \end{aligned} \tag{7.87}$$

We use the ansatz

$$V(X,t,T) = \frac{1}{2}\left[XG(t)X + V_0(t)\right] \tag{7.88}$$

with the symmetric $N \times N$ matrix G and obtain

$$\begin{aligned} 0 &= X\dot{G}X + \dot{V}_0 + XQX - XGBR^{-1}B^TGX \\ &\quad + XA^TGX + XGAX + \mathrm{tr}DG \,. \end{aligned} \tag{7.89}$$

All terms of order X^2 yield the Riccati equation

$$\dot{G} + A^TG + GA - GBR^{-1}B^TG = -Q \tag{7.90}$$

with the boundary condition $G(T) = \Omega$. The remaining relation is

$$\dot{V}_0 = -\mathrm{tr}DK \,. \tag{7.91}$$

The solution of the Ricatti equation (7.90) now allows us to formulate the complete control law from (7.86)

$$u^*(t) = -R(t)^{-1}B^T(t)G(t)X^*(t) \,, \tag{7.92}$$

while $X^*(t)$ is a solution of the linear differential equation (7.82) considering (7.92). Thus, the control law of the stochastic feedback control of a linear quadratic problem is completely equivalent to the control low of the deterministic control of linear quadratic problems. The effects of noise are only considered in the function $V_0(t)$ while $G(t)$ is not affected by $D(t)$, neither is $u^*(t)$. The only difference is the minimum expected performance $V(X,t,T)$ which differs from the minimum performance of the deterministic model by the term V_0.

References

1. B.J. Ford: Biologist **39**, 82 (1992)
2. A. Einstein, Annalen der Physik **17**, 132 (1905)
3. C. Bender, S.A. Orszag: *Advanced Mathematical Methods for Scientists and Engineers* (McGraw-Hill, New York, 1978)

4. J. Zinn-Justin: *Quantum Field Theory and Critical Phenomena* (Claredon Press, Oxford, 1990)
5. C. Grosche, F. Steiner: *Handbook of Feynman Path Integrals* (Springer, Berlin Heidelberg New York, 1998)
6. H. Kleinert: *Path Integrals in Quantum Mechanics, Statistics, Polymer Physics, and Financial Markets* (World Scientific Publishing, Singapore, 2004)
7. C. Holland: 'Small Noise Open Loop Control', SIAM J. Control **12**, 380 (1974).
8. C. Holland: 'Gaussian Open Loop Control Problems', SIAM J. Control **13**, 545 (1975).
9. V. Warfield: A stochastic maximum principle. PhD Thesis, Brown University, Providence, RI (1971)
10. A. Ranfagni, P. Moretti, D. Mugnai: *Trajectories and Rays: The Path-Summation in Quantum Mechanics and Optics* (World Scientific Publishing, Singapore, 1991)
11. W.H. Fleming: *Deterministic and Stochastic Optimal Control* (Springer, Berlin Heidelberg New York, 1975)
12. J.H. Davis: *Foundations of Deterministic and Stochastic Control* (Birkhäuser, Basel, 2002)
13. R. Gabasov, F.M. Kirillova, S.V. Prischepova: *Optimal Feedback Control* (Springer, Berlin Heidelberg New York, 1995)
14. T. Chen, B. Francis: *Optimal Sampled Data Control* (Springer, Berlin Heidelberg New York, 1995)

8

Filters and Predictors

8.1 Partial Uncertainty of Controlled Systems

Suppose we have a system under control, described by dynamical equations of motion for the N-dimensional state vector $X(t)$, and we have obtained an optimal deterministic control curve trajectory $X^*(t)$ and the corresponding optimum control $u^*(t)$ by the methods described in Chap. 7 by neglecting all noise terms, then we may write the desired evolution equation:

$$\dot{X}^*(t) = F(X^*(t), u^*(t), t) . \tag{8.1}$$

On the other hand, the real system considering the influence of the stochastic evolution equations may be described by the Ito stochastic differential equation (7.23). The noise terms always generate deviations of the real trajectory $X(t)$ from the nominal behavior, $Y(t) = X(t) - X^*(t)$, which require the control $u(t)$ instead of the nominal control $u^*(t)$ in order to keep the deviations $Y(t)$ small. Considering (7.23) and (8.1), we obtain the evolution equation:

$$\begin{aligned} \mathrm{d}Y &= [F(X^* + Y, u^* + w, t) - F(X^*, u^*, t)]\,\mathrm{d}t \\ &\quad + \sum_{k=1}^{R} \mathrm{d}_k(X^* + Y, u^* + w, t)\mathrm{d}W_k(t) . \end{aligned} \tag{8.2}$$

For small w and Y we may use the linearized stochastic evolution equation

$$dY(t) = [A(t)Y(t) + B(t)w(t)]\,dt + \sum_{k=1}^{R} d_k(t)dW_k(t) \tag{8.3}$$

with

$$A(t) = \frac{\partial F(X^*, u^*, t)}{\partial X^*} \quad B(t) = \frac{\partial F(X^*, u^*, t)}{\partial u^*} \tag{8.4}$$

and

$$\mathrm{d}_k(t) = \mathrm{d}_k(X^*, u^*, t) . \tag{8.5}$$

M. Schulz: *Control Theory in Physics and other Fields of Science*
STMP **215**, 213–264 (2006)

Thus, one obtains, together with the expansion of the performance up to the second-order in Y and w, a stochastic linear quadratic problem which we have discussed in Sect. 7.3.2. In particular, one obtains the control law (7.92) presenting the classical linear feedback relation.

However, the application of this theory to real problems causes some new problems. The first problem belongs to the stochastic sources which drive the system under control. It is often impossible to determine the coupling functions $\mathrm{d}_k(t)$, which connect the system dynamics with the noise processes. In addition, it cannot be guaranteed that a real system is described exclusively at the Markov level by pure diffusion processes related to several realizations of the Wiener process. In principle, the stochastic terms may also represent various jump processes or combined diffusion-jump processes.[1]

Since the majority of physical processes in complex systems consist of a sufficiently large number of different external noise sources, the estimation of the noise terms can be made in the framework of the limit distributions. This will be done in the following parts of this chapter.

The second problem belongs to the observability of a system. It means that we have the complete information about the stochastic dynamics of the system, given by the sum of the noise terms and the matrices $A(t)$ and $B(t)$, but we are not able to measure the state $X(t)$ or equivalently the difference $Y(t) = X(t) - X^*(t)$. Instead of this, we have only a reduced piece of information given by the observable output

$$Z(t) = C(t)X(t) + \eta(t) , \tag{8.6}$$

where the output $Z(t)$ is a vector of p components, usually with $p < N$, $C(t)$ is a matrix of type $p \times N$, and $\eta(t)$ represents the p-component noise process modelling the observation error. The problem occurs if the reduced information $Z(t)$ and all previous observations $Z(\tau)$ with $\tau < t$ can be used for the control of the system at the current time t. Such so-called filter problems will be considered in the two subsequent chapters.

Finally, it may be possible that the dynamics of the system is unknown. The only available information is the historical set of observations and control functions while the system itself behaves like a black box. In this case it is necessary to estimate the most probable evolution of the system under control.

[1] We remark that the stochastic Ito differential equation is related to a Fokker–Planck equation, which is only a special case of the differential Chapman–Kolmogorov equation (6.91). This equation is valid for all Markov processes and considers also jump processes. Thus we may reversely conclude that (8.3) can also be generalized to combined diffusion-jump processes.

8.2 Gaussian Processes

8.2.1 The Central Limit Theorem

Let us first analyze the properties of the stochastic contributions to a control problem if a detailed characterization of the noise terms is no longer possible. In other words, if we consider the N-component noise vector

$$\mathrm{d}\xi(t) = \sum_{j=1}^{R} \mathrm{d}_j(t)\mathrm{d}W_j(t) = \sum_{j=1}^{R} \mathrm{d}\xi_j(t) \tag{8.7}$$

or the discrete version

$$\xi(t) = \sum_{j=1}^{R} \int_{t}^{t+\Delta t} \mathrm{d}_j(t')\mathrm{d}W_j(t') = \sum_{j=1}^{R} \xi_j(t) \,, \tag{8.8}$$

then $\mathrm{d}\xi(t)$ and $\xi(t)$, respectively, can be interpreted as a sum of R independent random quantities $\mathrm{d}\xi_j(t)$ and $\xi_j(t)$, respectively. In future we consider the discrete representation (8.8). The extension to infinitesimal small changes $\mathrm{d}\xi_j(t)$ is always possible. Formally, each ξ_j represents an event of a stochastic process realized with the probability distribution $p^{(j)}(\xi_j)$. We remark that the events ξ_j must not be a weighted realization of the Wiener process. It is also possible to extend $\mathrm{d}W_j$ in (8.8) to arbitrary independent diffusion-jump processes [1, 27, 28] corresponding to the differential Chapman–Kolmogorov equation.

In order to characterize the stochastic processes driving the actual system, we need only the probability distribution of the sum $\xi(t)$ while the knowledge of the distribution functions of the single events is a secondary information. The number of these independent noise terms becomes very large for the majority of complex systems. On the other hand, the often-observed lack of distinguished features between the elementary noise processes and the absolute equality of the single noise terms $\xi_j(t)$ in (8.7) and (8.8) gives rise to the reasonable assumption that all the events $\xi_j(t)$ are realized with the same probability distribution function, $p^{(j)}(\xi_j) = p(\xi_j)$. For the sake of simplicity, we will use this physically motivated argument also for some of the following concepts.

In other words, we have the typical situation that the external sources produce a series of randomly distributed events

$$\{\xi_1, \xi_2, \ldots, \xi_R\} \,, \tag{8.9}$$

but the system utilizes only the sum

$$\xi = \sum_{j=1}^{R} \xi_j \tag{8.10}$$

for its further dynamical evolution. We assume for a moment that we know the probability distribution function $p(\xi_j)$ for the single events. In the following context we designate $p(\xi_j)$ also as an elementary probability distribution function. Because of the statistical independence the joint probability for the set (8.9) is given by

$$p(\xi_1, \xi_2, \ldots, \xi_R) = \prod_{j=1}^{R} p(\xi_j) \ . \tag{8.11}$$

Let us now determine the function $p_R(\xi)$ for the sum (8.10). We get

$$p_R(\xi) = \int \prod_{j=1}^{R} \mathrm{d}\xi_j \delta\left(\xi - \sum_{j=1}^{R} \xi_j\right) \prod_{j=1}^{R} p(\xi_j) \ . \tag{8.12}$$

The Markov property allows us to derive the complete functional structure of the probability distribution function $p_R(\xi)$ from the sole knowledge of the elementary probability density $p(\xi_j)$. It is convenient to use the characteristic function (6.50), which is defined as the Fourier transform of the probability density. Hence, we obtain

$$\begin{aligned} \hat{p}_R(k) &= \int \mathrm{d}\xi \ \exp\{ik\xi\} \, p_R(\xi) \\ &= \int \prod_{j=1}^{R} \mathrm{d}\xi_j \ \exp\left\{ik \sum_{j=1}^{R} \xi_j\right\} \prod_{j=1}^{L} p(\xi_j) \\ &= [\hat{p}(k)]^N \ . \end{aligned} \tag{8.13}$$

What can we learn from this approach? To this aim we provide a naive scaling procedure to (8.13). We start from the expansion of the characteristic function in terms of cumulants. For the sake of simplicity we focus for a short moment on single component event $\xi_j \in \mathbb{R}$. In this case we may write

$$\hat{p}(k) = \exp\left\{\sum_{n=1}^{\infty} \frac{c^{(n)}}{n!} (ik)^n\right\} , \tag{8.14}$$

and because of (8.13)

$$\hat{p}_R(k) = \exp\left\{\sum_{n=1}^{\infty} \frac{Rc^{(n)}}{n!} (ik)^n\right\} , \tag{8.15}$$

where k is now a simple scalar quantity instead a vector of a certain dimension. Obviously, when $R \to \infty$, the quantity ξ goes to infinity with the central tendency $\overline{\xi} = Rc^{(1)}$ the standard deviation $\sigma = \left(Rc^{(2)}\right)^{1/2}$. Since the drift can be zero or can be put to zero by a suitable shift $\xi \to \xi - \overline{\xi}$, we conclude that the relevant scale is that of the fluctuations, namely the variance σ. The corresponding range of k is simply its inverse, since ξ and k are conjugate in the Fourier transform. Thus, after rescaling $k \to \hat{k}R^{-1/2}$ the cumulant expansion reads

$$\hat{p}_R\left(\hat{k}\right) = \exp\left\{\sum_{n=1}^{\infty} \frac{c^{(n)}R^{1-n/2}}{n!}\left(i\hat{k}\right)^n\right\}. \tag{8.16}$$

Apart from the first cumulant, we find that the second cumulant remains invariant while all higher cumulants approach zero as $R \to \infty$. Thus, only the first and the second cumulants will remain for sufficiently large R and the probability distribution function $p_R(\xi)$ approaches a Gaussian function. The result of our naive argumentation is the central limit theorem. The precise formulation of this important theorem is:

> The sum, normalized by $R^{-1/2}$ of R random independent and identically distributed states of zero mean and finite variance, is a random variable with a probability distribution function converging to the Gaussian distribution with the same variance. The convergence is to be understood in the sense of a limit in probability, i.e., the probability that the normalized sum has a value within a given interval converges to that calculated from the Gaussian distribution.

We will now give a more precisely derivation of the central limit theorem. Formal proofs of the theorem may be found in probability textbooks such as Feller [18,29,30]. Here we follow a more physically motivated way by Sornette [31], using the technique of the renormalization group theory.

This powerful method [32] introduced in field theory and in critical phase transitions is a very general mathematical tool, which allows one to decompose the problem of finding the collective behavior of a large number of elements on large spatial scales and for long times into a succession of simpler problems with a decreasing number of elements, whose effective properties vary with the scale of observation. In the context of the central limit theorem, these elements refer to the elementary N-component events ξ_j.

The renormalization group theory works best when the problem is dominated by one characteristic scale which diverges at the so-called critical point. The distance to this criticality is usually determined by a control parameter which may be identified in our special case as R^{-1}. Close to the critical point, a universal behavior becomes observable, which is related to typical phenomena like scale invariance of self-similarity. As we will see below, the form stability of the Gaussian probability distribution function is such a kind of self-similarity.

The renormalization consists of an iterative application of decimation and rescaling steps. The first step is to reduce the number of elements to transform the problem in a simpler one. We use the thesis that under certain conditions the knowledge of all the cumulants is equivalent to the knowledge of the probability density. So we can write

$$p(\xi_j) = f\left(\xi_j, c^{(1)}, c^{(2)}, \ldots, c^{(m)}, \ldots\right), \tag{8.17}$$

where f is a unique function of ξ_j and the infinite set of all cumulants $\{c^{(1)}, c^{(2)}, \ldots\}$. Every distribution function can be expressed by the same function in this way, however with differences in the infinite set of parameters.

The probability distribution function $p_R(\xi)$ may be the convolution of $R = 2^l$ identical distribution functions $p(\xi_j)$. This specific choice of R is not a restriction since we are interested in the limit of large R and the way with which we reach this limit is irrelevant. We denote the result of the 2^l-fold convolution as

$$p_R(\xi) = f^{(l)}\left(\xi, c^{(1)}, c^{(2)}, \ldots, c^{(m)}, \ldots\right). \tag{8.18}$$

Furthermore, we can also calculate first the convolution between two identical elementary probability distributions

$$p_2(\xi) = \int p(\xi - \xi')\, p(\xi')\, d\xi', \tag{8.19}$$

which leads because of the general relation (8.13) to the formal structure

$$p_2(\xi) = f\left(\xi, 2c^{(1)}, 2c^{(2)}, \ldots, 2c^{(m)}, \ldots\right) \tag{8.20}$$

with the same function f as used in (8.17). With this knowledge we are able to generate $p_R(\xi)$ also from $p_2(\xi)$ by a 2^{l-1}-fold convolution

$$p_R(\xi) = f^{(l-1)}\left(\xi, 2c^{(1)}, 2c^{(2)}, \ldots, 2c^{(m)}, \ldots\right). \tag{8.21}$$

Here, we see the effect of the decimation. The new convolution considers only 2^{l-1} events. The decimation itself corresponds to the pairing due to the convolution (8.19) between two identical elementary probability distributions

The notation of the scale is inherent to the probability distribution function. The new elementary probability distribution function $p_2(\xi)$ obtained from (8.19) may display differences to the probability density we started from. We compensate for this by the scale factor λ^{-1} for ξ. This leads to the rescaling step $\xi \to \lambda^{-1}\xi$ of the renormalization group which is necessary to keep the reference scale.

With the rescaling of the components of the vector ξ, the cumulants are also rescaled and each cumulant of order m has to be multiplied by the factor λ^{-m}. This is a direct consequence of (6.55) because it demonstrates that the cumulants of order m have the dimension $|k|^{-m}$ and $|\xi|^m$, respectively. The conservation of the probabilities $p(\xi)\, d\xi = p(\xi')\, d\xi'$ introduces a prefactor λ^{-N} as a consequence of the change of the N-dimensional vector $\xi \to \xi'$. We thus obtain from (8.21)

$$p_R(\xi) = \lambda^{-N} f^{(l-1)}\left(\frac{\xi}{\lambda}, \frac{2c^{(1)}}{\lambda}, \frac{2c^{(2)}}{\lambda^2}, \ldots, \frac{2c^{(m)}}{\lambda^m}, \ldots\right). \tag{8.22}$$

The successive repeating of both decimation and the rescaling leads after l steps to

$$p_R(\xi) = \lambda^{-lN} f^{(0)} \left(\frac{\xi}{\lambda^l}, \frac{2^l c^{(1)}}{\lambda^l}, \frac{2^l c^{(2)}}{\lambda^{2l}}, \ldots, \frac{2^l c^{(m)}}{\lambda^{ml}}, \ldots \right) . \tag{8.23}$$

As mentioned above, $f^{(l)}(\xi, \ldots c^{(m)}, \ldots)$ is a function which is obtainable from a convolution of 2^l identical functions $f(\xi, \ldots c^{(m)}, \ldots)$. In this sense we obtain the matching condition $f^{(0)} \equiv f$ so that we arrive at

$$p_R(\xi) = \lambda^{-lN} f \left(\frac{\xi}{\lambda^l}, \frac{2^l c^{(1)}}{\lambda^l}, \frac{2^l c^{(2)}}{\lambda^{2l}}, \ldots, \frac{2^l c^{(m)}}{\lambda^{ml}}, \ldots \right) . \tag{8.24}$$

Finally we have to fix the scale λ. We see from (8.24) that the particular choice $\lambda = 2^{1/m_0}$ makes the prefactor of the m_0-th cumulant equal to 1 while all higher cumulants decrease to zero as $l = \log_2 R \to \infty$. The lower cumulants diverge with $R^{(1-m/m_0)}$, where $m < m_0$.

The only reasonable choice is $m_0 = 2$ because $\lambda = \sqrt{2}$ keeps the probability distribution function in a window with constant width. In this case, only the first cumulant may remain divergent for $R \to \infty$. As mentioned above, this effect can be eliminated by a suitable shift of ξ. Thus we arrive at

$$\lim_{R\to\infty} p_R(\xi) = R^{-N/2} f \left(\frac{\xi}{\sqrt{R}}, c^{(1)}\sqrt{R}, c^{(2)}, 0, \ldots, 0, \ldots \right) \tag{8.25}$$

In particular, if we come back to our original problem, we have thus obtained the asymptotic result that the probability distribution of the sum over incoming stochastic events has only its two first cumulant nonzero. Hence, the corresponding probability density is a Gaussian law.

If we return to the original scales, the final Gaussian probability distribution function $p_R(\xi)$ is characterized by the mean $\overline{\xi} = Rc^{(1)}$ and the covariance matrix $\tilde{\sigma} = Rc^{(2)}$, where $c^{(1)}$ and $c^{(2)}$ are the first two cumulants of the elementary probability density. Hence, we obtain

$$\lim_{R\to\infty} p_R(\xi) = \frac{1}{(2\pi)^{N/2}\sqrt{\det \tilde{\sigma}}} \exp\left\{ -\frac{1}{2} \left(\xi - \overline{\xi}\right) \tilde{\sigma}^{-1} \left(\xi - \overline{\xi}\right) \right\} \tag{8.26}$$

or with the rescaled and shifted states

$$\lim_{R\to\infty} p_R\left(\hat{\xi}\right) = \frac{1}{(2\pi)^{N/2}\sqrt{\det c^{(2)}}} \exp\left\{ -\frac{1}{2}\hat{\xi}\left[c^{(2)}\right]^{-1}\hat{\xi} \right\} . \tag{8.27}$$

The quantity $\hat{\xi}$ is simply the sum, normalized by $R^{-1/2}$ of R random independent and identically distributed events of zero mean and finite variance,

$$\hat{\xi} = \frac{\xi - \overline{\xi}}{\sqrt{R}} = \frac{1}{\sqrt{R}} \sum_{j=1}^{R} \left(\xi_j - c^{(1)} \right) . \tag{8.28}$$

In other words, (8.27) is the mathematical formulation of central limit theorem. The Gaussian distribution function itself is a fixed point of the convolution procedure in the space of functions in the sense that it is form stable under the renormalization group approach. Notice that form stability or alternatively self-similarity means that the resulting Gaussian function is identical to the initial Gaussian function after an appropriate shift and a rescaling of the variables.

We remark that the convergence to a Gaussian behavior also holds if the initially variables have different probability distribution functions with finite variance of the same order of magnitude. The generalized fixed point is now the Gaussian law (8.26) with

$$\overline{\xi} = \sum_{n=1}^{R} c_j^{(1)} \quad \text{and} \quad \tilde{\sigma} = \sum_{n=1}^{R} c_j^{(2)} , \tag{8.29}$$

where $c_j^{(1)}$ and $c_j^{(2)}$ are the mean trend vector and the covariance matrix, respectively, obtained from the now time-dependent elementary probability distribution function $p^{(j)}(\xi_j)$.

Finally, it should be remarked that the two conditions of the central limit theorem may be partially relaxed. The first condition under which this theorem holds is the Markov property. This strict condition can, however, be weakened, and the central limit theorem still holds for weakly correlated variables under certain conditions. The second condition that the variance of the variables be finite can be somewhat relaxed to include probability functions with algebraic tails $|\xi|^{-3}$. In this case, the normalizing factor is no longer $R^{-1/2}$ but can contain logarithmic corrections.

8.2.2 Convergence Problems

As a consequence of the renormalization group analysis, the central limit theorem is applicable in a strict sense only in the limit of infinite R. But, in practice, the Gaussian shape is a good approximation of the center of a probability distribution function if R is sufficiently large. It is important to realize that large deviations can occur in the tail of the probability distribution function $p_R(\xi)$, whose weight shrinks as R increases. The center is a region of width at least of the order of $\sqrt{R}$ around the average $\overline{\xi} = Rc^{(1)}$.

Let us make more precise what the center of a probability distribution function means. For the sake of simplicity we investigate events of only one component; i.e., ξ is now again a scalar quantity. As before, ξ is the sum of R identicales distributed variables ξ_j with mean $c^{(1)}$, variance $c^{(2)}$, and finite higher cumulants $c^{(m)}$. Thus, the central limit theorem reads

$$\lim_{R\to\infty} p_R(x) = \frac{1}{\sqrt{2\pi}} \exp\left\{-\frac{x^2}{2}\right\} , \tag{8.30}$$

where we have introduced the reduced variable

$$x = \frac{\hat{\xi}}{\sqrt{c^{(2)}}} = \frac{\xi - Rc^{(1)}}{\sqrt{Rc^{(2)}}} . \tag{8.31}$$

In order to analyze the convergence behavior for the tails [34], we start from the probability

$$P_{>}^{(R)}(z) = P^{(R)}(x > z) = \int_{z}^{\infty} p_R(x)\,\mathrm{d}x \tag{8.32}$$

and analyze the difference $\Delta P^{(R)}(z) = P_{>}^{(R)}(z) - P_{>}^{(\infty)}(z)$, where $P_{>}^{(\infty)}(z)$ is simply the complementary error function due to (8.30). If all cumulants are finite, one can develop a systematic expansion in powers of $R^{-1/2}$ of the difference $\Delta P^{(R)}(z)$ [33]:

$$\Delta P^{(R)}(z) = \frac{\exp\left\{-z^2/2\right\}}{\sqrt{2\pi}} \left[\frac{Q_1(z)}{R^{1/2}} + \frac{Q_2(z)}{R} \cdots + \frac{Q_m(z)}{R^{m/2}} \cdots\right], \tag{8.33}$$

where $Q_m(z)$ are polynomials in z, the coefficients of which depend on the first $m+2$ normalized cumulants of the elementary probability distribution function, $\lambda_k = c^{(k)}/[c^{(2)}]^{k/2}$. The explicit form of these polynomials can be obtained from the textbook of Gnedenko and Kolmogorov [34]. The two first polynomials are

$$Q_1(z) = \frac{\lambda_3}{6}\left(1 - z^2\right) \tag{8.34}$$

and

$$Q_2(z) = \frac{\lambda_3^2}{72} z^5 + \left(\frac{\lambda_4}{24} - \frac{5\lambda_3^2}{36}\right) z^4 + \left(\frac{5\lambda_3^2}{24} - \frac{\lambda_4}{8}\right) z^3 . \tag{8.35}$$

If the elementary probability distribution function has a Gaussian behavior, all its cumulants $c^{(m)}$ of order larger than 2 vanish identically. Therefore, all $Q_m(z)$ are also zero and the probability density $p_R(x)$ is a Gaussian.

For an arbitrary asymmetric probability distribution function, the skewness λ_3 is nonvanishing in general and the leading correction is $Q_1(z)$. The Gaussian law is valid if the relative error $\left|\Delta P^{(R)}(z)\right| / P_{>}^{(\infty)}(z)$ is small compared to 1. Since the error increases with z, the Gaussian behavior becomes observable at first close to the central tendency. The necessity condition $|\lambda_3| \ll R^{1/2}$ follows directly from $\left|\Delta P^{(R)}(z)\right| / P_{>}^{(\infty)}(z) \ll 1$ for $z \to 0$.

For large z, the approximation of $p_R(x)$ by a Gaussian law remains valid if the relative error remains small compared to 1. Here, we may replace the complementary error function $P_{>}^{(\infty)}(z)$ by its asymptotic representation $\exp\left\{-z^2/2\right\}/(\sqrt{2\pi}z)$. We thus obtain the inequality $|zQ_1(z)| \ll R^{1/2}$ leading to $\left|z^3\lambda_3\right| \ll R^{1/2}$. Because of (8.31), this relation is equivalent to the condition

$$\left|\xi - Rc^{(1)}\right| \ll |\lambda_3|^{-1/3} \sigma R^{2/3} . \tag{8.36}$$

It means that the Gaussian law holds in a region of an order of magnitude of $\left|\xi - Rc^{(1)}\right| \ll |\lambda_3|^{-1/3} \sigma R^{2/3}$ around the central tendency.

A symmetric probability distribution function has a vanishing skewness so that the excess kurtosis $\lambda_4 = c^{(4)}/\sigma^4$ provides the leading correction to the central limit theorem. The Gaussian law is now valid if $\lambda_4 \ll R$ and

$$\left|\xi - Rc^{(1)}\right| \ll |\lambda_4|^{-1/4} \sigma R^{3/4}, \tag{8.37}$$

i.e., the central region in which the Gaussian law holds is now of an order of magnitude $R^{3/4}$.

Another class of inequalities describing the convergence behavior with respect to the central limit theorem was found by Berry [35] and Esséen [36]. The Berry–Esséen theorems [37] provide inequalities controlling the absolute difference $\left|\Delta P^{(R)}(z)\right|$. Suppose the variance $c^{(2)}$ and the average

$$\eta = \int \left|\xi - c^{(1)}\right|^3 p(\xi)\, \mathrm{d}\xi \tag{8.38}$$

are finite quantities, then the first theorem reads

$$\left|\Delta P^{(R)}(z)\right| \leq \frac{3\eta}{\left[c^{(2)}\right]^{3/2} \sqrt{R}}. \tag{8.39}$$

The second theorem is the extension to not identically by distributed variables. Here, we have to replace the constant values of $c^{(2)}$ and η by

$$\overline{c^{(2)}} = \frac{1}{R} \sum_{j=1}^{R} c_j^{(2)} \tag{8.40}$$

and

$$\overline{\eta} = \frac{1}{R} \sum_{j=1}^{R} \eta_j, \tag{8.41}$$

where $c_j^{(2)}$ and η_j are obtained from the individual elementary probability distribution functions $p^{(j)}(\xi_j)$. Then, the following inequality holds

$$\left|\Delta P^{(R)}(z)\right| \leq \frac{6\overline{\eta}}{\left[c^{(2)}\right]^{3/2} \sqrt{R}}. \tag{8.42}$$

Notice that the Berry–Esséen theorems are less stringent than the results obtained from the cumulant expansion (8.33). We see that the central limit theorem gives no information about the behavior of the tails for finite R. Only the center is well-approximated by the Gaussian law. The width of the central region depends on the detailed properties of the elementary probability distribution functions.

The Gaussian probability distribution function is the fixed point or the attractor of a well-defined class of functions. This class is also denoted as the basin of attraction with respect to the corresponding functional space. When R increases, the functions $p_R(\xi)$ become progressively closer to the Gaussian attractor. As discussed above, this process is not uniform. The convergence is faster close to the center than in the tails of the probability distribution function.

8.3 Lévy Processes

8.3.1 Form-Stable Limit Distributions

While we had derived the central limit theorem, we saw that the probability density function $p_R(\xi)$ of the accumulated events could be expressed as a generalized convolution (8.12) of the elementary probability distribution functions $p(\xi)$. We want to use this equation in order to determine the set of all form-stable probability distribution functions. A probability density $p_R(\xi)$ is called a form-stable function if it can be represented by a function g, which is independent from the number R of convolutions,

$$p_R(\xi)\mathrm{d}\xi = g(\xi')\mathrm{d}\xi' , \tag{8.43}$$

where the variables are connected by the linear relation $\xi' = \alpha_R\xi + \beta_R$. Because the vector ξ has the dimension N, the $N \times N$ matrix α_R describes an appropriate rotation and dilation of the coordinates while the N-component vector β_R corresponds to a global translation of the coordinate system. Within the formalism of the renormalization group, a form-stable probability density law corresponds to a fixed point of the convolution procedure. The Fourier transform of g is given by

$$\begin{aligned}\hat{g}(k) = \int g(\xi')\mathrm{e}^{ik\xi'}\mathrm{d}\xi' &= \int p_R(\xi)\mathrm{e}^{ik(\alpha_R\xi+\beta_R)}\mathrm{d}\xi \\ &= \mathrm{e}^{ik\beta_R}\hat{p}_R(\alpha_R k) ,\end{aligned} \tag{8.44}$$

where we have used definition (6.50) of the characteristic function. The form stability requires that this relation must be fulfilled for all values of R. In particular, we obtain

$$\hat{p}_R(k) = \hat{g}(\alpha_R^{-1}k)\mathrm{e}^{-i\beta_R\alpha_R^{-1}k} \quad \text{and} \quad \hat{p}(k) = \hat{g}(\alpha_1^{-1}k)\mathrm{e}^{-i\beta_1\alpha_1^{-1}k} . \tag{8.45}$$

Without any restriction, we can choose $\alpha_1 = 1$ and $\beta_1 = 0$. The substitution of (8.45) into the convolution formula (8.13) yields now

$$\hat{g}(\alpha_R^{-1}k)\mathrm{e}^{-i\beta_R\alpha_R^{-1}k} = \hat{g}^R(k) . \tag{8.46}$$

Let us write

$$\hat{g}(k) = \exp\{\Phi(k)\} , \tag{8.47}$$

where $\Phi(k)$ is the cumulant generating function. Thus (8.46) can be written as

$$\Phi(\alpha_R^{-1}k) - i\beta_R\alpha_R^{-1}k = R\Phi(k) \tag{8.48}$$

and after splitting off the contributions linearly in k

$$\Phi(k) = iuk + \varphi(k) , \tag{8.49}$$

we arrive at the two relations,

$$\beta_R = \alpha_R u\left[\alpha_R^{-1} - R\right] \tag{8.50}$$

and

$$\varphi(\alpha_R^{-1} k) = R\varphi(k) \,. \tag{8.51}$$

The first equation gives simply the total shift of the center of the probability distribution function resulting from R convolution steps. As discussed in the context of the central limit theorem, the drift term can be put to zero by a suitable linear change of the variables ξ. Thus, β_R is no object of the further discussion. Second equation (8.51) is the true key for our analysis of the form stability. In the following investigation we restrict ourselves again to the one-variable case. The mathematical handling of the multidimensional case is similar, but the large number of possible degrees of freedom complicates the discussion.

The relation (8.51) requires that $\varphi(k)$ is a homogeneous function, $\varphi(\lambda k) = \lambda^\gamma \varphi(k)$ with the homogeneity coefficient γ. Considering that α_R must be a real quantity, we obtain $a_R = R^{-1/\gamma}$. Consequently, the function φ has the general structure

$$\varphi(k) = c_+ |k|^\gamma + c_- k |k|^{\gamma - 1} \tag{8.52}$$

with the three parameters c_+, c_-, and $\gamma \neq 1$.

A special solution occurs for $\gamma = 1$, because in this case $\varphi(k)$ merges with the separated linear contributions. Here, we obtain the special structure $\varphi(k) = c_+ |k| + c_- k \ln |k|$. The rescaling $k \to \lambda k$ leads then to $\varphi(\lambda k) = \lambda\varphi(k) + c_- k \ln \lambda$ and the additional term $c_- \ln \lambda$ may be absorbed in the shift coefficient β_R.

It is convenient to use the more common representation [38, 39]

$$\hat{g}(k) = \mathcal{L}_{a,b}^{\gamma}(k) = \exp\left\{ -a |k|^\gamma \left[1 + ib \tan\left(\frac{\pi\gamma}{2}\right) \frac{k}{|k|} \right] \right\} \tag{8.53}$$

with $\gamma \neq 1$. For $\gamma = 1$, $\tan(\pi\gamma/2)$ must be replaced by $(2/\pi) \ln |k|$. A more detailed analysis [38, 40] shows that $\hat{g}(k)$ is a characteristic function of a probability distribution function if and only if a is a positive scale factor, γ is a positive exponent, and the asymmetry parameter satisfies $|b| \leq 1$.

Apart from the drift term, (8.53) is the representation of any characteristic function corresponding to a probability density which is form-invariant under the convolution procedure. The set of these functions is known as the class of Lévy functions. Obviously, the Gaussian law is a special subclass. The Lévy functions are fully characterized by the expression of their characteristic functions (8.53). Thus, the inverse Fourier transform of (8.53) should lead to the real Lévy functions $\mathcal{L}_{a,b}^{\gamma}(\xi)$.

Unfortunately, there are no simple analytic expressions of the Lévy functions except for a few special cases, namely the Gaussian law ($\gamma = 2$), the Lévy–Smirnow law ($\gamma = 1/2$, $b = 1$)

$$\mathcal{L}_{a,1}^{1/2}(\xi) = \frac{2a}{\sqrt{\pi}\,(2\xi)^{3/2}} \exp\left\{ -\frac{a^2}{2\xi} \right\} \quad \text{for} \quad \xi > 0 \tag{8.54}$$

and the Cauchy law ($\gamma = 1$, $b = 0$)

$$\mathcal{L}^{1}_{a,0}\left(\xi\right) = \frac{a}{\pi^2 a^2 + \xi^2}\,, \tag{8.55}$$

which is also known as Lorentzian. One of the most important properties of the Lévy functions is their asymptotic power law behavior. A symmetric Lévy function ($b = 0$) centered at zero is completely defined by the Fourier integral

$$\mathcal{L}^{\gamma}_{a,0}\left(\xi\right) = \frac{1}{\pi}\int\limits_{0}^{\infty} \exp\left\{-a\left|k\right|^{\gamma}\right\}\cos(k\xi)\mathrm{d}k\,. \tag{8.56}$$

This integral can be written as a series expansion valid for $|\xi| \to \infty$

$$\mathcal{L}^{\gamma}_{a,0}\left(\xi\right) = -\frac{1}{\pi}\sum_{n=1}^{\infty}\frac{\left(-a\right)^n}{\left|\xi\right|^{\gamma n+1}}\frac{\Gamma\left(\gamma n+1\right)}{\Gamma\left(n+1\right)}\sin\left(\frac{\pi\gamma n}{2}\right)\,. \tag{8.57}$$

The leading term defines the asymptotic dependence

$$\mathcal{L}^{\gamma}_{a,0}\left(\xi\right) \sim \frac{C}{\left|\xi\right|^{1+\gamma}}\,. \tag{8.58}$$

Here, $C = a\gamma\Gamma\left(\gamma\right)\sin\left(\pi\gamma/2\right)/\pi$ is a positive constant called the tail and the exponent γ is between 0 and 2. The condition $\gamma < 2$ is necessary because a Lévy function with $\gamma > 2$ is unstable and converges to the Gaussian law. We will discuss this behavior below.

Lévy laws can also be asymmetric. Then we have the asymptotic behavior $\mathcal{L}^{\gamma}_{a,b}\left(\xi\right) \sim C_{+}/\xi^{1+\gamma}$ for $\xi \to \infty$ and $\mathcal{L}^{\gamma}_{a,b}\left(\xi\right) \sim C_{-}/\left|\xi\right|^{1+\gamma}$ for $\xi \to -\infty$ and the asymmetry is quantified by the asymmetry parameter b via

$$b = \frac{C_{+} - C_{-}}{C_{+} + C_{-}}\,. \tag{8.59}$$

The completely antisymmetric cases correspond to $b = \pm 1$. For $b = +1$ and $\gamma < 1$ the variable ξ takes only positive values while for $b = -1$ and $\gamma < 1$ the variable ξ is defined to be negative. For $1 < \gamma < 2$ and $b = 1$ the Lévy distribution is a power law $\xi^{-\gamma-1}$ for $\xi \to \infty$ while the function converges to zero for $\xi \to -\infty$ as $\exp\left(-\left|\xi\right|^{\gamma/(\gamma-1)}\right)$. The inverse situation occurs for $b = -1$.

All Lévy functions with the same exponent γ and the same asymmetry coefficient b are related by the scaling law

$$\mathcal{L}^{\gamma}_{a,b}\left(\xi\right) = a^{-1/\gamma}\mathcal{L}^{\gamma}_{1,b}\left(a^{-1/\gamma}\xi\right)\,. \tag{8.60}$$

Therefore we obtain

$$\overline{\left|\xi\right|^{\theta}} = \int\left|\xi\right|^{\theta}\mathcal{L}^{\gamma}_{a,b}\left(\xi\right)\mathrm{d}\xi = a^{\theta/\gamma}\int\left|\xi'\right|^{\theta}\mathcal{L}^{\gamma}_{1,b}\left(\xi'\right)\mathrm{d}\xi' \tag{8.61}$$

if the integrals in (8.61) exist. An important property of all Lévy distributions is that the variance is infinite. This behavior follows directly from the substitution of (8.53) into (6.52). Roughly speaking, the Lévy law does not decay sufficiently rapidly at $|\xi| \to \infty$ as it will be necessary for the integral (6.49) to converge. However, the absolute value of the spread (6.46) exists and suggests a characteristic scale of the fluctuations $D_{\mathrm{sp}}(t) \sim a^{1/\gamma}$. When $\gamma \leq 1$ even the mean and the average of the absolute value of the spread diverge. The characteristic scale of the fluctuations may be obtained from (8.61) via $\left[\overline{|\xi|^{\theta}}\right]^{1/\theta} \sim a^{1/\gamma}$ for a sufficiently small exponent θ. We remark that also for $\gamma \leq 1$ the median and the most probable value still exist.

8.3.2 Convergence to Stable Lévy Distributions

The Gaussian probability distribution function is not only a form-stable distribution, it is also the fixed point of the classical central limit theorem. In particular, it is the attractor of all the distribution functions having a finite variance. On the other hand, the Gaussian law is a special distribution of the form-stable class of Lévy distributions. It is then natural to ask if all other Lévy distributions are also attractors in the functional space of probability distribution functions with respect to the convolution procedure (Fig. 8.1).

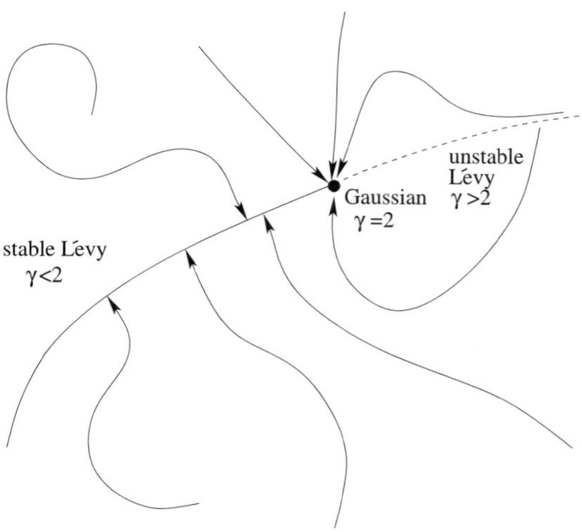

Fig. 8.1. The schematic convergence behavior of probability distribution functions in the functional space. The Gaussian law separates stable and unstable Lévy laws

There is a bipartite situation. Upon R convolutions, all probability distribution functions $p(\xi)$ with an asymptotic behavior $p(\xi) \sim C_{\pm} |\xi|^{-1-\gamma_{\pm}}$ and with $\gamma_{\pm} < 2$ are attracted to a stable Lévy distribution. In case of asymptotically symmetric functions, $C_+ = C_- = C$ and $\gamma_+ = \gamma_- = \gamma$, the fixed point is the symmetric Lévy law with the exponent γ and the scale parameter $a \sim RC$.

If the initial probability distribution functions have different tails, $C_+ \neq C_-$ but equal exponents, $p_R(\xi)$ converges to the asymmetric Lévy distribution with the exponent γ, the asymmetry parameter (8.59) and $a \sim R(C_+ + C_-)/2$ If the asymptotic exponents $\gamma_\pm$ of the elementary probability density $p(\xi)$ are different but $\min(\gamma_+, \gamma_-) < 2$, the convergence is to a completely asymmetric Lévy distribution with an exponent $\gamma = \min(\gamma_+, \gamma_-)$ and $b = 1$ for $\gamma_- < \gamma_+$ or $b = -1$ for $\gamma_- > \gamma_+$.

Finally, upon a sufficiently large number of convolutions, the Gaussian distribution attracts also all the probability distribution functions decaying as or faster than $|\xi|^{-3}$ at large $|\xi|$. Therefore, Lévy laws with $\gamma < 2$ are sometimes denoted as true Lévy laws.

Unfortunately, all Lévy distributions with $\gamma < 2$ have infinite variances. That limits its physical, but not its mathematical, meaning. Physically, Lévy distributions are meaningless with respect to finite systems. But in complex systems with an almost unlimited reservoir of hidden irrelevant degree of freedom, such probability distribution functions are quite possible at least over a wide range of the stochastic variables. Well-known examples of such wild distributions [13, 41] have been found to quantify the velocity-length distribution of the fully developed turbulence (Kolmogorov law) [14,20,21], the size–frequency distribution of earthquakes (Gutenberg–Richter law) [25, 26], or the destruction losses due to storms [22]. Further examples related to social and economic problems are the distribution of wealth [23, 24] also known as Pareto law, the distribution of losses due to business interruption resulting from accidents [15, 16] in the insurance business, or the distribution of losses caused by floods worldwide [17] or the famous classical St. Petersburg paradox discussed by Bernoulli [18, 19]

8.3.3 Truncated Lévy Distributions

As we have seen, Lévy laws obey scaling relations but have an infinite variance. A real Lévy distribution is not observed in finite physical systems. However, a stochastic process with finite variance and characterized by scaling relations in a large but finite region close to the center is the truncated Lévy distribution [43]. For many realistic problems, we have to ask for a distribution which in the tails is a power law multiplied by an exponential

$$p(\xi) \sim \frac{C_\pm}{|\xi|^{\gamma+1}} \exp\left\{-\frac{|\xi|}{\xi_0}\right\}. \tag{8.62}$$

The characteristic function of Lévy laws truncated by an exponential as in (8.62) can be written explicitly as [42, 43]

$$\begin{aligned} \ln \hat{p}(k) = {} & a\frac{\left(1 + k^2\xi_0^2\right)^{\gamma/2} \cos\left(\gamma \arctan\left(k\xi_0\right)\right) - 1}{\xi_0^\gamma \cos\left(\pi\gamma/2\right)} \\ & \times \left[1 + ib \tan\left(\gamma \arctan\left(|k|\, \xi_0\right)\right) \frac{k}{|k|}\right]. \end{aligned} \tag{8.63}$$

After R convolutions we get the characteristic distribution function

$$\ln \hat{p}_R(k) = -Ra\frac{\left(1+k^2\xi_0^2\right)^{\gamma/2}\cos\left(\gamma\arctan\left(k\xi_0\right)\right)-1}{\xi_0^{\gamma}\cos\left(\pi\gamma/2\right)} \times \left[1+ib\tan\left(\gamma\arctan\left(|k|\,\xi_0\right)\right)\frac{k}{|k|}\right]. \tag{8.64}$$

It can be checked that (8.63) recovers (8.53) for $\xi_0 \to \infty$. The behavior of $p_R(\xi)$ can be obtained from an inverse Fourier transform (6.51). In order to determine the characteristic scale of the probability distribution $p_R(\xi)$, we have to consider the main contributions to the inverse Fourier transform. This condition requires that the characteristic wave-number k_{char} is of an order of magnitude satisfying $\ln \hat{p}_R(k_{\text{char}}) \simeq 1$. This relation is equivalent to

$$Ra\left[\left(k^2+\xi_0^{-2}\right)^{\gamma/2}-\xi_0^{-\gamma}\right] \simeq 1. \tag{8.65}$$

For $R \ll \xi_0^{\gamma}$, (8.65) is satisfied if $k_{\text{char}}^2\xi_0^2 \gg 1$. Thus we obtain immediately $k_{\text{char}} \sim (Ra)^{-1/\gamma}$ and therefore the characteristic scale $\xi_{\text{char}} \sim (Ra)^{1/\gamma}$, which characterizes an ideal Lévy distribution.

When, on the contrary, $R \gg \xi_0^{\gamma}$, the characteristic value of k_{char} becomes much smaller than ξ_0^{-1}, and we find now the relation $k_{\text{char}} \sim (Ra)^{-1/2}\,\xi_0^{\gamma/2-1}$. The characteristic scale $\xi_{\text{char}} \sim (Ra)^{1/2}\,\xi_0^{1-\gamma/2}$ corresponding to what we expect from the Gaussian behavior.

Hence, as expected, a truncated Lévy distribution is not stable. It flows to an ideal Lévy probability distribution function for small R and then to the Gaussian distribution for large R. The crossover from the initial Lévy-like regime to the final Gaussian regime occurs if the characteristic scale of the Lévy distribution reaches the truncation scale $\xi_{\text{char}} \sim \xi_0$, i.e., if $Ra \sim \xi_0^{\gamma}$.

8.4 Rare Events

8.4.1 The Cramér Theorem

The central limit theorem states that the Gaussian law is a good description of the center of the probability distribution function $p_R(\xi)$ for sufficiently large R. We have demonstrated that the range of the center increases with increasing R but it is always limited for finite R. A similar statement is valid for the generalized version of the central limit theorem regarding the convergence behavior of Lévy laws. Fluctuations exceeding the range of the center are denoted as large fluctuations.

Of course, large fluctuations are rare events. The behavior of possible large fluctuations is not, or is only partially, affected by the predictions of the central limit theorem so that we should ask for an alternative description. We start our investigation from the general formulae (8.12) for a one-component event.

The characteristic function can also be calculated for an imaginary $k \to iz$ so that the Fourier transform becomes a Laplace transform

$$\hat{p}(z) = \int d\xi p(\xi) \exp\{-z\xi\} , \tag{8.66}$$

which holds under the assumption that the probability distribution function decays faster than an exponential for $|\xi| \to \infty$. We obtain again an algebraic relation for R convolution of the elementary probability distribution function $p(\xi)$,

$$\hat{p}_R(z) = [\hat{p}(z)]^R . \tag{8.67}$$

On the other hand, we assume that for sufficiently large R the probability density $p_R(\xi)$ may be written as

$$p_R(\xi) = \exp\left\{-R\mathcal{C}\left(\frac{\xi}{R}\right)\right\} , \tag{8.68}$$

where $\mathcal{C}(x)$ is the Cramér function [44, 45]. We will check by a construction principle, whether such a function exists for the limit $R \to \infty$. To this aim we calculate the corresponding Laplace transform

$$\hat{p}_R(z) = R \int dx \exp\{-R[\mathcal{C}(x) + zx]\} \tag{8.69}$$

by using the method of steepest descent. This method approximates the integral by the value of the integrand in a small neighborhood around its maximum $\tilde{x}$. The value of $\tilde{x}$ depends not on R and is a solution of

$$\frac{\partial}{\partial \tilde{x}} \mathcal{C}(\tilde{x}) + z = 0 . \tag{8.70}$$

With the knowledge of $\tilde{x}$ we can expand the Cramér function in powers of x around $\tilde{x}$

$$\mathcal{C}(x) + zx = \mathcal{C}(\tilde{x}) + z\tilde{x} + \frac{1}{2}\frac{\partial^2}{\partial \tilde{x}^2}\mathcal{C}(\tilde{x})[x - \tilde{x}]^2 + \cdots . \tag{8.71}$$

Note that the first-order term vanishes because of (8.70). Substituting (8.71) into (8.69), we obtain the integral

$$\hat{p}_R(z) = R \exp\{-R[\mathcal{C}(\tilde{x}) + z\tilde{x}]\} \times \int dy \exp\left\{-R\left[\frac{1}{2}\frac{\partial^2 \mathcal{C}(\tilde{x})}{\partial \tilde{x}^2} y^2 + \cdots\right]\right\} \tag{8.72}$$

with $y = x - \tilde{x}$. The leading term in the remaining integral is a Gaussian law of width $\delta y \sim R^{-1/2}$. With respect to this width all other contributions of the series expansion can be neglected for $R \to \infty$. Therefore, we focus here in the second-order term. The corresponding Gaussian integral exists if $\partial^2 \mathcal{C} / \partial x^2 > 0$. In this case we obtain

$$\hat{p}_R(z) \sim \sqrt{R / \left(\partial^2 \mathcal{C}(\tilde{x}) / \partial \tilde{x}^2\right)} \exp\{-R[\mathcal{C}(\tilde{x}) + z\tilde{x}]\} . \tag{8.73}$$

For $R \to \infty$, the leading term of the characteristic function is given by

$$\hat{p}_R(z) \sim \exp\{-R[\mathcal{C}(\tilde{x}) + z\tilde{x}]\} \ . \tag{8.74}$$

Combining (8.67), (8.74), and (8.70), we obtain the equations

$$\mathcal{C}(\tilde{x}) + z\tilde{x} + \ln \hat{p}(z) = 0 \quad \text{and} \quad \frac{\partial}{\partial \tilde{x}}\mathcal{C}(\tilde{x}) + z = 0 \tag{8.75}$$

which allow the determination of $\mathcal{C}(x)$. These two equations indicate that the Cramér function is the Legendre transform of $\ln \hat{p}(z)$. Hence, in order to determine $\mathcal{C}(\tilde{x})$ we must find the value of z which corresponds to a given $\tilde{x}$. The differentiation of (8.75) with respect to $\tilde{x}$ leads to

$$\frac{\partial}{\partial \tilde{x}}\mathcal{C}(\tilde{x}) + z + \tilde{x}\frac{\partial z}{\partial \tilde{x}} + \frac{\partial \ln \hat{p}(z)}{\partial z}\frac{\partial z}{\partial \tilde{x}} = \left[\tilde{x} + \frac{\partial \ln \hat{p}(z)}{\partial z}\right]\frac{\partial z}{\partial \tilde{x}} = 0 \ . \tag{8.76}$$

Because of $\partial z/\partial \tilde{x} = -\partial^2 \mathcal{C}(\tilde{x})/\partial \tilde{x}^2 < 0$ (see above), we find the relation

$$\tilde{x} = -\frac{\partial \ln \hat{p}(z)}{\partial z} \tag{8.77}$$

from where we can calculate $z = z(\tilde{x})$. Having $\mathcal{C}(\tilde{x})$, the Cramér theorem reads

$$p_R(\xi) = \exp\left\{-R\mathcal{C}\left(\frac{\xi}{R}\right)\right\} \quad \text{for} \quad R \to \infty \, . \tag{8.78}$$

This theorem describes large fluctuations outside the central region of $\hat{p}_R(\xi)$. The central region is defined by the central limit theorem, which requires $\xi \sim R^{\alpha}$ with $\alpha < 1$ (see (8.36) and (8.37)). Thus, the central region collapses to the origin in the Cramér theorem.

But outside of the center we have $|\xi|/R > 0$. Obviously, the scaling of the variables differs between the central limit theorem and the Cramér theorem. While the rescaling $\xi \to \xi/\sqrt{R}$ leads to the form-stable Gaussian behavior of $p_R(\xi)$ in the limit $R \to \infty$, the rescaling ξ/R yields another kind of form stability concerning the expression $R^{-1} \ln p_R(\xi)$.

Furthermore, the properties of the initial elementary probability distribution disappear close to the center for $R \to \infty$. Therefore, the central limit theorem describes a universal phenomenon. The Cramér function conserves the properties of the elementary probability distribution functions due to (8.75) so that the large fluctuations show no universal behavior.

8.4.2 Extreme Fluctuations

The Cramér theorem provides a concept for the treatment of large fluctuation as a sum of an infinite number of successive events. This limit $R \to \infty$ corresponds to the fact that the rescaled accumulated fluctuations ξ/R remains finite.

Another important regime is the extreme fluctuation regime [46]. Here we have to deal with finite R but $\xi/R \to \infty$.

In order to quantify this class of fluctuations, we start again from (8.12) and consider one-component events.

We use the representation $p(\xi) = \exp\{-f(\xi)\}$ and obtain

$$p_R(\xi) = \int \prod_{j=1}^{R} d\xi_j \delta\left(\xi - \sum_{j=1}^{R} \xi_j\right) \exp\left\{-\sum_{j=1}^{R} f(\xi_j)\right\} . \tag{8.79}$$

In order to simplify, we restrict ourselves on the case of an extreme positive fluctuation $\xi \to +\infty$. We have now two possibilities. On the one hand, the asymptotic behavior of the function $f(\xi)$ can be concave. Then we have $f(x) + f(y) > f(x+y)$ so that the dominant contributions to (8.79) are obtained from configurations with all fluctuations are very small except of one extreme fluctuation being almost equal to ξ. Therefore, we get

$$\ln p_L(\xi) \sim \ln p(\xi) \sim -f(\xi) . \tag{8.80}$$

On the other hand, if the asymptotic behavior of $f(\xi)$ is convex, $f(x) + f(y) < f(x+y)$, the minimum of the exponentials is given by the symmetric configuration $\xi_j = \xi/R$ for all $j = 1, \ldots, R$. The convexity condition requires a global minimum of the sum of all exponentials in (8.79) so that

$$\sum_{j=1}^{R} f(\xi_j) \geq Rf\left(\frac{\xi}{R}\right) . \tag{8.81}$$

We apply again the method of the steepest descent. To this aim we introduce the deviations $\delta\xi_j = \xi_j - \xi/R$ and expand the sum in (8.81) around its minimum

$$\sum_{j=1}^{R} f(\xi_j) = Rf\left(\frac{\xi}{R}\right) + \frac{1}{2} f''\left(\frac{\xi}{R}\right) \sum_{j=1}^{R} (\delta\xi_j)^2 + o\left(|\delta\xi|^3\right) , \tag{8.82}$$

where we have used the constraint $\delta\xi_1 + \delta\xi_2 + \cdots + \delta\xi_R = 0$. We substitute this expression into (8.79). Then, with the assumption of convexity, $f''(\xi/R) > 0$, the integral (8.79) can be estimated. We get the leading term

$$p_R(\xi) \sim \exp\left\{-Rf\left(\frac{\xi}{R}\right)\right\} . \tag{8.83}$$

This approximate result approaches the true value for $\xi/R \to \infty$. Apparently, (8.83) and (8.68) are identical expressions. But we should reminder that (8.68) holds for $R \to \infty$ but finite ξ/R, while (8.83) requires $\xi/R \to \infty$. However, the Cramér function $\mathcal{C}(x)$ becomes equal to $f(x)$ for $x \to \infty$.

In summary, the knowledge of the tails of an elementary probability distribution $p(\xi)$ allows the determination of the tails of the probability distribution function $p_R(\xi)$ via

$$p_R(\xi) \sim \left[p\left(\frac{\xi}{R}\right)\right]^R \tag{8.84}$$

if $\ln p^{-1}(\xi)$ is a convex function in ξ. On the other hand, if $\ln p^{-1}(\xi)$ is concave, we get $p_R(\xi) \sim p(\xi)$ for $\xi/R \to \infty$.

8.5 Kalman Filter

8.5.1 Linear Quadratic Problems with Gaussian Noise

Let us now study a stochastic system under control which is described by the linear evolution equation of the type (8.3).

$$\dot{X}(t) = A(t)X(t) + B(t)u(t) + \xi(t)\,, \tag{8.85}$$

where $\xi(t)$ is the N-component noise vector modeling the uncertainty of the system. Because of the central limit theorem (see Sect. 8.2), the probability distribution functions of the components of $\xi(t)$ are assumed to be those of a Gaussian stochastic process. Furthermore, we introduce a p-component output

$$Y(t) = C(t)X(t) + \eta(t)\,, \tag{8.86}$$

where $C(t)$ is a matrix of type $p \times N$ and $\eta(t)$ represents the p-component Gaussian random observation error. Both noise vectors have zero mean

$$\overline{\xi(t)} = 0 \quad \text{and} \quad \overline{\eta(t)} = 0 \tag{8.87}$$

while the correlation functions are given by

$$\overline{\xi_\alpha(t)\xi_\beta(t')} = \Omega_{\alpha\beta}(t)\delta(t-t') \quad \overline{\eta_\alpha(t)\eta_\beta(t')} = \Theta_{\alpha\beta}(t)\delta(t-t') \tag{8.88}$$

and

$$\overline{\xi_\alpha(t)\eta_\beta(t')} = 0\,. \tag{8.89}$$

The initial value of the state vector, $X_0 = X(0)$, may have the mean $\overline{X}_0$ while the covariance matrix is given by

$$\overline{(X_0 - \overline{X}_0)_\alpha (X_0 - \overline{X}_0)_\beta} = \sigma_{\alpha\beta}\,. \tag{8.90}$$

Obviously, we have a double problem. The first part must be the reconstruction of the state $X(t)$ from the knowledge of the observations $Y(t)$ while the second problem is the control of the system.

8.5.2 Estimation of the System State

The problem of the optimal estimate of the state of a system from the available observations is also called a filtering procedure. To solve this problem, we split the state and the observation variable

$$X(t) = \widetilde{X}(t) + X_u(t) \quad \text{and} \quad Y(t) = \widetilde{Y}(t) + Y_u(t) \tag{8.91}$$

with

$$\dot{\widetilde{X}}(t) = A(t)\widetilde{X}(t) + \xi(t) \quad \text{and} \quad \dot{X}_u(t) = A(t)X_u(t) + B(t)u(t) \tag{8.92}$$

and

$$\widetilde{Y}(t) = C(t)\widetilde{X}(t) + \eta(t) \quad \text{and} \quad Y_u(t) = C(t)X_u(t) \tag{8.93}$$

while the initial conditions are

$$\widetilde{X}(0) = \delta X_0 \quad \text{and} \quad X_u(0) = \overline{X}_0\,. \tag{8.94}$$

Note that the initial fluctuations are given by $\delta X_0 = X_0 - \overline{X}_0$. Now we consider the evolution of $\widetilde{X}(t)$ and try to reconstruct this state at the current time t from the knowledge of $\widetilde{Y}(t')$ with $t' < t$. To this aim we define a certain basic $\{e_1, e_2, \ldots, e_N\}$ spanning the phase space $\mathbb{P}$. Then the projections of the current state onto this basic

$$x_k(t) = \widetilde{X}(t)e_k \tag{8.95}$$

$(k = 1, \ldots, N)$ represent the dynamics of the system completely. On the other hand, we may introduce the scalar quantities

$$\theta_k(t) = \int_0^t \mathrm{d}t' \Lambda_k(t')\widetilde{Y}(t') \tag{8.96}$$

$(k = 1, \ldots, N)$ and ask for certain p-component vector functions $\Lambda_k(t)$ satisfying the N minimum problems

$$J_k(t) = \frac{1}{2}\overline{(x_k(t) - \theta_k(t))^2} \to \min \tag{8.97}$$

at the current time[2] $t > 0$. It means that we have decomposed the filtering problem into N separate minimum problems leading to N optimal p-component filter functions Λ_k. In order to solve these minimum problems, we consider the differential equations $(k = 1, \ldots, N)$

$$\dot{Z}_k(t') = -A^T(t')Z_k(t') + C^T(t')\Lambda_k(t') \tag{8.98}$$

for $t' \in [0, t]$ with matrices $A(t')$ and $C(t')$ from (8.85) and (8.86), respectively, and the final condition

$$Z(t) = e_k\,. \tag{8.99}$$

We transform this equation by the application of (8.85) and (8.86):

$$\begin{aligned} \frac{d\widetilde{X}Z_k}{dt} &= \widetilde{X}C^T\Lambda_k + \xi Z_k \\ &= \widetilde{Y}\Lambda_k - \eta\Lambda_k + \xi Z_k\,. \end{aligned} \tag{8.100}$$

By integrating both sides of this equation between 0 and t, we get with (8.99), (8.95), and (8.96)

[2] Note that the initial time is $t = 0$.

$$x_k(t) - \theta_k(t) = \delta X_0 Z_k(0) + \int_0^t dt' \left[\xi(t') Z_k(t') - \eta(t') \Lambda_k(t')\right] . \tag{8.101}$$

Hence, by squaring both sides, performing the average and considering (8.88), we obtain

$$\begin{aligned}
J_k(t) &= \frac{1}{2} Z_k(0) \overline{\delta X_0 \delta X_0^T} Z_k(0) \\
&+ \frac{1}{2} \int_0^t dt' \int_0^t dt'' Z_k(t') \overline{\xi(t') \xi^T(t'')} Z_k(t'') \\
&+ \frac{1}{2} \int_0^t dt' \int_0^t dt'' \Lambda_k(t') \overline{\eta(t') \eta^T(t'')} \Lambda_k(t'') \\
&= \frac{1}{2} Z_k(0) \sigma Z_k(0) \\
&+ \frac{1}{2} \int_0^t dt' \left[Z_k(t') \Omega Z_k(t') + \Lambda_k(t') \Theta \Lambda_k(t')\right] .
\end{aligned} \tag{8.102}$$

Thus, the filtering problem is reduced to a deterministic linear quadratic control problem with performance $J_k(t)$, the constraints (8.98), and the final conditions (8.99). However, the roles of the final and initial times have been interchanged in the performance functional. This fact can be managed by the reflection of the time direction. Then the comparison of the problem with the results of Sect. 3.1.4 requires now the solution

$$\Lambda_k = \Theta^{-1} C G Z_k , \tag{8.103}$$

where the symmetric $N \times N$ matrix G is a solution of the Ricatti equation[3]

$$\dot{G} - G A^T - A G + G C^T \Theta^{-1} C G = \Omega \tag{8.104}$$

with the initial condition

$$G(0) = \sigma \tag{8.105}$$

while the function Z_k is the solution of

$$\dot{Z}_k = \left[C^T \Theta^{-1} C G - A^T\right] Z_k \tag{8.106}$$

with the final condition

$$Z_k(t) = e_k . \tag{8.107}$$

Thus the wanted estimation $\widehat{x}_k(t)$ of the state vector with respect to the basic vector e_k is given by

[3] The changed sign is also a consequence of the mentioned time reflection.

$$\begin{aligned}\widehat{x}_k(t) = \theta_k(t) &= \int_0^t \mathrm{d}t' \Lambda_k(t')\widetilde{Y}(t') \\ &= \int_0^t \mathrm{d}t' Z_k(t') G(t') C^T(t') \Theta^{-1} \widetilde{Y}(t') \\ &= e_k \int_0^t \mathrm{d}t' \Gamma^T(t',t) G(t') C^T(t') \Theta^{-1} \widetilde{Y}(t') \end{aligned} \tag{8.108}$$

and therefore

$$\widehat{X}(t) = \int_0^t \mathrm{d}t' \Gamma^T(t',t) G(t') C^T(t') \Theta^{-1} \widetilde{Y}(t') . \tag{8.109}$$

Here, $\Gamma(t',t)$ is the Green's function solving the differential equation (8.106). This solution may be formally written as

$$\Gamma(t',t) = \exp\left\{ \int_t^{t'} \left[C^T(\tau)\,\Theta^{-1} C(\tau)\, G(\tau) - A^T(\tau) \right] \mathrm{d}\tau \right\} . \tag{8.110}$$

Thus we get

$$\begin{aligned}\Gamma^T(t',t) &= \exp\left\{ -\int_{t'}^{t} \left[G(\tau)\, C^T(\tau)\,\Theta^{-1} C(\tau) - A(\tau) \right] \mathrm{d}\tau \right\} , \\ &= \widetilde{\Gamma}(t,t') , \end{aligned} \tag{8.111}$$

where $\widetilde{\Gamma}(t,t')$ is Green's function associated to $A - GC^T\Theta^{-1}C$. In other words, the optimal estimation $\widehat{X}(t)$ satisfies the differential equation

$$\dot{\widehat{X}}(t) = \left[A - GC^T\Theta^{-1}C \right] \widehat{X}(t) + G(t) C^T(t) \Theta^{-1} \widetilde{Y}(t) \tag{8.112}$$

with the initial condition $\widehat{X}(0) = 0$. We remark that the optimal estimation depends essentially on the strength of the noise concerning the state evolution of the system, the observation error, and the uncertainties of the initial state (see (8.112) and (8.104)).

The estimation $\widehat{X}(t)$ is taken for $u = 0$. In order to obtain the estimation $\widehat{X}_u(t)$ for the presence of a finite control, we must add the deterministic solution $X_u(t)$, obtained from the solution of the second group of equations of (8.92) and (8.93), $\widehat{X}_u(t) = \widehat{X}(t) + X_u(t)$. Because of (8.92) and (8.93), the complete estimation fulfills the differential equations

$$\dot{\widehat{X}}_u = \left[A - GC^T\Theta^{-1}C \right] \widehat{X}_u + Bu + GC^T\Theta^{-1}Y , \tag{8.113}$$

where we have taken into account (8.91) for the elimination of $\widetilde{Y}(t)$. In order to complete the estimation equation (8.113), we have to consider the initial condition

$$\widehat{X}_u(0) = \overline{X}_0 \,. \tag{8.114}$$

Equation (8.113) is the famous Kalman filter equation [90, 91]. Since the estimation of a state from the knowledge of continuous observations is a very important problem for a large class of physical experiments, we will illustrate the algorithm with a simple example which belongs to the standard classical measurement. Let us assume that a one-component quantity $X(t)$ follows a well-defined deterministic law $\dot{X}(t) = u(t)$, where $u(t)$ is the expected time-dependent trajectory while the observations, $Y(t) = X(t) + \eta(t)$, are flawed with an error $\eta(t)$ of the variance Θ. Thus, we have $A = 0$, $B = 1$, $C = 1$, and $\Omega = 0$. The Ricatti equation (8.104) becomes now

$$\dot{G} + \Theta^{-1}G^2 = 0 \quad \text{with} \quad G(0) = \sigma \,, \tag{8.115}$$

where σ is the variance of the initial data. The solution of this equation is simply $G = \sigma\Theta/(\sigma t + \Theta)$. Thus, we get the estimation equation

$$\dot{\widehat{X}}_u + \frac{\sigma}{\sigma t + \Theta}\widehat{X}_u = u + \frac{\sigma}{\sigma t + \Theta} Y \tag{8.116}$$

and therefore the optimal estimated state

$$\widehat{X}_u(t) = \frac{\int_0^t \mathrm{d}t' \left[\sigma t' u(t') + \Theta u(t') + \sigma Y(t')\right] + \overline{X}_0 \Theta}{\sigma t + \Theta} \,. \tag{8.117}$$

This formula gives the optimal estimation of state on the basis of the observations $Y(t)$. In principle, the Kalman filter can be interpreted as a special regression model. In order to obtain the deviations of the estimated state from the real state, we substitute $Y = X + \eta$ and $u = \dot{X}$ in (8.117) and obtain after integration by parts

$$\widehat{X}_u(t) = X(t) + \frac{1}{\sigma t + \Theta}\left[\Theta\left(\overline{X}_0 - X_0\right) + \int_0^t \mathrm{d}t' \sigma \eta\right] . \tag{8.118}$$

Obviously, the estimation error, $\left|\widehat{X}_u(t) - X(t)\right|$, is dominated by the integral over the noise process. Hence, we get for a Gaussian noise the relative error

$$\left|\frac{\widehat{X}_u(t) - X(t)}{X(t)}\right| \sim t^{-1/2} \left|X(t)\right|^{-1/2} . \tag{8.119}$$

Thus, the Kalman filter estimation leads to an asymptotically convergence to the true behavior if $X(t)$ decays not faster than t^{-1} for $t \to \infty$.

Finally, we will define the meaning of the function $G(t)$. To this aim we make use of (8.106) and (8.104) and evaluate the derivative

$$\frac{\mathrm{d}Z_k(t)G(t)Z_k(t)}{\mathrm{d}t} = Z_k(t)G(t)C^T(t)\Theta^{-1}C(t)G(t)Z_k(t) + Z_k(t)\Omega Z_k(t)\,. \tag{8.120}$$

Thus, the performance functional (8.102) can be written as

$$\begin{aligned} J_k(t) &= \frac{1}{2}Z_k(0)\sigma Z_k(0) + \frac{1}{2}\int_0^t dt' \frac{dZ_k(t')G(t')Z_k(t')}{dt'} \\ &= \frac{1}{2}Z_k(t)G(t)Z_k(t) \\ &= \frac{1}{2}e_k G(t) e_k\,. \end{aligned} \tag{8.121}$$

The last stage of this relation is a consequence of (8.107). On the other hand, the performance may also be written as (8.97)

$$\begin{aligned} J_k(t) &= \frac{1}{2}\overline{(x_k(t) - \widehat{x}_k(t))^2} \\ &= \frac{1}{2}\overline{e_k\left(\widetilde{X}(t) - \widehat{X}(t)\right) \circ \left(\widetilde{X}(t) - \widehat{X}(t)\right)e_k} \\ &= \frac{1}{2}\overline{e_k\left(X(t) - \widehat{X}_u(t)\right) \circ \left(X(t) - \widehat{X}_u(t)\right)e_k}\,. \end{aligned} \tag{8.122}$$

The comparison of (8.121) with (8.122) yields

$$G_{\alpha\beta}(t) = \overline{\left[X_\alpha(t) - \widehat{X}_{u,\alpha}(t)\right]\left[X_\beta(t) - \widehat{X}_{u,\beta}(t)\right]} \tag{8.123}$$

where we have used the component representation. Thus, the matrix G is the variance of the optimal estimation error. Thus, $G(t)$ gives a direct information about how good is the estimate performed on the basis of the data available up to the current time t.

8.5.3 Ljapunov Differential Equation

Let us now introduce the difference $Z(t) = X(t) - \widehat{X}_u(t)$ between the current state and its estimation. Because of (8.113), (8.85), and (8.86), we obtain the evolution equation

$$\dot{Z} = \left[A - GC^T\Theta^{-1}C\right]Z - GC^T\Theta^{-1}\eta(t) + \xi(t) \tag{8.124}$$

and therefore

$$\dot{\overline{Z}} = \left[A - GC^T\Theta^{-1}C\right]\overline{Z}\,. \tag{8.125}$$

Because of[4] $\overline{Z}(0) = \overline{X(0)} - X_0 = 0$, the last equation requires $\overline{Z}(t) = 0$. On the other hand, we know from (8.123) that

$$\overline{Z_\alpha(t)Z_\beta(t)} = G_{\alpha\beta}(t)\,. \tag{8.126}$$

[4] Recall that $\widehat{X}_u(0) = X_0$ (see (8.114)).

The second quantity we may analyze is the fluctuation of the optimal estimation, $W(t) = \widehat{X}_u(t) - \overline{\widehat{X}_u(t)}$ with $\overline{W(t)} = 0$. From (8.113), (8.85), and (8.86) we obtain the evolution equation

$$\begin{aligned}\dot{W} &= \left[A - GC^T\Theta^{-1}C\right] W \\ &\quad + GC^T\Theta^{-1}C\left[X(t) - \overline{X}\right] + GC^T\Theta^{-1}\eta(t) \\ &= AW + GC^T\Theta^{-1}CZ + GC^T\Theta^{-1}\eta(t)\,,\end{aligned} \tag{8.127}$$

where we have used in the last stage the identity $\overline{X} = \overline{\widehat{X}}_u$ (corresponding to $\overline{Z}(t) = 0$). We are now interested in the correlations between both quantities, $Z(t)$ and $W(t)$. Thus, we may combine (8.125) and (8.127) to

$$\frac{d}{dt}\overrightarrow{\Psi} = M\overrightarrow{\Psi} + H\overrightarrow{\xi} \tag{8.128}$$

with

$$\overrightarrow{\Psi} = \begin{pmatrix} Z \\ W \end{pmatrix} \quad M = \begin{pmatrix} A - GC^T\Theta^{-1}C & 0 \\ GC^T\Theta^{-1}C & A \end{pmatrix} \tag{8.129}$$

and

$$H = \begin{pmatrix} I & -GC^T\Theta^{-1} \\ 0 & GC^T\Theta^{-1} \end{pmatrix} \quad \overrightarrow{\xi} = \begin{pmatrix} \xi \\ \eta \end{pmatrix}. \tag{8.130}$$

The formal solution of (8.128) is given by

$$\overrightarrow{\Psi}(t) = U(t,0)\overrightarrow{\Psi}(0) + \int_0^t dt' U(t,t')H(t')\overrightarrow{\xi}(t') \tag{8.131}$$

with

$$\frac{\partial}{\partial t}U(t,t') = M(t)U(t,t') \quad \text{and} \quad U(t,t) = 1\,. \tag{8.132}$$

Thus we obtain

$$\begin{aligned}F(t) &= \overline{\overrightarrow{\Psi}(t)\overrightarrow{\Psi}^T(t)} \\ &= U(t,0)\overrightarrow{\Psi}(0)\overrightarrow{\Psi}^T(0)U^T(t,0) \\ &\quad + \int_0^t dt' U(t,t')H(t')KH^T(t')U^T(t,t')\,,\end{aligned} \tag{8.133}$$

where we have introduced the correlation matrix K via

$$\overline{\overrightarrow{\xi}(t')\overrightarrow{\xi}^T(t)} = K\delta(t-t')\,. \tag{8.134}$$

The derivative of (8.133) with respect to the time yields the so-called differential Ljapunov equation

$$\frac{\mathrm{d}}{\mathrm{d}t}F(t) = M(t)F(t) + F(t)M^T(t) + H(t)KH^T(t)\,. \tag{8.135}$$

Because of (8.88), the correlation matrix has the form

$$K = \begin{pmatrix} \Omega & 0 \\ 0 & \Theta \end{pmatrix} . \tag{8.136}$$

Hence, we obtain from (8.135) the relations

$$\begin{aligned}\frac{\mathrm{d}}{\mathrm{d}t}\overline{ZZ^T} &= \left[A - GC^T\Theta^{-1}C\right]\overline{ZZ^T} \\ &\quad + \overline{ZZ^T}\left[A^T - C^T\Theta^{-1}CG\right] + \Omega + GC^T\Theta^{-1}CG\end{aligned} \tag{8.137}$$

and

$$\begin{aligned}\frac{\mathrm{d}}{\mathrm{d}t}\overline{ZW^T} &= \left[A - GC^T\Theta^{-1}C\right]\overline{ZW^T} \\ &\quad + \overline{ZZ^T}C^T\Theta^{-1}CG + \overline{ZW^T}A^T - GC^T\Theta^{-1}CG\end{aligned} \tag{8.138}$$

as well as

$$\begin{aligned}\frac{\mathrm{d}}{\mathrm{d}t}\overline{WW^T} &= (A + GC^T\Theta^{-1}C)\overline{ZW^T} + A\overline{WW^T} \\ &\quad + \overline{WZ^T}C^T\Theta^{-1}CG + \overline{WW^T}A^T + GC^T\Theta^{-1}CG^T .\end{aligned} \tag{8.139}$$

The first equation is because of (8.126) equivalent to the Ricatti equation (8.104) and corresponds to the above-derived identity (8.126). The second equation, (8.138), has the initial condition $\overline{Z(0)W^T(0)} = 0$ due to (8.114). On the other hand, (8.138) is a homogeneous differential equation because of (8.126). Because of the initial condition, we get the unique solution

$$\overline{Z(t)W^T(t)} = 0 . \tag{8.140}$$

Thus, the third equation reduces again to a differential equation of the Ljapunov type

$$\frac{\mathrm{d}}{\mathrm{d}t}\overline{WW^T} = A\overline{WW^T} + \overline{WW^T}A^T + GC^T\Theta^{-1}CG^T . \tag{8.141}$$

8.5.4 Optimal Control Problem for Kalman Filters

We come now to the second point of our problem, namely the control of a system on the basis of the filtered data. We consider a quadratic functional of type (7.83)

$$\begin{aligned}&J[X', \tau, u, T] \\ &= \frac{1}{2}\int_\tau^T dt \left[\overline{X(t)Q(t)X(t)}\Big|_{X(\tau)=X'} + \overline{u(t)R(t)u(t)}\Big|_{X(\tau)=X'} \right] ,\end{aligned} \tag{8.142}$$

which we will minimize. We replace $X(t)$ by $Z(t) + \widehat{X}_u(t)$ and obtain

$$\begin{aligned}\overline{X(t)Q(t)X(t)} &= \overline{\widehat{X}_u(t)Q(t)\widehat{X}_u(t)} \\ &\quad + 2\overline{Z(t)Q(t)\widehat{X}_u(t)} + \overline{Z(t)Q(t)Z(t)} .\end{aligned} \tag{8.143}$$

The second term can be rewritten as

$$\overline{Z(t)Q(t)\widehat{X}_u(t)} = \overline{Z(t)Q(t)W(t)} = 0 \tag{8.144}$$

because of (8.140) as well as $\overline{Z}(t) = 0$ and $W(t) = \widehat{X}_u(t) - \overline{\widehat{X}_u(t)}$. The third term becomes

$$\begin{aligned}\overline{Z(t)Q(t)Z(t)} &= Q_{\alpha\beta}(t)\overline{Z_\alpha(t)Z_\beta(t)} \\ &= \sum_{\alpha,\beta} Q_{\alpha\beta}(t)G_{\alpha\beta}(t) \\ &= \operatorname{tr}\, Q(t)G(t)\,.\end{aligned} \tag{8.145}$$

Hence, the performance can now be written as

$$\begin{aligned}&J[X',\tau,u,T] \\ &= \frac{1}{2}\int_\tau^T dt\left[\overline{\widehat{X}_u(t)Q(t)\widehat{X}_u(t)}\Big|_{X(\tau)=X'} + \overline{u(t)R(t)u(t)}\Big|_{X(\tau)=X'}\right] \\ &+ \frac{1}{2}\int_\tau^T dt\,[\operatorname{tr}\, Q(t)G(t)]\end{aligned} \tag{8.146}$$

and we get together with the evolution equation (8.113), a linear quadratic problem for the estimated state. This equation may be written as

$$\dot{\widehat{X}}_u = A\widehat{X}_u + Bu + GC^T\Theta^{-1}(Y - C\widehat{X}_u)\,, \tag{8.147}$$

where

$$Y - C\widehat{X}_u = C\left(X - \widehat{X}_u\right) = CZ \tag{8.148}$$

is a random quantity with zero mean (see Sect. 8.5.3). Hence, the optimal control law is given by

$$u^*(t) = -R^{-1}(t)B^T P(t)\widehat{X}_u^* \tag{8.149}$$

with $P(t)$ as a solution of the Ricatti equation

$$\dot{P} + A^T P + PA - PBR^{-1}B^T P = -Q \tag{8.150}$$

with the final condition $P(T) = 0$ and the optimal controlled estimation

$$\dot{\widehat{X}}_u^* = A\widehat{X}_u^* - BR^{-1}(t)B^T P(t)\widehat{X}_u^* + GC^T\Theta^{-1}(Y - C\widehat{X}_u^*)\,. \tag{8.151}$$

The result is again a feedback control with a stochastic input at the right-hand side of (8.151).

Let us illustrate the idea of a controlled filter by a simple example. We consider the one-dimensional motion of a Brownian particle, which should be localized at a certain point in space. Without any restriction, this may be the origin of our coordinate system. Then we have the stochastic evolution equation

$$\dot{X} = u + \xi\,, \tag{8.152}$$

where $u(t)$ is the control force and ξ is the noise. As localization functional we use the quadratic form (8.142) with constant coefficients Q and R. Then, the optimal feedback control law (7.92) yields

$$u^* = -R^{-1}G(t)X^*(t) \tag{8.153}$$

with $G(t)$ a solution of the Ricatti equation (8.150)

$$\dot{G} - R^{-1}G^2 = -Q \tag{8.154}$$

with the final condition $G(T) = 0$. Hence, we get

$$G(t) = \sqrt{QR}\ \tanh \frac{\sqrt{Q}(T-t)}{\sqrt{R}} \tag{8.155}$$

and the optimal control trajectory is described by an effective Ornstein-Uhlenbeck process

$$\dot{X}^* = -\sqrt{\frac{Q}{R}}\tanh\sqrt{\frac{Q}{R}}(T-t)\ X^* + \xi\,. \tag{8.156}$$

The main problem is that this control requires the precise determination of the particle position. In fact, each observation has an intrinsic error. Thus, we measure not the current position $X(t)$ but

$$Y(t) = X(t) + \eta\,. \tag{8.157}$$

If we now interpret this observation as the true position, we have the apparently feedback control law

$$u = -\sqrt{\frac{Q}{R}}\ \tanh\sqrt{\frac{Q}{R}}(T-t)\ Y \tag{8.158}$$

and therefore the trajectory

$$\dot{X} = -\sqrt{\frac{Q}{R}}\ \tanh\sqrt{\frac{Q}{R}}(T-t)\ \ (X+\eta) + \xi\,. \tag{8.159}$$

Now we replace the observation data Y in the control law (8.158) by the estimated state $\widehat{X}$

$$u = -\sqrt{\frac{Q}{R}}\ \tanh\sqrt{\frac{Q}{R}}(T-t)\ \widehat{X}\,. \tag{8.160}$$

The quantity $\widehat{X}$ is coupled to the observations Y via the law (8.113). In our special case we have the corresponding evolution equation

$$\dot{\widehat{X}} = -g\Theta^{-1}\widehat{X} - \sqrt{\frac{Q}{R}}\tanh\sqrt{\frac{Q}{R}}(T-t)\ \widehat{X} + g\Theta^{-1}Y\,, \tag{8.161}$$

where we have replaced u by the control law (8.160). The function g is the solution of the Ricatti equation (8.104)

$$\dot{g} + \Theta^{-1} g^2 = \Omega \tag{8.162}$$

with the initial condition[5] $g(0) = 0$. Thus, we obtain the solution

$$g(t) = \sqrt{\Theta\Omega} \tanh\sqrt{\frac{\Omega}{\Theta}}t \tag{8.163}$$

and the estimated state is then given by

$$\begin{aligned}\dot{\widehat{X}} &= -\widehat{X}\sqrt{\frac{\Omega}{\Theta}} \tanh\sqrt{\frac{\Omega}{\Theta}}t - \sqrt{\frac{Q}{R}} \tanh\sqrt{\frac{Q}{R}}(T-t)\ \widehat{X} \\ &\quad +Y\sqrt{\frac{\Omega}{\Theta}} \tanh\sqrt{\frac{\Omega}{\Theta}}t\end{aligned} \tag{8.164}$$

or with (8.157)

$$\begin{aligned}\dot{\widehat{X}} &= -\widehat{X}\sqrt{\frac{\Omega}{\Theta}} \tanh\sqrt{\frac{\Omega}{\Theta}}t - \sqrt{\frac{Q}{R}} \tanh\sqrt{\frac{Q}{R}}(T-t)\ \widehat{X} \\ &\quad +(X+\eta)\sqrt{\frac{\Omega}{\Theta}} \tanh\sqrt{\frac{\Omega}{\Theta}}t\end{aligned} \tag{8.165}$$

while the real state is given by

$$\dot{X}^* = -\sqrt{\frac{Q}{R}} \tanh\sqrt{\frac{Q}{R}}(T-t)\ \ \widehat{X} + \xi\,. \tag{8.166}$$

The different behavior of the three-control mechanism is presented in Figs. 8.2 and 8.3. We remark that all data presented here are generated with the same sets of randomly distributed variables ξ and η. Thus, the uncontrolled mechanism corresponds to the standard Brownian motion. Of course, the best control result occurs for the full information about the current state. This case corresponds to the optimal feedback control. We get the standard feedback control law (8.153) and the trajectory is a random process described by (8.156).

The second and third control regimes correspond to the situation that the measurement process shows some uncertainties. In the second case, these disturbances are not really considered, i.e., we assume that the variable $Y(t)$ is the real state. As a consequence, the system overreacts to the control and the fluctuation of the particle around the origin increases in comparison to the optimal feedback control. Much better as the second case is the third regime, where the current state is estimated from the observation data via the Kalman filter procedure. In fact, this control regime produces nearly the same results as the optimal feedback control.

[5] We assume that the particle was initially injected at the position $X_0 = 0$ without any uncertainty.

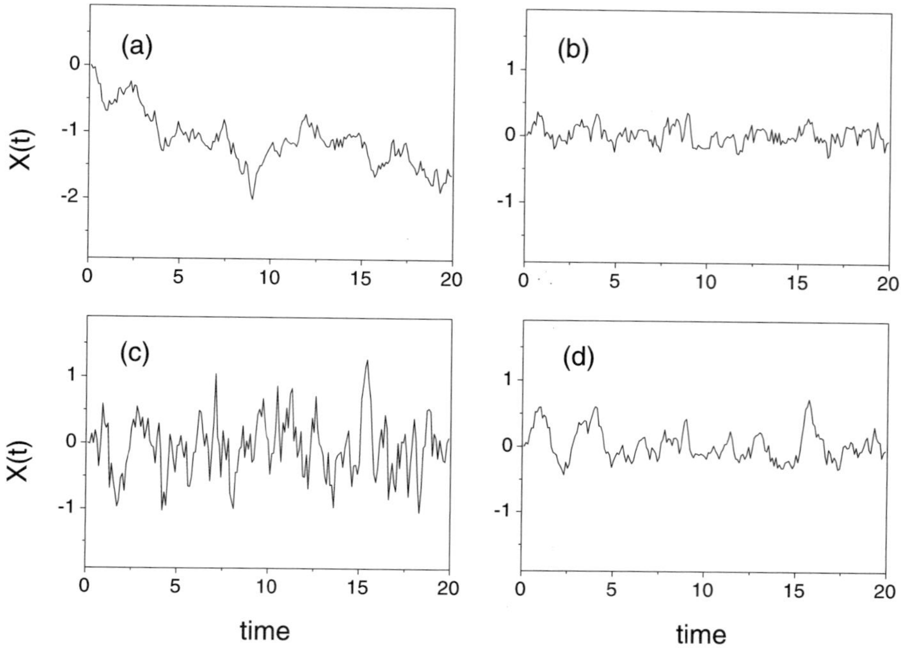

Fig. 8.2. The trajectory X for various control regimes corresponding to the same sample of noise and the parameters $Q = 10$, $R = 1$, $\Omega = 1$, $\Theta = 1$, and $T = 20$. (**a**) Without control, i.e., $u = 0$. The behavior of X is a free diffusion. (**b**) Under optimal control. The fluctuations of X are minimal compared with all other control regimes. The behavior is similar to an Ornstein–Uhlenbeck process. (**c**) With a control (8.158). The additional noise terms contained in the observations Y destabilize the trajectory X in comparison to the optimal controlled system. (**d**) With Kalman filter. The fluctuations of X are of the order of magnitude of the fluctuations of the optimal case

8.6 Filters and Predictors

8.6.1 General Filter Concepts

Filters play a double role in the context of stochastic control mechanisms. The first meaning is the preparation and transformation of disturbed observation data into reasonable estimations $\widehat{X}$ describing the current state of the system better than the pure observations. As we have seen above, the Kalman filter is able to reduce essentially the influence of intrinsic noise effects which inevitably occur during the measurement processes.

The second role is the application of the filtered data for the construction of an appropriate control $u = -K\widehat{X}$. The determination of the feedback K is a standard procedure leading to the stochastic optimal feedback control (7.92).

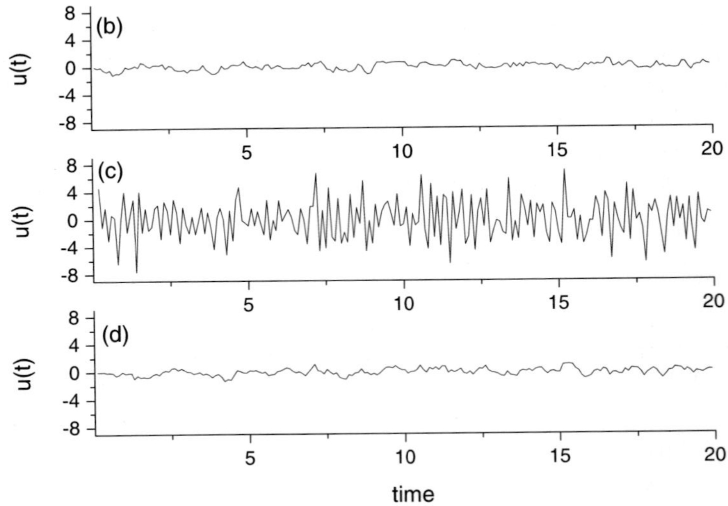

Fig. 8.3. The control function u for the three non-trivial control regimes (b-d) presented in Fig. 8.2. The optimal control (**b**) and the control with Kalman filter (**d**) are of the same order of magnitude while the control on the basis of the observations (**c**) shows essential stronger fluctuations

This procedure depends not essentially on the filtering process. Also, in case we are not able to solve the stochastic control equations, we can always find by empirical methods a suitable matrix K defining the feedback.

Thus, the main problem of filtering is the selection and preparation of available observations in order to compute a reasonable input $\widehat{X}$ for the control law. A filter is a special case of a larger class of estimation procedures which may be characterized as the determination of an estimate $\widehat{X}(t)$ from a given noise output observation $Y(t')$ with $t' < \tau$. We speak about a filtering problem, if $\tau = t$, i.e., if the end of the observation record corresponds to the current time. The problem is called a prediction problem if $t > \tau$ and a smoothing problem if $t < \tau$.

Smoothing problems are only of partial interest in the framework of control theory, whereas the solution of filtering and prediction problems is often helpful for the determination of a powerful control.

8.6.2 Wiener Filters

For simplicity, we consider now time discrete processes with $t_n = n\delta t$. A generalization to time-continuous processes is always possible. The original Wiener filtering process concerns the problem of linear causal estimation of a process [92]. The observation data are again modeled as the sum of a deterministic term mapping the state vector X onto the observation vector Y and an independent zero mean white noise. As for the Kalman filter, we are interested in the elimination of the error from the observations Y, i.e., we ask for an optimal estimated state which may be used for a subsequent control of the system. Furthermore, we assume that the system state X has the same dimension N as the observation state Y. The statement that a causal estimation is desired means that the estimated quantity at the current time depends only on the past of the observation data. The linearity requires the ansatz

$$\widehat{X}_n = \sum_{k=-\infty}^{n} K_{n-k} Y_k = \sum_{k=0}^{\infty} K_k Y_{n-k} \,. \tag{8.167}$$

The filter coefficients K_k are assumed to be such that the expression (8.167) is convergent with respect to the mean square. The problem is now to determine these coefficients. The appropriate criterion used for the Wiener filter is the averaged orthogonality between the observed states Y_n and the errors of the estimation, $X_n - \widehat{X}_n$, i.e.,

$$\overline{\left(X_n - \widehat{X}_n\right) Y_k} = 0 \quad \text{for} \quad k = -\infty, \ldots, n \,. \tag{8.168}$$

Thus we obtain

$$\overline{X_n Y_{n-j}} = \overline{\widehat{X}_n Y_{n-j}} \quad \text{for} \quad j = 0, \ldots, \infty \tag{8.169}$$

and therefore with (8.167)

$$\overline{X_n Y_{n-j}} = \sum_{k=0}^{\infty} K_k \overline{Y_{n-k} Y_{n-j}} \tag{8.170}$$

or

$$C_{XY}(j) = \sum_{k=0}^{\infty} K_k C_{YY}(j-k) \tag{8.171}$$

with the correlation functions

$$C_{XY}(j) = \overline{X_n Y_{n-j}} \quad \text{and} \quad C_{YY}(j) = \overline{Y_n Y_{n-j}} \,. \tag{8.172}$$

Both correlation functions are well defined for an arbitrary linear system following the dynamics given by (8.85) and (8.86). It is always possible to calculate these matrices following the same procedure presented in Sect. 6.9.1. From here, one obtains straightforwardly the wanted filter coefficients.

8.6.3 Estimation of the System Dynamics

The uncertainty of a system under control increases essentially if we have no information about the true system dynamics, i.e., the evolution functions $F(X, u, t)$ or equivalent in case of a linear problem the matrices $A(t)$ and $B(t)$, are unknown. The only information which is available is several observation records.

In contrast to the above-discussed Wiener and Kalman filters, we must now estimate also the system dynamics from the observation records. This means that we must solve a prediction problem because the knowledge of the system dynamics is equivalent to the knowledge of the future evolution and vice versa. Since we have no information about the real system dynamics and we obtain also in future no more information as the continuation of the observation records, it is no longer necessary to estimate the complete state evolution of the system. The present situation allows us not to see more than the observations, i.e., neither it can be proven a certain assumption about the intrinsic dynamics nor this assumption can be disproved. From this point of view, the treatment of such black box systems is an application of the principle of Occam's razor [47,48]. This idea is attributed to the 14th-century Franciscan monk William of Occam, which states that entities should not be multiplied unnecessarily. The most useful statement of this principle is that the better theory of two competing theories which make exactly the same predictions is the simpler one. Occam's razor is used to cut away unprovable concepts.

In principle, each forecasting concept about the observations belonging to a system with hidden intrinsic dynamics defines also a more or less suitable model connecting the current and historical observations with an estimation of the future evolution. In so far, this models represent a substitute system from which we may obtain substitute evolution equations which are the necessary constraints for a successful control. The uncertainties of such models are considered in appropriable noise terms. Thus, if we have estimated the evolution of the underlying system, we come back to the classical stochastic control problems. The system output, the observations, is automatically the input of the control function while all functions defining the control law are obtainable from the estimated evolution equations, i.e., the forecasting equations.

In the subsequent sections we will give some few ideas which may be helpful for the characterization and application of several prediction methods. Since these techniques do not belong to the central topics of the control theory, we restrict our carrying out to a brief discussion of the main features.

8.6.4 Regression and Autoregression

For simplicity, we consider again time discrete processes. At the beginning of the last century, standard predictions were undertaken by simply extrapolating a given time series through a global fit procedure. The principle is very simple. Suppose we have a time series of observations $\{Y_1, Y_2, \ldots, Y_L\}$ with the corresponding points in time $\{t_1, t_2, \ldots, t_L\}$ and Y_n vectors of the p-dimensional observation space. Then we can determine a regression function f in such a way that the distance between the observations Y_n and the corresponding values $f(t_n)$ becomes sufficiently small. There are two problems. The first one is the choice of a suitable parametrized regression function. This is usually an empirical step which depends often on the amount of experience. The second problem is the definition of a suitable measure for the distance.

Standard techniques as least mean square methods minimize a certain utility function, for example,

$$F = \sum_{n=1}^{L} (Y_n - f(t_n))^2 \tag{8.173}$$

by variation of the parameters of the function f. For instance, the well-known linear regression requires the determination of parameters A and B, which define the regression function f via $f(t) = A + Bt$. Obviously, the choice of the utility function is important for the determination of the parameters of the regression function. For example, the simple regression function $f(t) = Bt$ may be estimated by

$$F_1 = \sum_{n=1}^{L} (Y_n - Bt_n)^2 \quad \text{and} \quad F_2 = \sum_{n=1}^{L} \left(\frac{Y_n}{Bt_n} - 1 \right)^2 . \tag{8.174}$$

The first function stresses the absolute deviation between the observation and the regression function, while the second expression stresses the relative error. The first function leads to the estimation $B = \langle Yt \rangle_L / \langle t^2 \rangle_L$ while the second one yields $B = \langle Y^2 t^{-2} \rangle_L / \langle Y t^{-1} \rangle_L$, where we have used the definition

$$\langle g \rangle_L = \frac{1}{L} \sum_{n=1}^{L} g_n . \tag{8.175}$$

It is important to define both the regression function and the utility function in agreement with the present knowledge about the underlying system. After the determination of the regression parameters, the predictions are simply given by

$$\widehat{Y}_{L+k} = f(t_{L+k}) . \tag{8.176}$$

The beginning of modern time series prediction was in 1927, when Yule [9] introduced the autoregressive model to predict the annual number of sunspots. Such models are usually linear or polynomial and they are driven by white noise. In this context, predictions are carried out on the basis of parametric autoregressive (AR), moving-average (MA), or autoregressive moving-average (ARMA) models [10–12].

The autoregressive process AR(m) is defined by

$$Y(t_n) = a_0 + \sum_{k=1}^{m} a_k Y(t_{n-k}) + \eta(t_n) , \tag{8.177}$$

where a_k ($k = 0, \ldots, m$) are parametrized matrices of type $p \times p$ and η_n represents the current noise. We can use an appropriate method of estimation, such as ordinary least squares, to get suitable approximations $\hat{a}_k$ of the initially unknown parameters a_k. After the estimation of these model parameters, we get the fitted model

$$\widehat{Y}(t_n) = \hat{a}_0 + \sum_{k=1}^{m} \hat{a}_k Y(t_{n-k}) \,. \tag{8.178}$$

Clearly different regression methods give different estimates, but they are all estimates on the basis of the same more or less unknown, but true distribution of $Y(t_n)$. In this sense, $\widehat{Y}(t_n)$ is an estimation of the true conditional mean of $Y(t_n)$, which may be generally denoted as $E\left(Y(t_n) \mid \omega_{n-1}\right)$, where ω_{n-1} is the information set available at time t_{n-1}. In case of the above-introduced autoregressive process AR(m), we have $\omega_{n-1} = \{Y(t_{n-1}), \ldots, Y(t_{n-m})\}$. This notation makes explicit how the conditional mean and therefore the prediction is constructed on the assumption that all data up to that point are known, deterministic variables.

A natural way for the estimation of the coefficients a_k considers the Mori–Zwanzig equations (6.126). As pointed out, this equation is an exact, linear relation. In a discrete version, this equation reads

$$Y_\alpha(t_{n+1}) = Y_\alpha(t_n) + \sum_{\beta=1}^{p} \sum_{k=0}^{n} \Xi_{\alpha\beta}(t_n - t_k) Y_\beta(t_k) + \eta_\alpha(t_{n+1}) \,, \tag{8.179}$$

where we have used the component representation. Note that we have replaced the notations for the relevant quantities, $G_\alpha \to Y_\alpha$, and for the residual forces, $f_\alpha \to \eta_\alpha$, while the frequency matrix and the memory kernel are collected in the matrix $\Xi_{\alpha\beta}(t_n - t_k)$. Of course, the residual forces, the memory, and the frequency matrix contained in the original Mori–Zwanzig equations are implicitly dependent on the initial state at t_0. Thus, for a stationary system, the matrix $\Xi_{\alpha\beta}(t)$ is independent on the initial state and the residual forces may be interpreted as a stationary noise. In order to determine the matrix $\Xi_{\alpha\beta}(t)$, we remember that the correlation functions of the relevant quantities are exactly defined by (6.132). This equation reads in its discrete form

$$\begin{aligned} \overline{Y_\alpha(t_{n+1}) Y_\gamma(t_0)} &= \overline{Y_\alpha(t_n) Y_\gamma(t_0)} \\ &\quad + \sum_{\beta=1}^{p} \sum_{k=0}^{n} \Xi_{\alpha\beta}(t_n - t_k) \overline{Y_\beta(t_k) Y_\gamma(t_0)} \,. \end{aligned} \tag{8.180}$$

Besides the error due to the discretization, (8.180) is a exact relation. In case of a stationary system, (8.180) holds for all initial times t_0 with the same matrix function $\Xi_{\alpha\beta}(t)$. Thus, we can replace the correlation functions $\overline{Y_\alpha(t_n) Y_\gamma(t_0)}$ by the estimations

$$C_{\alpha\gamma}(t_n - t_0) = \left\langle Y_\alpha(t_n) Y_\gamma(t_0) \right\rangle_L = \frac{1}{L-n} \sum_{k=1}^{L-n} y_\alpha(t_{n+k}) y_\gamma(t_k) \tag{8.181}$$

(with $n < L$), which are obtainable from empirical observations. Thus, we arrive at the matrix equation

$$C_{\alpha\gamma}([n+1]\,\delta t) = C_{\alpha\gamma}(n\delta t) + \sum_{\beta=1}^{p} \sum_{k=0}^{n} \Xi_{\alpha\beta}([n-k]\,\delta t) C_{\beta\gamma}(k\delta t) \,, \tag{8.182}$$

where we have used $t_{n+1} = t_n + \delta t$. Equation (8.182) allows the determination of the matrix $\Xi_{\alpha\beta}(t)$ on the basis of the empirically estimated correlation functions $C_{\alpha\gamma}(t)$. After the estimation of the matrix functions $\Xi_{\alpha\beta}(t)$ we get the prediction formula

$$\widehat{Y}_{\alpha}(t_{n+1}) = Y_{\alpha}(t_n) + \sum_{\beta=1}^{p} \sum_{k=0}^{n} \Xi_{\alpha\beta}(t_n - t_k) Y_{\beta}(t_k) . \tag{8.183}$$

We remark that a repeated application of such prediction formulas allows also the forecasting of the behavior at later times, but of course, there is usually an increasing error.

The prediction formulas of moving averages and autoregressive processes are related. A moving average is a weighted average over the finite or infinite past. In general, a moving average can be written as

$$\overline{Y}(t_n) = \frac{\sum_{k=0}^{n-1} a_k Y(t_{n-k})}{\sum_{k=0}^{n-1} a_k} , \tag{8.184}$$

where the weights usually decrease with increasing k. The weight functions are often chosen heuristically under consideration of possible empirical investigations. The prediction formula is simply given by

$$\widehat{Y}(t_{n+1}) = \overline{Y}(t_n) . \tag{8.185}$$

The main difference between autoregressive processes and moving averages is the interpretation of the data with respect to the prediction formula. In an autoregressive process, the input is always understood as a deterministic series, in spite of the stochastic character of the underlying model. On the other hand, the moving average assumes that all observations are realizations of a stochastic process. Autoregressive moving averages (ARMA) are combinations of moving averages and autoregressive processes. Such processes play an important role for the analysis of modified ARCH and GARCH processes [49–52].

8.6.5 The Bayesian Concept

Decision Theory

Suppose we have several models F_i $(i = 1, \ldots, M)$ as possible candidates predicting the evolution of a given black box system. The problem is now to decide which model gives the best approach to the reality. This decision can be carried out on the basis of Bayes' theorem. We denote each model as a hypothesis B_i $(i = 1, \ldots, M)$. The possible hypotheses are mutually exclusive, i.e., in the language of set theory we have to write $B_i \cap B_j = \emptyset$, and exhaustive. The probability that Hypothesis B_i appears is $P(B_i)$. Furthermore, we consider an event A, which may be conditioned by the hypotheses. Thus, (6.59) can be written as

$$P(A \mid B_i)P(B_i) = P(B_i \mid A)P(A) \tag{8.186}$$

for all $i = 1, \ldots, M$. Furthermore, (6.64) leads to

$$P(B_i \mid A) = \frac{P(A \mid B_i)P(B_i)}{\sum_{i=1}^{M} P(A \mid B_i)\, P(B_i)} \,. \tag{8.187}$$

This is the standard form of Bayes' theorem. In the present context, we denote $P(B_i)$ as the "a priori" probability, which is available before the event A appears. The likelihood $P(A \mid B_i)$ is the conditional probability that the event A occurs under Hypothesis B_i. The quantity $P(B_i \mid A)$ may be interpreted as the probability that Hypothesis B_i was true under the condition that event A occurs. Therefore, $P(B_i \mid A)$ is also denoted as the "a posteriori" probability which may be empirically determined after the appearance of A.

Bayesian Theory and Forecasting

The above-discussed Bayesian theory of model or decision selection [53–57] generates insights not only into the theory of decision making, but also into the theory of predictions. The Bayesian solution to the model selection problem is well known: it is optimal to choose the model with the highest a posteriori probability. On the other hand, the knowledge of the a posteriori probabilities is not only important for the selection of a model, but it gives also an essential information for a reasonable combination of forecast results since the a posteriori probabilities are associated with the forecasting models F_i.

For the sake of simplicity, we consider only two models. Then, we have the a posteriori probabilities $P(F_1 \mid \omega)$ that model 1 is true, $P(F_2 \mid \omega)$ that model 2 is true under the condition, and that a certain event ω occurs. The estimation of these a posteriori probabilities is obtainable from the scheme discussed above. Furthermore, we have the mean square deviations

$$\left.\overline{(Y - \widehat{Y})^2}\right|_{F_1} = \int dy (Y - \widehat{Y})^2 p\,(Y \mid F_1) \tag{8.188}$$

and

$$\left.\overline{(Y - \widehat{Y})^2}\right|_{F_2} = \int dy (Y - \widehat{Y})^2 p\,(Y \mid F_2) \tag{8.189}$$

describing the expected square difference between an arbitrary forecast $\widehat{Y}$ and outcome Y of the model. Because of

$$p\,(Y \mid \omega) = p\,(Y \mid F_1)\, P(F_1 \mid \omega) + p\,(Y \mid F_2)\, P(F_2 \mid \omega)\,, \tag{8.190}$$

we get the total mean square deviation

$$\left.\overline{(Y - \widehat{Y})^2}\right|_{\omega} = \left.\overline{(Y - \widehat{Y})^2}\right|_{F_1} P(F_1 \mid \omega) + \left.\overline{(Y - \widehat{Y})^2}\right|_{F_2} P(F_2 \mid \omega)\,, \tag{8.191}$$

which is expected under the condition that the event ω appears. The prediction $\widehat{Y}$ is up to now a free value. We chose this value by minimization of total mean square deviation. We get

$$\frac{\partial}{\partial \widehat{Y}} \overline{(Y - \widehat{Y})^2}\Big|_\omega = 2\left[\overline{Y}\big|_{F_1} - \widehat{Y}\right] P(F_1 \mid \omega) + 2\left[\overline{Y}\big|_{F_2} - \widehat{Y}\right] P(F_2 \mid \omega) = 0 \tag{8.192}$$

and therefore the optimal prediction

$$\widehat{Y} = \overline{Y}\big|_{F_1} P(F_1 \mid \omega) + \overline{Y}\big|_{F_2} P(F_2 \mid \omega) . \tag{8.193}$$

This relation allows us to combine predictions of different models in order to obtain a likely forecast. For example, the averages $\overline{Y}\big|_{F_1}$ and $\overline{Y}\big|_{F_2}$ may be the results of two moving-average procedures. At least one of these forecasting models fails. The a posteriori probabilities $P(F_i \mid \omega)$ can be interpreted as the outcome of certain tests associated with the event ω, which should determine the correct moving-average model. The model selection theory requires that we have to consider only that model which has the largest a posteriori probability, i.e., we get either $\widehat{Y} = \overline{Y}\big|_{F_1}$ or $\widehat{Y} = \overline{Y}\big|_{F_2}$. However, the Bayesian forecast concept allows also the consideration of unfavorable models with small, but finite weights.

8.6.6 Neural Networks

Introduction

As discussed above, time series predictions have usually been performed by the use of parametric regressive, autoregressive, moving-average, or autoregressive moving-average models. The parameters of the prediction models are obtained from least mean square algorithms or similar procedures. A serious problem is that these techniques are basically linear. On the other hand, many time series are probably induced by strong nonlinear processes due to the high degree of complexity of the underlying system.

In this case, neural networks provide alternative methods for a forecasting of the further development of time series. Neural networks are powerful when applied to problems whose solutions require knowledge about a system or a model which is difficult or impossible to specify, but for which there is a large set of past observations available [58–60]. The neural network approach to time series prediction is parameter free in the sense that such methods do not need any information regarding the system that generates the signal. In other words, the system can be interpreted as a black box with certain inputs and outputs. The aim of a forecasting using neural networks is to determine the output with a suitable accuracy when only the input is known. This task is carried out by a process of learning from the so-called training patterns presented to the network and changing network structure and weights in response to the output error.

From a general point of view, the use of neural networks may be understood as a step back from rule-based models to data-driven methods [61].

Spin Glasses and Neural Networks

Let us discuss why neural networks are useful for the prediction of the evolution time series. Such systems can store patterns and they can recall these items on the basis of an incomplete input. A typical application is the evolution of the system state along a stable orbit. If a neural network detects similarities between a current time series and an older one, it may extrapolate the possible time evolution of the current time series on the basis of the historical experience. Usually, the similarities are often not very trivially recognizable. The weights of the stored properties used for the comparison of different pattern depend on the architecture of the underlying network. First of all, we will explain why neural networks have a so-called adaptive memory.

Neural networks have some similarities with a real nervous system consisting of interacting nerve cells [62, 63]. Therefore, let us start our investigation from a biological point of view. The human nervous system is very large. It consists of approximately 10^{11} highly interconnected nerve cells. Electric signals induce transmitter substances to be released at the synaptic junctions where the nerves almost touch (Fig. 8.4). The transmitters generate a local flow of sodium and potassium cations which raises or lowers the electrical potential. If the potential exceeds a certain threshold, a soliton-like excitation propagates from the cell body down to the axon. This then leads to the release of transmitters at the synapses to the next nerve cell. Obviously, the nervous system may be interpreted as a large cellular automaton [83–86] of identical cells but with complicated topological connections. In particular, each cell has effectively just two states, an active one and a passive one. We adopt a spin analogy: the state of the cell α ($\alpha = 1, \ldots, N$) may be given by $S_\alpha = \pm 1$, where $+1$ characterizes the active state and -1 the passive state. The electrical potential may be a weighted sum of the activity of the neighbored nerve cells

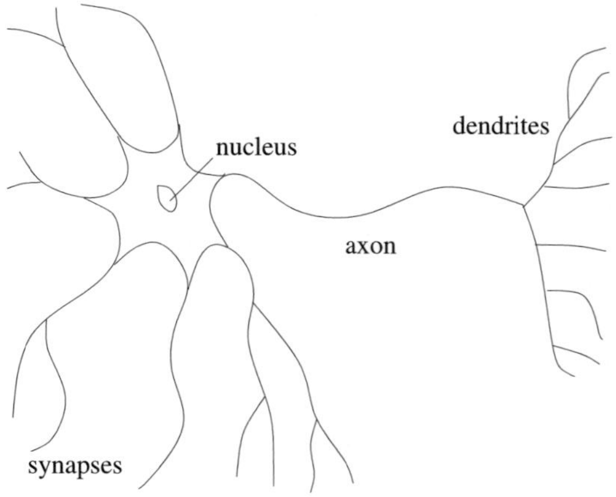

Fig. 8.4. Schematic representation of a nerve cell

$$V_\alpha = \sum_\beta J_{\alpha\beta} S_\beta \,. \tag{8.194}$$

The coupling parameters $J_{\alpha\beta}$ describe the influence of cell β on cell α. We remark that there is usually no symmetry, i.e., $J_{\alpha\beta} \neq J_{\beta\alpha}$. Of course, the absolute value and the sign of the parameters $J_{\alpha\beta}$ depend on the strength of the biochemically synaptic junction from cell β to cell α. The transition rule of this cellular automaton reads

$$S_\alpha(t_{n+1}) = \operatorname{sgn}\left(V_\alpha(t_n) - \theta_\alpha\right) = \operatorname{sgn}\left(\sum_\beta J_{\alpha\beta} S_\beta(t_n) - \theta_\alpha\right) , \tag{8.195}$$

where θ_α is the specific threshold of the cell [87–89]. Let us now transform this deterministic cellular automaton model in a probabilistic one. To this aim, we introduce the probability that the cell α becomes active at t_{n+1}

$$p_\alpha^+(t_{n+1}) = \psi(V_\alpha(t_n) - \theta_\alpha) \,, \tag{8.196}$$

where ψ is a sigmoidal function with the boundaries $\psi(-\infty) = 0$ and $\psi(\infty) = 1$. Equation (8.196) implies $p_\alpha^- = 1 - p_\alpha^+$. This generalization is really observed in nervous systems. The amount of transmitter substance released at a synapse can fluctuate so that a cell remains in the passive state even though $V_\alpha(t_n)$ exceeds the threshold θ_α. For the sake of simplicity, we focus on the symmetric case $J_{\alpha\beta} = J_{\beta\alpha}$. The special choice

$$\psi(x) = \frac{1}{1 + \exp\{-2x/T\}} \tag{8.197}$$

is particularly convenient because it corresponds to an Ising model with a so-called Glauber dynamics. It means that a cell changes its state independently from possible changes of other cells. For symmetric $J_{\alpha\beta}$, the succession of these changes drives the system to the states with low energy, and the system reaches after a sufficiently long relaxation time the thermodynamical equilibrium characterized by the stationary Gibb's distribution $\exp\{-H/T\}$ with the Hopfield–Hamiltonian [68–70]

$$H = -\frac{1}{2}\sum_{\alpha\beta} J_{\alpha\beta} S_\alpha S_\beta + \sum_\alpha \theta_\alpha S_\alpha \tag{8.198}$$

and the temperature T. From here, we can reproduce (8.196) and (8.197) in a very simple manner. The cell α can undergo the transitions $+1 \to +1$, $-1 \to -1$, $-1 \to +1$, and $+1 \to -1$ with the corresponding energy differences $\Delta H_{+,+} = \Delta H_{-,-} = 0$ and $\Delta H_{-,+} = -\Delta H_{+,-} = 2\left(V_\alpha - \theta_\alpha\right)$, which follow directly from (8.198). Thus, Gibb's measure requires the conditional probabilities

$$p_\alpha\left(+ \mid +\right) = \frac{\exp(-\Delta H_{+,+}/T)}{\exp(-\Delta H_{+,+}/T) + \exp(-\Delta H_{-,+}/T)} \tag{8.199}$$

and

$$p_\alpha(+\mid -) = \frac{\exp(-\Delta H_{+,-}/T)}{\exp(-\Delta H_{+,-}/T) + \exp(-\Delta H_{-,-}/T)} \,. \tag{8.200}$$

Considering the values of the energy differences, we get $p_\alpha^+ = p_\alpha(+\mid +) = p_\alpha(+\mid -)$, where p_α^+ satisfies (8.196) and (8.197). Obviously, our special model of a neural network is nothing other than a spin glass, i.e., an Ising model with stochastic, but symmetric interaction constants $J_{\alpha\beta}$ and the set of spin variables $S = \{S_1, \ldots, S_N\}$.

Now we come back to the question how a neural network can store items and how it can recall the items on the basis of an incomplete input. We restrict ourselves to the above-introduced simple spin glass model [64–66]. A pattern may be defined by a particular configuration $\sigma = \{\sigma_1, \sigma_2, ...\}$. Such a pattern is called a training pattern. Usually, we have to deal with more than one training pattern $\sigma^{(m)}$ with $m = 1, 2, \ldots, M$. Let us define the coupling constants as [67–70]

$$J_{\alpha\beta} = \frac{1}{N} \sum_{m=1}^{M} \sigma_\alpha^{(m)} \sigma_\beta^{(m)} \,. \tag{8.201}$$

The prefactor N^{-1} is just a convenient choice for defining the scale of the couplings. (8.201) is known as the Hebb rule. In the following discussion we set $\theta_\alpha = 0$, although the theory can also be worked without this simplification. Thus, because of (8.201), the Hamiltonian (8.198) becomes

$$H = -\frac{N}{2} \sum_{m=1}^{M} \left(\sigma^{(m)}, S\right)^2 \,, \tag{8.202}$$

where we have introduced the scalar product

$$(\sigma, \sigma') = \frac{1}{N} \sum_{\alpha=1}^{N} \sigma_\alpha \sigma'_\alpha \,. \tag{8.203}$$

In case of only one pattern, $M = 1$, the Hamiltonian can be written as $H = -N\left(\sigma^{(1)}, S\right)^2/2$. In other words, the configurations with the lowest energy ($H = -N/2$) are given by $S = \sigma^{(1)}$ and by $S = -\sigma^{(1)}$. Both states are visited with the highest probability in course of the random motion of the system through its phase space. Nevertheless, the dynamics of the system shows another remarkable feature. If the system has reached one of these ground states, say $\sigma^{(1)}$, it will occupy for a large time several states in the nearest environment of $\sigma^{(1)}$. The possibility that the system escapes from this basin of attraction and arrives the environment of the opposite ground state, $-\sigma^{(1)}$, is very small and decreases rapidly with decreasing temperature. It means that an initially given pattern $S(0)$ approaches for low temperatures T relatively fastly the nearest environment of that ground state $\sigma^{(1)}$ or $-\sigma^{(1)}$ which is the same basin of attraction as $S(0)$. Here, it will be present for a long time before a very rare set of suitable successive steps drives the system close to the opposite ground state. In other words, the system finds in a finite time with a very high probability that ground state and therefore that training pattern which is close to the initial state.

If we have a finite number $M \ll N$ of statistically independent training patterns, every one of them is a locally stable state. We remark that two patterns $\sigma^{(m)}$ and $\sigma^{(n)}$ are completely independent if the scalar product $\left(\sigma^{(m)}, \sigma^{(n)}\right)$ vanishes, $\left(\sigma^{(m)}, \sigma^{(n)}\right) = 0$. Statistic independence means that $\sigma^{(m)}$ and $\sigma^{(n)}$ represent two random series of values ± 1. Thus, we find the estimation $\left|\left(\sigma^{(m)}, \sigma^{(n)}\right)\right| \sim N^{-1/2}$. Let us set $S = \sigma^{(k)}$. Then we obtain from (8.202)

$$\begin{aligned} H &= -\frac{N}{2} \sum_{m=1}^{M} \left(\sigma^{(m)}, \sigma^{(k)}\right)^2 \\ &= -\frac{N}{2} \left[1 + \sum_{m \neq k} \left(\sigma^{(m)}, \sigma^{(k)}\right)^2\right] \\ &\approx -\frac{N}{2} + o\left(M\right) . \end{aligned} \tag{8.204}$$

It is simple to show that the training patterns $\sigma^{(m)}$ (and the dual patterns $-\sigma^{(m)}$) define the ground states of the Hamiltonian. It means that the thermodynamic evolution at sufficiently low temperatures of the neural network with a finite number of training patterns finds after a moderate period again the ground state which most resembles the initial state $S(0)$. That is the main property of an adaptive memory. Each configuration learned by the neural network is stored in the coupling constants (8.201). A given initial configuration $S(0)$ of the network is now interpreted as disturbed training pattern. The neural network acts to correct these errors in the input just by following its dynamics to the nearest stable state. Hence, the neural network assigns an input pattern to the nearest training pattern.

The neural network can still recall all M patterns (and the M dual patterns) as long as the temperature is sufficiently low and $M/N \to 0$ for $N \to \infty$. We remark that in case of $N \to \infty$ the system can no longer escape from the initially visited basin of attraction if the temperature is below a critical temperature T_c. It means that the system now always finds the ground state which is close to the initial state. The critical temperature is given by $T_c = 1$, i.e., for $T > 1$ the system always reaches the thermodynamic equilibrium. In other words, for $T > T_c$ the neural network behaves in a manner similar to a paramagnetic lattice gas and the equilibrium state favors no training patterns. On the other hand, for very low temperatures and a sufficiently large distance between the input pattern $S(0)$ and the training pattern, the dynamics of the system may lead the evolution $S(t)$ into spurious ghost states other than the training states. These ghost states are also minima of the free energy which occurs because of the complexity of the Hamiltonian (8.202). But it turns out that these ghost states are unstable above $T_0 = 0.46$. Hence, by choosing the temperature slightly above T_0, we can avoid these states while still keeping the training patterns stable.

Another remarkable situation occurs for $c = M/N > 0$. Here, the training states remain stable for a small enough c. But beyond a critical value $c^{\star}(T)$, they suddenly lose their stability and the neural network behaves like a real spin glass [71, 72]. Especially, the typical ultrametric structure of the spin glass states occurs in this phase. At $T = 0$, the curve $c^{\star}(T)$ reaches its maximum value of $c^{\star}(0) \approx 0.138$. For the completeness we remark that above a further curve, $c_p(T)$, the spin glass phase melts to a paramagnetic phase. However, both the spin glass phase and the paramagnetic phase are useless for an adaptive memory. Only the phase capturing the training patterns is meaningful for the application of neural networks.

Topology of Neural Networks

The above-discussed physical approach to neural networks is only a small contribution to the main stream of the mathematical and technical efforts concerning the development in this discipline. Beginning in the early sixties [73,74], the degree of scientific development of neural networks and the number of practical applications grow exponentially [68, 75–77, 80, 93].

In neural networks, computational models or nodes are connected through weights that are adapted during use to improve performance. The main idea is equivalent to the concept of cellular automata: a high performance occurs because of interconnection of the simple computational elements. A simple node labelled by α provides a linear combination of Γ weights $J_{\alpha 1}$, $J_{\alpha 2}, \ldots$, $J_{\alpha \Gamma}$ and Γ input values S_1, $S_2, \ldots$, S_Γ, and passes the result through a usually nonlinear transition or activation function ψ

$$\widehat{S}_\alpha = \psi\left(\sum_{\beta=1}^{\Gamma} J_{\alpha\beta} S_\beta\right) . \tag{8.205}$$

The function ψ is monotone and continuous, most commonly of a sigmoidal type. In this representation, the output of the neuron is a deterministic result $\widehat{S}_\alpha$, which may be a part of the input for the next node. In general, the output can be formulated also on the basis of probabilistic rules (see above).

The neural network not only consists of one node but is usually an interconnected set of many nodes as well. There is the theoretical experience that massively interconnected neural networks provide a greater degree of robustness than weakly interconnected networks. By robustness we mean that small perturbations in parameters and in the input data will result in small deviations of the output data from their nominal values.

Besides their node characteristics, neural networks are characterized by the network topology. The topology can be determined by the connectivity matrix Θ with the components $\Theta_{\alpha\beta} = 1$ if a link from the node α to the node

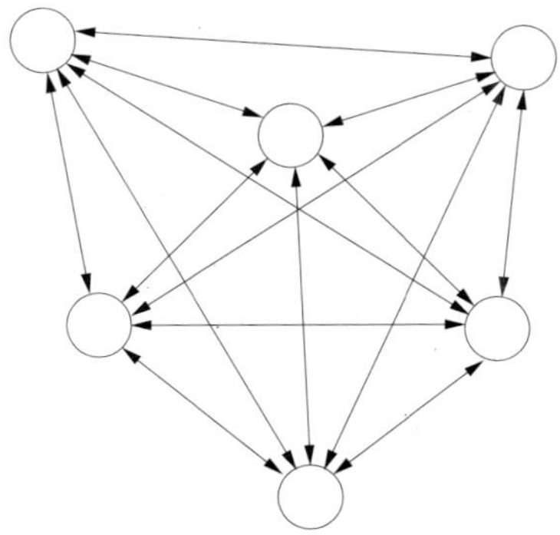

Fig. 8.5. The graph of a Hopfield network with 6 nodes

β exists, and $\Theta_{\alpha\beta} = 0$ otherwise. A link from α to β means that the output of α is the input of β.

Only such weights $J_{\alpha\beta}$ can have nonzero values which corresponds to the connectivity $\Theta_{\alpha\beta} = 1$. In other words, we may write

$$J_{\alpha\beta} = \Theta_{\alpha\beta} g_{\alpha\beta} , \tag{8.206}$$

where $\Theta_{\alpha\beta}$ is fixed by the respective network architecture and remains unchanged during the learning process, while the $g_{\alpha\beta}$ should capture the training patters.

Obviously, the connectivity matrix is not necessarily a symmetric one. We may describe this matrix symbolically by a corresponding network graph which consists of arrows and nodes. In particular, each arrow stands for an existing link, and the direction of the arrow indicates the flow of information. The above-discussed Hopfield network has the ideal connectivity $\Theta_{\alpha\beta} = 1$ for all $\alpha \neq \beta$. Thus, the topology of the Hopfield network represented by a graph in which each node is connected to each other node by a double arrow (Fig. 8.5). The dilution of such a topology by a random pruning procedure leads to a stochastic neural network or a so-called neural cluster. From the topological point of view, both types of neural networks distinguish not at all or only very weakly between input neurons and output neurons. The only exception is the case of a diluted network containing nodes with only outgoing arrows or only incoming arrows so that these nodes can be classified as input nodes or output nodes. Usually, these nodes are defined by the underlying program structure, but not by the topology of these networks.

Another versions of neural networks show a so-called layer structure, where the input nodes and output nodes can be identified on the basis of the topological structure. Formally, these networks consist of an input layer, several hidden layers, and an output layer (Fig. 8.6). Topologically, these neural networks contain no loops. Therefore, layer networks are sometimes denoted as filters or feedforward networks. The input pattern is transformed by determin-

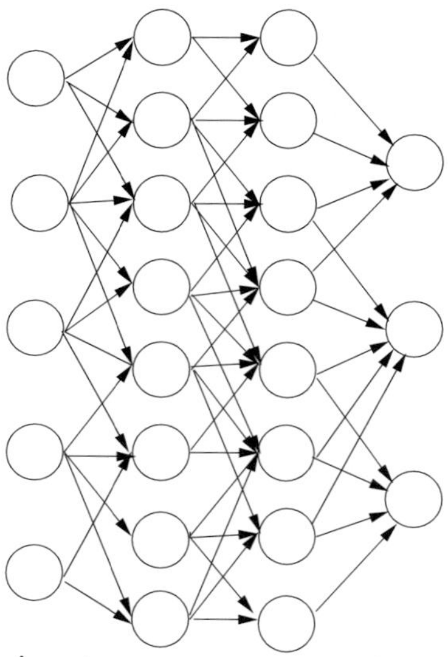

Fig. 8.6. Typical graph of a layer network

istic or, more rarely, by probabilistic rules into several intermediate patterns at the hidden layers and the final pattern at the output layer.

Modern layer networks imply several feedback mechanism between subsequent and previous layers. Therefore, we distinguish between two categories of neural networks: feedforward networks or filters without any loops and recurrent networks, where loops occur because of feedback connections. In other words, subsequent layers have the possibility to send data to previous layers which may be used for the change of the weights or of the activation functions of the previous layer in order to obtain an improved treatment of the next input. Another frequently used version consists in multiple restarts of the computation using the output of subsequent layers as a new input of previous layers. Such a technique can be used to stabilize the final output.

Between the Hopfield network and the feedforward network exist a lot of intermediate levels. The so-called Kohonen network [80] or feature map consists of a regular d-dimensional lattice and an input layer. Each node of the regular lattice is bidirectional connected with all nodes of a neighborhood shell, and each node of the input layer is connected by directed links with all nodes of the Kohonen layer. The important property of such a network is that at the end of the computation steps the node with the largest output is set to 1 while all other nodes are defined to be zero. Thus, a Kohonen network can be used for the classification of incoming patterns.

The bidirectional associative memory [81] consists of two layers, the input and the output layers. All components of the connectivity matrix corresponding to links between both layers have the value 1, while all other coefficients vanish. Thus, the network topology of such a network is characterized by a symmetric matrix. In a manner similar to the Hopfield model, the bidirectional associative memory approaches a stationary state after a sufficiently large number of iterative computation steps with the difference that for odd steps the data flow from the input to the output nodes while a data backflow from the output nodes to the input nodes occurs for even computation steps.

Other neural networks, for instance, the adaptive resonance network [75] or the learning vector quantizers [82], are further realizations of combinations of layer structures.

Training of Neural Networks

A neural network is characterized by its topology and its node characteristics and the training patterns captured in the values of the weights $J_{\alpha\beta}$. The remaining question is, how can a neural network store the training patterns? As discussed above, the problem can be solved straightforwardly for a Hopfield network. A similar situation occurs for the bidirectional adaptive memory. But other networks with complicated loops and asymmetric connectivity matrices need a special learning procedure in order to prepare the originally nonspecified system for the subsequent working phase. The training requires a sufficiently strong adaptability of the network.

In general, adaptability may be interpreted as the ability to react to changes in their environment through a learning process [79]. In our case, the environment of a neural network is given by a real system, for example, a market, the internal dynamics of which is widely unknown. In order to use a neural network for predictions, the neural network is fed with all (or a limited set of the) historical observations $Y(t_1)$, $Y(t_2)$, ...$Y(t_L)$, which we know from the dynamics of the real system at every discrete time step t_n. The output of the neural system may be $\widehat{Y}(t_{n+1})$ while $Y(t_{n+1})$ is the response of the unknown system. The error signal $e(t_{n+1})$ is formed as the difference of both output signals, $e(t_{n+1}) = \widehat{Y}(t_{n+1}) - Y(t_{n+1})$, and the parameters of the weights of the neural network are adjusted using this error information.

The aim of a learning procedure is to update iteratively the weights $J_{\alpha\beta}(t_n)$ of an adaptive system at each time step t_n so that a nonnegative error measure $\mathcal{E}$ is reduced at each time step t_n, $\mathcal{E}(J(t_{n+1})) \leq \mathcal{E}(J(t_n))$. This will generally ensure that after the training process, the neural network has captured the relevant properties of the unknown system that we are trying to model. Using $\Delta J(t_n) = J(t_{n+1}) - J(t_n)$, we obtain

$$\begin{aligned}\Delta\mathcal{E}(J(t_n)) &= \mathcal{E}(J(t_{n+1})) - \mathcal{E}(J(t_n)) \\ &= \sum_{\alpha\beta} \left.\frac{\partial\mathcal{E}(J)}{\partial J_{\alpha\beta}}\right|_{J=J(t_n)} \Delta J_{\alpha\beta}(t_n)\end{aligned} \tag{8.207}$$

and therefore

$$\sum_{\alpha\beta} \left.\frac{\partial \mathcal{E}(J)}{\partial J_{\alpha\beta}}\right|_{J=J(t_n)} \Delta J_{\alpha\beta}(t_n) \leq 0 . \tag{8.208}$$

This equation is always fulfilled for the special choice

$$\Delta J_{\alpha\beta}(t_n) = -\Lambda \left.\frac{\partial \mathcal{E}(J)}{\partial J_{\alpha\beta}}\right|_{J=J(t_n)} , \tag{8.209}$$

where Λ is a small positive scalar called the learning rate or the adaptation parameter. A learning procedure controlled by (8.209) is also denoted as gradient-descent–based learning process. We remark that gradient-based algorithms inherently forget old data, which have particular importance for performance of the learning procedure.

The quasi-Newton learning algorithm bases on the second-order derivative of the error function. If we expand the error function in a Taylor series, we have

$$\begin{aligned}\Delta\mathcal{E}(J(t_n)) = &\sum_{\alpha\beta} \left.\frac{\partial \mathcal{E}(J)}{\partial J_{\alpha\beta}}\right|_{J=J(t_n)} \Delta J_{\alpha\beta}(t_n) \\ &+\frac{1}{2}\sum_{\alpha\beta\gamma\delta} \left.\frac{\partial^2 \mathcal{E}(J)}{\partial J_{\alpha\beta}\partial J_{\gamma\delta}}\right|_{J=J(t_n)} \Delta J_{\alpha\beta}(t_n)\Delta J_{\gamma\delta}(t_n) .\end{aligned} \tag{8.210}$$

Using the extremum condition $\partial\Delta\mathcal{E}(J(t_n))/\partial\Delta J_{\alpha\beta}(t_n) = 0$, we get the changes

$$\Delta J_{\alpha\beta}(t_n) = -\sum_{\gamma\delta} \left[\left(\frac{\partial}{\partial J} \circ \frac{\partial}{\partial J}\mathcal{E}(J)\right)^{-1}\right]_{\alpha\beta\gamma\delta} \left.\frac{\partial \mathcal{E}(J)}{\partial J_{\gamma\delta}}\right|_{J=J(t_n)} . \tag{8.211}$$

As a simple example, let us calculate the changes $\Delta J_{\alpha\beta}(t_n)$ for a neural network with only one node and an input vector of dimension Γ. Such a simple neural network is denoted as perceptron. The error function may be given by

$$\mathcal{E} = e^2(t_n) = \left[r(t_n) - \psi\left(\sum_{\beta=1}^{\Gamma} J_\beta(t_n) Y_\beta(t_n)\right)\right]^2 \tag{8.212}$$

with $J_\beta = J_{1\beta}$. Therefore, we obtain

$$\frac{\partial \mathcal{E}}{\partial J_\alpha(t_n)} = -2e(t_n)\psi'\left(\sum_{\beta=1}^{\Gamma} J_\beta(t_n) Y_\beta(t_n)\right) Y_\alpha(t_n) , \tag{8.213}$$

and the gradient-descent–based learning process is defined by the equation

$$J_\alpha(t_{n+1}) = J_\alpha(t_n) + 2\Lambda\psi'\left(\sum_{\beta=1}^{\Gamma} J_\beta(t_n) Y_\beta(t_n)\right) e(t_n) Y_\alpha(t_n) . \tag{8.214}$$

When deriving a learning algorithm for a general neural network, the network architecture should be taken into account. This leads, of course, to relative complicated nonlinear equations, which must be treated during the training procedure of a network.

In principle, the above-introduced learning algorithms are special procedures referring to the class of adaptive learning. Roughly speaking, the idea behind this concept is to forget the past when it is no longer relevant and adapt to the changes in the environment. We remark that the term gear-shifting is sometimes used for the above-discussed gradient-descent–based learning when the learning rate is changed during training. Another popular learning algorithm is deterministic and stochastic learning methods [2–4]. Finally, we mention another learning procedure which is called the constructive learning. This modern version deals with the change of architecture or topological interconnections in the network during training. Neural networks for which the topology can change in course of the learning procedure are called ontogenic neural networks [5]. The standard procedures of constructive learning are network growing and network pruning. The growing mechanism begins with a very simple network, and if the error is too big, new subnetwork units or single network units are added to the network [6]. In contrast, network pruning starts from a large neural network and if the error is smaller as a lower limit, the size of the network is reduced [7, 8].

References

1. B. Øksendal, A. Sulem: *Applied Stochastic Control of Jump Diffusion* (Springer, Berlin Heidelberg New York, 2005)
2. S. Kirkpatrick, C.D. Gelatt Jr, M.P. Vecchi: Science **220**, 671 (1983)
3. K. Rose: Proc. IEEE **86**, 2210 (1998)
4. H. Szu, R. Harley: Proc. IEEE **75**, 1538 (1987)
5. E. Fiesler, R. Beale: *Handbook of Neural Computation* (Oxford University Press, Oxford, 1997)
6. M. Hochfeld, S.E. Fahlman: IEEE Trans. Neural Networks **3**, 603 (1992)
7. R. Reed: IEEE Trans. Neural Networks **4**, 740 (1993)
8. J. Sum, C.S. Leung, G.H. Young, W.K. Kan: IEEE Trans. Neural Networks **10**, 161 (1999)
9. G.U. Yule: Phil. Trans. R. Soc. London A **226**, 267 (1927)
10. G.E.P. Box, G.M. Jenkins: *Time Series Analysis: Forcasting and Control* (Holden-Day, New York, 1976)
11. L. Ljung, T. Soderstrom: IEEE Trans. Neural Networks **5**, 803 (1983)
12. J. Makhoul: Proc. IEEE **63**, 561 (1995).
13. B.B. Mandelbrot: *The Fractal Geometry of Nature* (W.H. Freeman, San Francisco, CA, 1982)
14. A.N. Kolmogorov: Dokl. Akad. Nauk. SSSR **30**, 9 (1941)
15. D. Zajdenweber: Fractals **3**, 601 (1995)
16. D. Zajdenweber: Risk and Insurance **63**, 95 (1996)
17. V.F. Pisarenko, Hydrol. Proc. **12**, 461 (1998)
18. W. Feller: *An Introduction to Probability Theory and Its Applications*, vol. 1, 3rd edn (Wiley, New York, 1968)

19. P.A. Samuelson: J. Econ. Literature **15**, 24 (1977)
20. A.N. Kolmogorov: Dokl. Akad. Nauk. SSSR **31**, 9538 (1941)
21. A.N. Kolmogorov: Dokl. Akad. Nauk. SSSR **32**, 16 (1941)
22. S.P. Nishenko, C.C. Barton: Geol. Soc. Am., Abstracts with Programs **25**, 412 (1993)
23. D. Zajdenweber: *Hasard et Prévision* (Economica, Paris, 1976)
24. D. Zajdenweber: Scale invariance in economics and finance. In: *Scale Invariance and Beyond* ed by B. Dubrulle, F. Graner, D. Sornette (EDP Sciences and Springer, Berlin Heidelberg New York, 1997)
25. D. Sornette, C. Vanneste, L. Knopoff: Phys. Rev. A **45**, 8351 (1992)
26. D. Sornette, A. Sornette: Bull. Seism. Soc. Am. **89**, 1121 (1999)
27. P. Protter: Stochastic Integration and Differential Equations, 2nd edn (Springer, Berlin Heidelberg New York, 2003)
28. K. Sato: *Lévy Processes and Infinitely Divisible Distributions* (Cambridge University Press, Cambridge, 1999)
29. V.K. Vijay: *An Introduction to Probability Theory and Mathematical Statistics* (Wiley, New York, 1976)
30. E.J. Dudewicz: *Modern Mathematical Statistics* (Wiley, New York, 1988)
31. D. Sornette: *Critical Phenomena in Natural Sciences* (Springer, Berlin Heidelberg New York, 2000)
32. J. Zinn-Justin: *Quantum Field Theory and Critical Phenomena* (Claredon Press, Oxford, 1990)
33. P.L. Chebyshev: Acta Math. **14**, 305 (1890)
34. B.V. Gnedenko, A.N. Kolmogorov: *Limit Distribution for Sum of Independent Random Variables* (Addison-Wesley, Reading, MA, 1954)
35. A.C. Berry: Trans. Am. Soc. **49**, 122 (1941)
36. C.G. Esséen: Acta Math. **77**, 1 (1945)
37. W. Feller: *An Introduction to Probability Theory and Its Applications*, vol 2, 2nd edn (Wiley, New York, 1971)
38. P. Lévy: *Caleul des prohabilités* (Gauthier-Villars, Paris, 1925)
39. A.Ya. Khintchine, P. Lévy: C. R. Acad. Sci. Paris **202**, 374 (1936)
40. G. Samorodnitsky, M.S. Taqqu: *Stable Non-Gaussian Random Processes: Stochastic Models with Infinite Variance* (Chapman and Hall, New York, 1994)
41. B.B. Mandelbrot: Science **279**, 783 (1998)
42. I. Koponen: Phys. Rev. E **52**, 1197 (1995)
43. R.N. Mantegna, H.E. Stanley: Phys. Rev. Lett. **73**, 2946 (1994)
44. O.E. Lanford: Entropy and equilibrium states in classical mechanics. In: *Statistical Mechanics and Mathematical Problems*, Lecture Notes in Physics, vol. 20, ed by A. Lenard (Springer, Berlin Heidelberg New York), p. 1
45. U. Frisch: *Turbulence, The Legacy of A.N. Kolmogorov* (Cambridge University Press, Cambridge, 1995)
46. U. Frisch, D. Sornette: J. Phys. I France **7**, 1155 (1997)
47. W.M. Troburn: Mind **23**, 297 (1915)
48. W.M. Troburn: Mind **26**, 345 (1918)
49. R.F. Engle: Econometrica **50**, 987 (1982)
50. T. Bollerslev: J. Econometrics **31**, 307 (1986)
51. T. Bollerslev, R.Y. Chou, K.F. Kroner: J. Econometrics **52**, 5 (1992)
52. T. Bollerslev, R.F. Engle, D.B. Nelson: ARCH models. In: *Handbook of Econometrics*, vol. 4, ed by R.F. Engle, D.L. McFadden (Elsevier, North-Holland, 1994)

53. M.S. Geisel: Bayesian comparisons of simple macroeconomic models. In: *Studies in Bayesian Econometrics and Statistics*, ed by S. Feinberg, A. Zellner (North-Holland, Amsterdam, 1974)
54. M.S. Geisel: Comparing and choosing among parametric statistical models: a Bayesian analysis with macroeconomic applications. PhD dissertation, University of Chicago (1970)
55. A. Zellner: *An Introduction to Bayesian Inference in Econometrics* (Wiley, New York, 1971)
56. A. Zellner: *Basic Issues in Econometrics* (University of Chicago Press, Chicago, 1984)
57. J. Picard: Statistical Learning Theory and Stochastic Optimization (Springer, Berlin Heidelberg New York, 2000)
58. R.M. Dillon, C.N. Manikopoulos: Electron. Lett. **27**, 824 (1991)
59. C.R. Gent, C.P. Sheppard: Comput. Control Eng. J. 109 (1992)*Au: Pelase supply a volume number for this reference.*
60. B. Townshend: Signal-Processing ICASSP **91**, 429 (1991)
61. N.A. Gershenfeld, A.S. Weigend: The future of time series: learning ansd understanding. In: *Time Series Prediction: Forecasting the Future and Understanding the Past*, ed by A.S. Weigend, N.A. Gershenfeld (Addison-Wesley, Reading, MA, 1993)
62. S.W. Kuffler, J.G. Nichols, A.R. Martin: *From Neuron to Brain* (Sinauer Associates, Sunderland, MA, 1984)
63. E.R. Kandel, J.H. Schwartz: *Principles of Neural Science* (Elsevier, Amsterdam, 1985)
64. G. Parisi: Phys. Rev. Lett. **43**, 1754 (1979)
65. K.H. Fischer, J.A. Hertz: *Spin Glasses* (Cambridge University Press, Cambridge, 1991)
66. G. Parisi: J. Phys. A **13**, 1101 (1980)
67. D.O. Hebb: *The Organization of Behavior* (Wiley, New York, 1949)
68. J.J. Hopfield: Proc. Natl Acad. Sci. USA **79**, 2554 (1982)
69. J.J. Hopfield: Proc. Natl Acad. Sci. USA **81**, 3088 (1984)
70. D.J. Amit, H. Gutsfreund, H. Sompolinsky: Phys. Rev. A **32**, 1007 (1985)
71. D.J. Amit, H. Gutsfreund, H. Sompolinsky: Phys. Rev. Lett. **55**, 1530 (1985)
72. D.J. Amit, H. Gutsfreund, H. Sompolinsky: Ann. Phys. (NY) **173**, 30 (1987)
73. F. Rosenblatt: *Principles of Neurodynamics* (Spartan, Washington, DC, 1962)
74. B. Widrow, M.E. Hoff: Proc. WESCON Convention **4**, 96 (1960)
75. S. Grossberg: Prog. Theor. Biol. **3**, 51 (1974)
76. D.E. Rumelhart, G.E. Hinton, R. Williams: Nature **323**, 533 (1986)
77. B. Widrow, M.E. Hoff: Proc. IEEE **78**, 1415 (1990)
78. T. Kohonen: Biol. Cybernet. **43**, 59 (1982)
79. S. Haykin: IEEE Signal Proces. Mag. **15**, 66 (1999)
80. T. Kohonen: Biol. Cybernet. **43**, 59 (1982)
81. U. Blien, H.-G. Lindner: Jahrbücher für Nationalökonomie und Statistik **212**, 497 (1993)
82. M. Pytlik: *Diskriminierungsanalyse und künstliche Neuronale Netze zur Klassifizierung von Jahresabschlüssen* (Peter Lang GmbH, Frankfurt, 1995)
83. N. Metropolis, S. Ulam: J. Am. Statist. Assoc. **44**, 335 (1949)
84. G. Peng, H.J. Heermann: Phys. Rev. E **49**, 1796 (1994)
85. M. Schulz, S. Trimper: J. Phys. A: Math. Gen. **33**, 7289 (2000)

86. G.B. Ermentrout, L. Edlestein-Keshet: J. Theoret. Biol. **160**, 97 (1993)
87. W.S. McCullough, W. Pitts: Bull. Math. Biophys. **5**, 115 (1943)
88. E.R. Caianiello: J. Theor. Biol. **1**, 204 (1961)
89. W.A. Little: Math. Biosci. **109**, 101 (1974)
90. A. Saberi, P. Sannuti, B.M. Chen: *H_2 Optimal Control* (Prentice-Hall, New York, 1995)
91. P. Colaneri, J.C. Geromel, A. Locatelli: *Control Theory and Design* (Academic, London, 1997)
92. J.H. Davis: *Foundations of Deterministic and Stochastic Control* (Birkhäuser, Basel, 2002)
93. D.J. Burr: Artificial neural networks: a decade of progress. In: *Artificial Neural Networks for Speech and Vision*, ed by R.J. Mammone (Chapman and Hall, New York, 1993)

9

Game Theory

9.1 Unpredictable Systems

All systems analyzed up to now were more or less predictable. That means, we had some information about the initial state and the dynamics of the system. But also in case of insufficient information about the internal dynamics of the system, we have always supposed that the knowledge of the history allows us to conclude at least partially the future evolution of the system. The key for these previous considerations was the suggestion that the evolution of a certain system is always determined by a set of deterministic coupled, but partially hidden degrees of freedom. The strength of the interaction between the measurable relevant quantities and the nonobservable, but always present irrelevant degrees of freedom determines which control concept is appropriate. In case of no irrelevant variables, we expect to deal with a deterministic system.

However, there exist also nondeterministic systems without an intrinsic dynamics of hidden variables. In principle, all quantum mechanical systems are suitable candidates for this class of problems. The outcome of a quantum mechanical experiment has, in combination with the measurement instruments, often a pronounced random character. The Einstein–Rosen–Podolski–Gedanken experiment [1], specified by a practicable realization [6], and the application of Bell's inequality [7–9], lead to the intensively experimentally [10] proved statement that a local deterministic theory using hidden parameters is not able to reproduce the observed quantum mechanical results.

Let us now give some concepts, how such an unpredictable system can be controlled. For the sake of simplicity, we assume that the system has a discrete number of outcomes which we denote as system states. Furthermore, the system may be embedded in an environment which may be characterized also by a finite set of different states. These states are also called channels. The controller may now open one arbitrary channel while all other channels are closed. In other words, the controller is able to fix the environment of the system in a certain sense. The aim of the controller is to choose such a channel

M. Schulz: *Control Theory in Physics and other Fields of Science*
STMP **215**, 265–277 (2006)

that the nonpredictable outcome of the system leads to the best result for the control problem in mind. In other words, the controller must make a certain decision, or the controller takes a certain action. The control concept requires that this action is done before the system realizes the output.

On the other hand, we may also interpret the system as a decision maker. It has its own set of actions, the outcomes, and can choose them in a way that interferes with the achievements of the controller. That is because the action of the system follows the action of the controller which may have changed the constraints of the system under control.

A typical quantum mechanical example illustrating this situation is the traditional double slit experiment (Fig. 9.1).

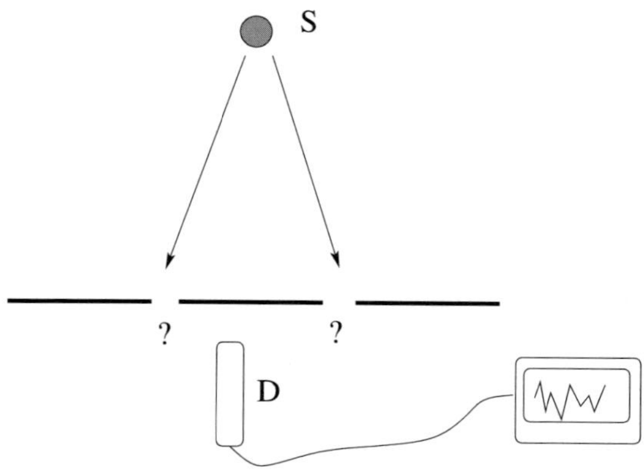

Fig. 9.1. A quantum mechanical problem in the language of the game theory: at which slide should one locate the detector D in order to measure a quantum particle from the source S?

There is a possibility for the controller to position a detector at the right or left slit while the system takes the action that a particle passes through either the right or the left slit. If the actions of the controller and the system match, we observe a detector signal, otherwise we do not.

This allows us to introduce two action spaces, namely:

- the controller action space $\mathbb{U}$, where each $u \in \mathbb{U}$ is referred to as a control action or as an open channel,
- the system action space $\mathbb{S}(u)$, where each $X \in \mathbb{S}(u)$ is referred to as a system action or a system state.

Note that the designation $\mathbb{S}(u)$ takes into account that the system can "know" the action of the controller so that its action space may depend on the actual configuration of the controller. This is a very natural situation for many quantum mechanical processes. The setting of a certain channel corresponds to a change of the common system composed of the actual quantum system and the controller. Thus, the quantum system may have different outcomes for different states of the controller.

The last ingredient we need for a control is the formulation of the control aim. This is again defined by performance or costs. We suppose that the cost depends on which actions were chosen by both the controller and the system. Thus we have a performance function $J(u, X)$ with

$$J : \mathbb{U} \times \mathbb{S}(u) \to \mathbb{R} \tag{9.1}$$

Since we have discrete sets, the performance is often represented by a matrix. For example, suppose that $\mathbb{U}$ and $\mathbb{S}$ each contains three actions. This results in nine possible outcomes, which can be specified by the following cost matrix:

$$\begin{array}{cc} \mathbb{U} & \begin{pmatrix} J_{11} & J_{12} & J_{13} \\ J_{21} & J_{22} & J_{23} \\ J_{31} & J_{32} & J_{33} \end{pmatrix} \\ & \mathbb{S} \end{array} \tag{9.2}$$

The controller action, u, selects a row and the system action, X, selects a column of the cost matrix. The resulting cost is given by the corresponding matrix element.

From this point of view, the control problem can be interpreted as a game of the controller against the system. Therefore, it seems reasonable to solve the optimal control problem by minimizing the performance in the framework of the game theory. From this point of view, we suggest that the game theoretical control concept is slightly different from the ideas presented in the previous chapters. Obviously, the controller does not modify the system dynamics as was the case up to now. Rather, the controller cooperates with the system in a special manner. In fact, the controller chooses its decisions in such a manner that these together with the more or less open dynamics of the system yield the expected control aim.

9.2 Optimal Control and Decision Theory

9.2.1 Nondeterministic and Probabilistic Regime

What is the best decision for the controller in its game against the system? There are two general possibilities: either the controller knows the probability with which the system answers its action or it does not. The first case defines the probabilistic regime while the second case is the complete nondeterministic regime [11]. The latter case occurs especially if only few or no observations about the system actions are available so that we cannot estimate the probability distribution of the system outcomes.

Under the non-deterministic regime, there is no additional information other than the knowledge of the actions and the cost matrix. The only reasonable approach for the controller is to make a decision by assuming the worst case. This pessimistic position is often humorously referred to as Murphy's law[1] [12]. Hence, the optimal decision of the controller is given by

$$u^* = \arg\min_{u \in \mathbb{U}} \left(\max_{X \in \mathbb{S}(u)} J(u, X) \right) \tag{9.3}$$

The optimal action u^* may be interpreted as the lowest-cost choice under a worst-case assumption.

The probabilistic regime is applicable if the controller has gathered enough data to reliably estimate the conditional probability $P(X \mid u)$ of a system action X under the condition that the controller has taken the action u. This formulation implies that we consider a stationary system. We use the expected case assumption and conclude

$$u^* = \arg\min_{u \in \mathbb{U}} \left(\left. \overline{J(u, X)} \right|_u \right) \tag{9.4}$$

with the conditional average

$$\left. \overline{J(u, X)} \right|_u = \sum_{X \in \mathbb{S}(u)} J(u, X) P(X \mid u) \tag{9.5}$$

For an illustration, let us consider a 3×3 cost matrix

$$\mathbb{U} \left\{ \begin{matrix} 1 \\ 2 \\ 3 \end{matrix} \right. \underbrace{\begin{pmatrix} 1 & -1 & 5 \\ 4 & 0 & -2 \\ 2 & 0 & -1 \end{pmatrix}}_{\substack{1 \; 2 \; 3 \\ \mathbb{S}}} \tag{9.6}$$

The worst-case analysis requires

$$\max_{X \in \mathbb{S}} J(1, X) = 5 \qquad \max_{X \in \mathbb{S}} J(2, X) = 4 \qquad \max_{X \in \mathbb{S}} J(3, X) = 2 \tag{9.7}$$

and therefore $u^* = 3$. On the other hand, the probabilistic regime requires the knowledge of probabilities. Let us assume that the actions of the system and the controller are independent of each other. Thus we have $P(X \mid u) = P(X)$. With the special choice $P(X = 1) = 0.1$, $P(X = 2) = 0.6$, and $P(X = 3) = 0.3$, we obtain

$$\left. \overline{J(1, X)} \right|_{u=1} = 1.0 \qquad \left. \overline{J(2, X)} \right|_{u=2} = -0.2 \qquad \left. \overline{J(3, X)} \right|_{u=3} = -0.1 \tag{9.8}$$

so that $u^* = 2$. The best decision in case of the probabilistic regime depends on the probability distribution. For instance, in case of $P(X = 1) = P(X = 2) = P(X = 3) = 1/3$, our example yields $u^* = 3$.

[1] If anything can go wrong, it will.

9.2.2 Strategies

Suppose the controller has the possibility to receive information characterizing the current state of the system immediately before opening a channel. These observations may allow the controller to improve its decision with respect to minimization of the costs.

For convenience, we suppose that the set $\mathbb{O}$ of possible observations Y is finite. The set $\mathbb{O}(X) \subseteq \mathbb{O}$ indicates the possible observations $Y \in \mathbb{O}(X)$ under the consideration that the subsequent system action is X. Furthermore, in the case of the probabilistic regime the conditional probabilities $P(Y \mid X)$ are available. The likelihood $P(Y \mid X)$ suggests the observation Y before the system action X occurs.

A strategy is a function θ connecting a given observation Y of the system with the controller decision, i.e., $u = \theta(Y)$. In other words, for each observation Y the strategy θ provides an action to the controller in order to minimize the costs.

Our aim is now to find the optimal strategy. In the case of the nondeterministic model, the sets $\mathbb{O}(X)$ must be used to determine the allowed system actions. That means, we have to determine the sets

$$\mathbb{S}(Y) = \{X \in \mathbb{S} \mid Y \in \mathbb{O}(X)\} \tag{9.9}$$

Then, the optimal strategy is

$$\theta^*(Y) = \arg\min_{u \in \mathbb{U}} \left(\max_{X \in \mathbb{S}(Y)} J(u, X) \right) \tag{9.10}$$

Obviously, the advantage of having the observation Y is that the set of available system states is reduced to $\mathbb{S}(Y) \subseteq \mathbb{S}$. The probabilistic regime requires the considerations of the above mentioned conditional probabilities. For the sake of simplicity, we restrict ourselves to the case that the system action depend does not on the controller action[2], i.e., $P(X \mid u) = P(X)$. Using the Bayes theorem (8.187), we get

$$P(X \mid Y) = \frac{P(Y \mid X)P(X)}{\sum\limits_{X' \in \mathbb{S}} P(Y \mid X')P(X')} \tag{9.11}$$

Note that $P(X \mid Y)$ is again an "a posteriori" probability in the sense of Bayesian statistics, which represents the probability that the system takes the action X after we have measured the observation Y. In the same context, the $P(X)$ are the corresponding "a priori" probabilities. The optimal strategy is then

$$\theta^*(Y) = \arg\min_{u \in \mathbb{U}} \left(\overline{J(u, X)}\Big|_Y \right) \tag{9.12}$$

[2] Otherwise, we need further information about the probability $P(X \mid u, Y)$ that a system takes the state X after the observation Y and the subsequent opening of channel u by the controller.

with the conditional Bayes' risk

$$\overline{J(u,X)}\Big|_Y = \sum_{X\in\mathbb{S}} J(u,X)P(X \mid Y) \tag{9.13}$$

Using (9.11), we may also write

$$\overline{J(u,X)}\Big|_Y = \frac{\sum_{X\in\mathbb{S}} J(u,X)P(Y \mid X)P(X)}{\sum_{X'\in\mathbb{S}} P(Y \mid X')P(X')} \tag{9.14}$$

and therefore

$$\begin{aligned}\theta^*(Y) &= \arg\min_{u\in\mathbb{U}} \left(\frac{\sum_{X\in\mathbb{S}} J(u,X)P(Y \mid X)P(X)}{\sum_{X'\in\mathbb{S}} P(Y \mid X')P(X')} \right) \\ &= \arg\min_{u\in\mathbb{U}} \left(\sum_{X\in\mathbb{S}} J(u,X)P(Y \mid X)P(X) \right)\end{aligned} \tag{9.15}$$

The problem can be extended to the case of multiple observations before the controller opens a certain channel and the system answers with its action. In this case the controller measures L observations, $Y_1,\ldots,Y_L$; each is assumed to belong to an observation space $\mathbb{O}_i$ $(i = 1,\ldots,L)$. The strategies now depend on all observations

$$\theta : \ \mathbb{O}_1 \times \mathbb{O}_2 \times \ldots \times \mathbb{O}_L \to \mathbb{U} \tag{9.16}$$

The nondeterministic regime requires the selection of all admissible X which belong to the observation set. This requires the knowledge of the subsets

$$\mathbb{S}(Y_i) = \{X \in \mathbb{S} \mid Y_i \in \mathbb{O}_i(X)\} \tag{9.17}$$

which may be used to construct

$$\mathbb{S}(Y_1,Y_2,\ldots,Y_L) = \mathbb{S}(Y_1) \cap \mathbb{S}(Y_2) \cap \cdots \cap \mathbb{S}(Y_L) \tag{9.18}$$

Thus, the optimal strategy for the nondeterministic regime is given by

$$\theta^*(Y_1,Y_2,\ldots,Y_L) = \arg\min_{u\in\mathbb{U}} \left(\max_{X\in\mathbb{S}(Y_1,Y_2,\ldots,Y_L)} J(u,X) \right) \tag{9.19}$$

The probabilistic regime can be extended in the same way. For simplicity, we assume that the observations are conditionally independent events. That means we have

$$P(Y_1,Y_2,\ldots,Y_L \mid X) = \prod_{k=1}^{L} P(Y_k \mid X) \tag{9.20}$$

$$P(X \mid Y_1,Y_2,...,Y_L) = \frac{\prod_{k=1}^{L} P(Y_k \mid X)P(X)}{\sum_{X'\in\mathbb{S}} \prod_{k=1}^{L} P(Y_k \mid X')P(X')} \tag{9.21}$$

Following the same steps which led to (9.15), we now arrive at

$$\theta^*(Y) = \arg\min_{u \in \mathbb{U}} \left(\sum_{X \in \mathbb{S}} J(u, X) \prod_{k=1}^{L} P(Y_k \mid X) P(X) \right) \tag{9.22}$$

We remark that the conditional independence between the observations is an additional assumption which we have used for a better illustration. However, this simplification is often used in practice, since the estimation of the complete conditional probabilities $P(Y_1, Y_2, \ldots, Y_L \mid X)$ requires a large record of observations which is not always available.

Finally, we stress again on the specific feature of a control on the basis of game theoretical concepts. In contrast to the concepts discussed above, the controller does not force the system to a special (optimal) dynamics. The controller chooses its actions in such a way that these decisions together with the free, but nearly unknown dynamics of the system lead to the expected control aim.

9.3 Zero-Sum Games

9.3.1 Two-Player Games

Many-player games are not often used for the control of a more or less passive system. But these games are often used for modeling the intrinsic, partially competing control mechanisms of systems with a very high degree of complexity. In this sense, what follows may be understood as a part of control theory. We now focus on two-player games. For the case of many players we refer to the special literature [13–16].

Suppose there are two players making their own decisions. Each player has a finite set of actions $\mathbb{U}_1$and $\mathbb{U}_2$. Furthermore, each player has a cost function $J_i(u_1, u_2)$ with $u_i \in \mathbb{U}_i$ $(i = 1, 2)$. A zero-sum game is then given by

$$J_1(u_1, u_2) + J_2(u_1, u_2) = 0 \tag{9.23}$$

That means, the cost for one player is a reward for the other. Obviously, in zero-sum games the interests of the players are completely opposed. Controllers with such properties are relatively rare. Because the theory of zero-sum game is very clear, this concept is often used also when it is partially incorrect, just to exploit the known results.

The goal of both players is to minimize their costs under symmetric conditions. That means, both players make their decisions simultaneously. Furthermore, it is assumed that the players know the cost functions[3] and that both opponents follow a reasonable concept. The latter condition requires that each player is interested to obtain the best cost whenever possible.

[3] This implies that each player knows the intentions of the opponent.

9.3.2 Deterministic Strategy

In order to obtain a solution of the two-player zero-sum game, we use the worst-case concept. From the viewpoint of player 1, the opponent is assumed to act similar to the passive system under the nondeterministic regime. Thus, we have

$$u_1^* = \arg\min_{u_1 \in \mathbb{U}_1} \left(\max_{u_2 \in \mathbb{U}_2} J_1(u_1, u_2) \right) \tag{9.24}$$

Because of the symmetry of the game, we obtain immediately

$$u_2^* = \arg\min_{u_2 \in \mathbb{U}_2} \left(\max_{u_1 \in \mathbb{U}_1} J_2(u_1, u_2) \right) \tag{9.25}$$

or equivalently

$$u_2^* = \arg\max_{u_2 \in \mathbb{U}_2} \left(\min_{u_1 \in \mathbb{U}_1} J_1(u_1, u_2) \right) \tag{9.26}$$

The optimal actions $u^{*(1)}$ and $u^{*(2)}$ are also called security strategies. The solution of this deterministic strategy problem must not lead to a unique solution. For instance, the cost matrix J_1

$$\mathbb{U}_1 \left\{ \begin{matrix} 1 \\ 2 \\ 3 \end{matrix} \right. \underbrace{\begin{pmatrix} 1 & 0 & -1 \\ -2 & 1 & 0 \\ 1 & 0 & 2 \end{pmatrix}}_{\substack{1 \quad 2 \quad 3 \\ \mathbb{U}_2}} \tag{9.27}$$

has the solutions $u_1^* = 1$ and $u_1^* = 2$ while the same problem gives a unique solution for the second player, $u_2^* = 2$. This uncertainty cannot be solved in the context of a deterministic strategy. Here, we need the probabilistic concept presented below. However, we can define the estimate of the upper value of the game from the viewpoint of player 1. This is simply the border J_+ defined by

$$J_+ = \max_{u_2 \in \mathbb{U}_2} J_1(u_1^*, u_2) \tag{9.28}$$

while the lower value is given by

$$J_- = \min_{u_1 \in \mathbb{U}_1} J_1(u_1, u_2^*) \tag{9.29}$$

In our example, we have $J_+ = 1$ and $J_- = 0$. Then, we have the inequalities

$$J_- \leq J_1(u_1^*, u_2^*) \leq J_+ \tag{9.30}$$

A unique solution for both players requires $J_- = J_+$. In this case, the security strategies are denoted as a saddle point of the game. A saddle point always requires

$$J_1(u_1^*, u_2^*) = \min_{u_1 \in \mathbb{U}_1} \left(\max_{u_2 \in \mathbb{U}_2} J_1(u_1, u_2) \right)$$
$$= \max_{u_2 \in \mathbb{U}_2} \left(\min_{u_1 \in \mathbb{U}_1} J_1(u_1, u_2) \right) \tag{9.31}$$

A saddle point is sometimes interpreted as an equilibrium of the game, because both players have no interest to change their choices. We remark that a system can have multiple saddle points. A simple example is given by the cost matrix J_1

$$\mathbb{U}_1 \begin{cases} 1 \\ 2 \\ 3 \\ 4 \end{cases} \begin{pmatrix} \mathbf{1} & 0 & -3 & \mathbf{1} \\ 2 & 2 & 0 & 2 \\ \mathbf{1} & 0 & -1 & \mathbf{1} \\ 1 & -1 & 3 & 2 \end{pmatrix} \atop \underbrace{\;1 \quad 2 \quad 3 \quad 4\;}_{\mathbb{U}_2} \tag{9.32}$$

with four saddle points (u_1^*, u_2^*), namely $(1,1)$, $(1,4)$, $(3,1)$, and $(3,4)$. From the necessary condition for each saddle point, $J_+ = J_-$, and from (9.30) it follows immediately that all saddle points must have the same costs.

9.3.3 Random Strategy

The main problem of the deterministic strategy was the uncertainty in the behavior of a player if no saddle point exists. To overcome this critical point, we now introduce stochastic rules. That means, for each game each player chooses randomly a certain action. Under the assumption that the same game is repeatedly played over a sufficiently large number of trials, the costs per game tend to their expected value. Suppose that player i ($i = 1, 2$) has m_i actions, given by u_i with $u_i = 1, \ldots, m_i$. Then, the probability that the player i chooses the action u_i is $p^{(i)}(u_i)$. The normalization requires

$$\sum_{u_i=1}^{m_i} p^{(i)}(u_i) = 1 \quad \text{for} \quad i = 1, 2 \tag{9.33}$$

The two sets of probabilities are written as two vectors

$$p^{(1)} = \left\{ p^{(1)}(1), \ldots, p^{(1)}(m_1) \right\} \quad \text{and} \quad p^{(2)} = \left\{ p^{(2)}(1), \ldots, p^{(2)}(m_2) \right\} \tag{9.34}$$

Because of $p^{(i)}(u_i) \geq 0$ and (9.33), each vector $p^{(i)}$ lies on a $(m_i - 1)$-dimensional simplex of the $\mathbb{R}^{m_i}$; see also Chap. 10. The expected costs[4] for given probability vectors $p^{(1)}$ and $p^{(2)}$ are

$$\overline{J}_1(p^{(1)}, p^{(2)}) = \sum_{u_1=1}^{m_1} \sum_{u_2=1}^{m_2} p^{(1)}(u_1) p^{(2)}(u_2) J_1(u_1, u_2) \tag{9.35}$$

or in a more compact form

[4] from the view of player 1.

$$\overline{J}_1(p^{(1)}, p^{(2)}) = p^{(1)} J_1 p^{(2)} \tag{9.36}$$

where J_1 is the cost matrix with respect to player 1. Following the above discussed concepts, the random security strategies are obtainable from an appropriate choice of the probability vectors $p^{(1)}$ and $p^{(2)}$ through

$$p^{*(1)} = \arg\min_{p^{(1)}} \left(\max_{p^{(2)}} \overline{J}_1(p^{(1)}, p^{(2)}) \right) \tag{9.37}$$

and

$$p^{*(2)} = \arg\max_{p^{(2)}} \left(\min_{p^{(1)}} \overline{J}_1(p^{(1)}, p^{(2)}) \right) \tag{9.38}$$

Furthermore, the upper value of the expected cost function is given by

$$\overline{J}_+ = \max_{p^{(2)}} \overline{J}_1(p^{*(1)}, p^{(2)}) \tag{9.39}$$

while the lower value is defined by

$$\overline{J}_- = \min_{p^{(1)}} \overline{J}_1(p^{(1)}, p^{*(2)}) \tag{9.40}$$

The most fundamental result in the zero-sum game theory, namely the equivalence between upper and lower values

$$\overline{J}_+ = \overline{J}_- = J_0 \tag{9.41}$$

was shown by von Neumann [17,18]. The quantity J_0 is the expected value of the game. The necessary existence of a saddle point under a random strategy demonstrates the importance of a probabilistic concept when making decisions against an intelligent player. Of course, when playing the game over a sufficiently long time with a deterministic strategy, the opponent could learn the strategy and would win every time. But if the player uses a random strategy, the second player has no concrete idea about the next strategy used by the first player and vice versa.

9.4 Nonzero-Sum Games

9.4.1 Nash Equilibrium

We focus again on two-player games. But now we allow arbitrary cost functions $J_1(u_1, u_2)$ and $J_2(u_1, u_2)$ for both players, i.e., the condition (9.23) is no longer valid. That means, both players are not always in strict competition. There may exist situations in which both players have similar interests, i.e., there is the possibility for both players to win. This is a general concept of redundant control systems. Both controllers, i.e., both players, have the same or nearly the same intention in controlling the output of a system.

Each player would like to minimize its cost. Firstly, we consider again deterministic strategies to solve the nonzero-sum game. Because of the independence of the two players, each applies its security strategy without making reference to the cost function of the other player. Because of the assumed deterministic character, the strategy of the opponent may be fixed. Then we say a pair of actions u_1^* and u_2^* is a Nash equilibrium if

$$J_1^* = J_1(u_1^*, u_2^*) = \min_{u_1 \in \mathbb{U}_1} J_1(u_1, u_2^*) \tag{9.42}$$

and

$$J_2^* = J_2(u_1^*, u_2^*) = \min_{u_2 \in \mathbb{U}_2} J_2(u_1^*, u_2) \tag{9.43}$$

Obviously, a Nash equilibrium can be detected in a pair of matrices J_1 and J_2 by finding a matrix position (u_1, u_2) such that the corresponding element $J_1(u_1, u_2)$ is the lowest among all elements in the column u_2 of J_1 and the element $J_2(u_1, u_2)$ is the lowest among all elements in row u_1 of J_2. Let us illustrate this procedure by a simple example. We consider the matrices

$$J_1 = \begin{pmatrix} \mathbf{1} & \mathbf{-1} & 0 & \mathbf{2} \\ 3 & 0 & 1 & 4 \\ 2 & 1 & \mathbf{-2} & 3 \end{pmatrix} \qquad J_2 = \begin{pmatrix} \mathbf{-1} & \mathbf{-1} & 1 & 2 \\ 2 & \mathbf{0} & 2 & 1 \\ 4 & 2 & \mathbf{1} & 2 \end{pmatrix} \tag{9.44}$$

It is simple to check that the Nash equilibria exist for the positions $(1,1)$, $(1,2)$, and $(3,3)$. It is a typical feature that a nonzero game has multiple Nash equilibria. From a first glance, the Nash equilibrium at $(1,2)$ seems to be the best choice because it yields negative costs for both players. However, the general decision as to which Nash equilibrium is the optimal choice is by no means a trivial procedure.

The simplest case occurs if both players do not have the same rights. Then, one player, say player 1, is the master player while the second one is the slave. Under this consideration a lexicographic order of the Nash equilibria defines the optimal solution. That means, firstly we search for the pair (J_1^*, J_2^*) which has the lowest value of J_1^*. If two or more pairs satisfy this condition, we consider that pair of these candidates which has also the lowest value of J_2^*. In our example such a concept would lead to the decision that the nash equilibrium at $(3,3)$ is the optimal choice. That should be the typical situation for possible intrinsic control mechanisms of complex systems. The first decision comes from the main controller, and only if it cannot give a unique answer, does the second controller decide the common strategy.

The situation becomes much more complicated if both players have equal rights. Then the definition of the best solution implies a suitable ordering of the Nash equilibria. It is often only a partial ordering procedure because some pairs (J_1^*, J_2^*) are incomparable[5]. In the last case, the player must communicate or collaborate in order to avoid higher costs. If the players do not find an agreement, the possibility of higher costs is often unavoidable. For example, this is the case if both players favor actions which are related to

[5] For example the pairs $(0,1)$ and $(1,0)$ cannot be ordered under the assumption of players with equal rights.

different Nash equilibria. Two well-known standard examples of such incomplete ordered Nash equilibria are the "Battle of the Sexes" and the "Prisoner's Dilemma" [19]. Finally, we remark that a nonzero-sum game also can have no Nash equilibrium.

9.4.2 Random Nash Equilibria

Let us now analyze nonzero-sum games with random strategies. Similar to Sect. 9.3.3, we introduce the probability vectors (9.34). Then we define the expected costs

$$\overline{J}_i(p^{(1)}, p^{(2)}) = \sum_{u_1=1}^{m_1} \sum_{u_2=1}^{m_2} p^{(1)}(u_1) p^{(2)}(u_2) J_i(u_1, u_2) \tag{9.45}$$

($i = 1, 2$). Then a pair of probability vectors $p^{*(1)}$ and $p^{*(2)}$ is said to be a mixed Nash equilibrium if

$$\overline{J}_1(p^{*(1)}, p^{*(2)}) = \min_{p^{(1)}} \overline{J}_1(p^{(1)}, p^{*(2)}) \tag{9.46}$$

and

$$\overline{J}_2(p^{*(1)}, p^{*(2)}) = \min_{p^{(2)}} \overline{J}_2(p^{*(1)}, p^{(2)}) \tag{9.47}$$

It was shown by Nash that every nonzero-sum game has a mixed Nash equilibrium [2]. Unfortunately, it cannot be guaranteed that multiple mixed Nash equilibria appear. That means there is no reliable way to avoid higher costs at least for one player unless the players collaborate. The determination of a mixed Nash equilibrium is a bilinear problem [3]. This requires usually numerical investigations using nonlinear programming concepts [4, 5].

References

1. A. Einstein, B. Podolski, N. Rosen: Phys. Rev. **47**, 777 (1935)
2. J. Nash: Ann. Math. **54**, 286 (1951)
3. T. Basar, G.J. Olsder: *Dynamic Noncooperative Game Theory*, 2nd edn (Academic, London, 1995)
4. D.G. Luenberger: *Introduction to Linear and Nonlinear Programming* (Wiley, New York, 1973)
5. S.G. Nash, A. Sofer: *Linear and Nonlinear Programming* (McGraw-Hill, New York, 1996)
6. D. Bohm: *Quantum Theory* (Prentice-Hall, New York, 1951)
7. J.S. Bell: Physics **1**, 195 (1965)
8. J.F. Clauser, A. Shimony: Rep. Prog. Phys. **41**, 1881 (1978)
9. B. d'Espagnat: Scientific Am. **241**, 128 (1979)
10. H. Kleinpoppen: Phys. Rev. Lett. **54**, 1790 (1985)
11. M.A. Erdmann: on probabilistic strategies for robot tasks. PhD thesis, Massachusetts Institute of Technology, Cambridge, MA (1989)

12. A. Bloch: *Murphy's Law and Other Reasons Why Things Go Wrong* (PSS Adult, 1977)
13. H. Dawid, A. Mehlmann: Complexity **1**, 51 (1996)
14. J.C. Harsanyi, R. Selten: *A General Theory of Equilibrium Selection in Games* (MIT Press, Cambridge, 1988)
15. R. Isaacs: *Differential Games* (Wiley, New York, 1965)
16. D.M. Kreps: *Game Theory and Economic Modelling* (Oxford University Press, New York, 1990)
17. J.V. Neumann: Mathematische Annalen, **100** 295 (1928)
18. J.V. Neumann, O. Morgenstern: *Theory of Games and Economic Behavior* (Princeton University Press, Princeton, NJ, 1944)
19. A. Mehlmann: *Wer gewinnt das Spiel?* (Vieweg, Braunschweig, 1997)

10

Optimization Problems

10.1 Notations of Optimization Theory

10.1.1 Introduction

Several problems, for example Pontryagin's maximum principle or the minimax problems of game theoretical approaches, require the determination of an extremum of a given function. These are typical optimization problems. Most of the instructive problems which have been presented in the previous chapters were relatively simple. Since only few degrees of freedom are considered, these problems were solvable by empirical concepts or by standard analytical methods.

However, the treatment of sufficiently complex structures often requires specific techniques. In this case it will be helpful to know some basic considerations of modern optimization theory.

Optimization methods are not unknown in physics. A standard example is free energy problems of thermodynamics. Here the equilibrium state of a system coupled with a well-defined heat bath is obtainable by the minimization of free energy. But also many other physical applications have turned out to be in fact optimization problems, for example the determination of quantum mechanical ground states using variational principles, the investigation of systems in random environments, or the folding principles of proteins.

The main link between control theory and classical optimization theory is due to Pontryagin's maximum principle. Considering the state X and the adjoint variables, the generalized momenta P as free parameters, the maximum principle requires the maximization of the Hamiltonian

$$H = H(X, P, u) = H(u) \to \max \tag{10.1}$$

with respect to the n-component control u. The standard way of solving this problem is a search for extremal solutions

$$\frac{\partial H(u^*)}{\partial u^*} = 0 \tag{10.2}$$

M. Schulz: *Control Theory in Physics and other Fields of Science*
STMP **215**, 279–293 (2006)

and, as a subsequent step, the decision if one of these solutions corresponds to the global maximum or not. Unfortunately, this problem becomes much more complicated if the control u should satisfy several constraints. For instance, the control vector u can be restricted to a region G of the complete control space U, or u has only discrete values.

10.1.2 Convex Objects

Convex Sets

The convexity of sets plays an important role in the optimization theory. In the above introduced case we have to check if the region $\mathbb{G}$ forms a convex set. The convexity is a very helpful property for many optimization problems. In particular, the theory of optimization on convex sets is well established [1]. The convexity of a region $\mathbb{G}$ requires that for each set of P points $\{u^{(1)}, \ldots, u^{(P)}\}$ with $u^{(i)} \in \mathbb{G}$ $(i = 1, \ldots, P)$, the linear form

$$v = \sum_{i=1}^{P} \lambda_i u^{(i)} \tag{10.3}$$

with the real numbers $\lambda_i \geq 0$ and

$$\sum_{i=1}^{P} \lambda_i = 1 \tag{10.4}$$

is also an element of $\mathbb{G}$, i.e., $v \in \mathbb{G}$, see Fig. 10.1. The verification of if a region is convex or not is not trivial. A special situation occurs if the region is described by a set of L linear inequalities of the type

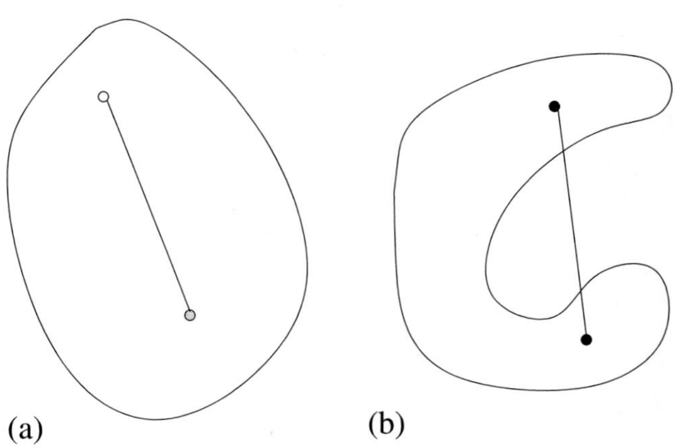

Fig. 10.1. Convex (**a**) and nonconvex (**b**) sets. Each line between two points of a convex set is also a subset of the convex set, while a line between two points of a nonconvex set is not necessarily a subset of the nonconvex set

$$\sum_{\beta=1}^{n} G_{\alpha\beta} u_\beta \leq g_\alpha \quad \text{with} \quad \alpha = 1, \ldots, L \tag{10.5}$$

or in a more compact form, by

$$Gu \leq g \tag{10.6}$$

with the $L \times n$ matrix G and the L-component vector g. In this case, we may replace u by a linear combination

$$u = \lambda u^{(1)} + (1 - \lambda) u^{(2)} \quad \text{with} \quad 0 \leq \lambda \leq 1 \tag{10.7}$$

of two points $u^{(1)}$ and $u^{(2)}$ both satisfying (10.6). Thus, we obtain

$$\begin{aligned} Gu &= G\left[\lambda u^{(1)} + (1 - \lambda) u^{(2)}\right] \\ &= \lambda G u^{(1)} + (1 - \lambda) G u^{(2)} \\ &\leq \lambda g + (1 - \lambda) g = g \end{aligned} \tag{10.8}$$

Regions which are defined by (10.6) are called convex polyhedrons.

Convex Functions

The decision if a local extremum u^* of a function $H(u)$ is also a global minimum (or maximum) often needs special investigations. A helpful situation occurs if the function is convex. A function $H(u)$ over a region $\mathbb{G}$ is denoted to be convex if for each pair $\left(u^{(1)}, u^{(2)}\right)$ of points with $u^{(i)} \in \mathbb{G}$ $(i = 1, 2)$ and for each λ with $0 \leq \lambda \leq 1$ the relation

$$H(\lambda u^{(1)} + (1 - \lambda) u^{(2)}) \leq \lambda H(u^{(1)}) + (1 - \lambda) H(u^{(2)}) \tag{10.9}$$

holds. Obviously, this definition requires that the convex function[1] must be declared over a convex region. A sufficient condition that a function is convex over a certain region $\mathbb{G}$ is that the Hesse matrix

$$\mathcal{H} = \left(\frac{\partial^2 H(u)}{\partial u_\alpha \partial u_\beta}\right) \tag{10.10}$$

is positive definite for all points $u \in \mathbb{G}$. Unfortunately, this condition requires the computation of all eigenvalues or equivalently of all submatrices of $\mathcal{H}$ and the subsequent proof that these quantities are positive. That is a very expansive procedure, especially in the case of higher dimensional variables u. An important property of convex function is the relation to its tangent planes. It is simple to check by using (10.9) that

$$H(u) \geq H(u^{(0)}) + \frac{\partial H(u^{(0)})}{\partial u^{(0)}} (u - u^{(0)}) \quad \text{for} \quad u, u^{(0)} \in \mathbb{G} \tag{10.11}$$

---

[1] Convex functions correspond to a global minimum. In case we are interested in a local maximum, we may consider concave functions or we can change the sign of the function. The latter step implies an exchange of minimum and maximum points.

i.e., a convex function always lies above its tangent planes. A local minimum of a convex function $H(u)$ over a convex region $\mathbb{G}$ is always the global minimum. This statement follows directly from (10.11) by identifying $u^{(0)}$ with the position of the minimum. Thus we have $\partial H(u^{(0)})/\partial u^{(0)} = 0$ and therefore $H(u) \geq H(u^{(0)})$ for all $u \in \mathbb{G}$.

Linear functions $H(u) = cu + d$ with the n-dimensional vector c and the scalar d are always convex[2]. Quadratic functions

$$H(u) = \frac{1}{2}uCu + cu + d \tag{10.12}$$

are convex if the symmetric matrix C of type $n \times n$ is positive definite. Although these classes of functions seem to be very special, they play an important role in control theoretical problems. Recall that the Hamiltonian of many control problems is often a linear function of the control variable u, especially if the performance does not depend on u. Furthermore, linear quadratic problems also lead to functions $H(u)$ which are elements of this special set of functions.

10.2 Optimization Methods

10.2.1 Extremal Solutions Without Constraints

The simplest case of an optimization problem occurs if the function $H(u)$ is a continuous function declared over a certain region $\mathbb{G}$ of the control space. Then, either the possible candidates for the global minimum (maximum) are the solutions of the extremal equation

$$\frac{\partial H(u)}{\partial u} = 0 \tag{10.13}$$

or the minimum (maximum) is located at the boundary $\partial\mathbb{G}$ of the region $\mathbb{G}$. The solution of the n usually nonlinear equations can be realized by analytic methods only if the dimension of the problem, n, is very low or if the function $H(u)$ is very simple. Especially the two classes of linear and quadratic functions are of special interest. Since a linear function with the exception of $H(u) = \text{const.}$ has no extremal points, i.e., no solutions of (10.13), the optimum is always located at the border of $\mathbb{G}$. Here, we need the techniques of linear optimization. The quadratic function (10.12) requires the solution of a linear equation

$$Cu = -c \tag{10.14}$$

This equation has for $\det C \neq 0$ a unique solution, $u^* = -C^{-1}c$. If C is positive definite and $u^* \in \mathbb{G}$, the optimization problem is solved.

[2] And simultaneously also concave.

If the analytical solution of (10.13) fails, the application of numerical methods seems to be an alternative approach. Here, we distinguish between deterministic methods and random techniques. An example for a deterministic technique is Newton's procedure. Here, we assume that $u^{(k)}$ is an approximation of the wanted extremum. Then, the expansion around this point up to the second-order gives

$$H(u) \approx H(u^{(k)}) + \frac{\partial H(u^{(k)})}{\partial u^{(k)}}(u - u^{(k)}) + \frac{1}{2}(u - u^{(k)}) \frac{\partial^2 H(u^{(k)})}{\partial u^{(k)} \partial u^{(k)}}(u - u^{(k)}) \quad (10.15)$$

This is a quadratic form from which we can calculate straightforwardly the corresponding extremum $\widetilde{u}^*$. However, because the right-hand side is only an approximation of the complete function $H(u)$, the solution $\widetilde{u}^*$ is also only an approximation of the true extremal point. On the other hand, $\widetilde{u}^*$ is usually a better approximation of the extremal point as $u^{(k)}$. Thus, we may identify the solution $\widetilde{u}^*$ with $u^{(k+1)}$ and start the iteration procedure again with the new input $u^{(k+1)}$. Repeated application of this procedure may lead to a continuous approach of $u^{(k)}$ to the true extremum u^* for $k \to \infty$.

Other traditional deterministic methods (see also [46]) are steepest descent algorithms, subgradient methods [5], the Fletcher–Reeves algorithm [6] and the Polak–Ribiere algorithm [7], trust region methods [6], or the coordinate method in Hooke and Jeves [8,9,46].

Stochastic optimization methods [45] work usually without derivatives. The idea is very simple. First, we choose a point $\overline{u} \in \mathbb{G}$ and determine $\overline{H} = H(\overline{u})$. Then, the region $\mathbb{G}$ is embedded in a hypercube $\mathbb{C} \supset \mathbb{G}$ of dimension n and unit length l. A randomly chosen set of n real numbers $\xi_i \in [0, l]$ (with $i = 1, \ldots, n$) is used to determine a point u' of the hypercube. If $u \in \mathbb{G}$ and $H(u) < \overline{H}$, we set $\overline{u} = u'$ and $\overline{H} = H(u')$; otherwise $\overline{u}$ and $\overline{H}$ remain unchanged. Repeated application of this algorithm then leads to a successive improvement of the estimation $u^* = \overline{u}$ and $H(u^*) = \overline{H}$.

Such algorithms require a random generator which produces uniformly distributed random numbers. Unfortunately, computer-generated random numbers are not really stochastic, since computer programs are deterministic algorithm. But, given an initial number (generally called the seed) a number of mathematical operations can be performed on the seed so as to generate apparently unrelated pseudorandom numbers. The output of random number generators is usually tested with various statistical methods to ensure that the generated number series are really random in relation to one another. There is an important caveat: if we use a seed more than once, we will get identical random numbers every time. However, several commercial programs pull the seed from somewhere within the system, so the seed is unlikely to be the same for two different simulation runs.

A given random number algorithm generates a series of random numbers $\{\eta_1, \eta_2, \ldots, \eta_N\}$ with a certain probability distribution function. If we know this distribution function $p_{\mathrm{rand}}(\eta)$, we now have from the rank ordering statistics [10, 11] that the likely rank of a random number η in a series of N numbers is

$$n = NP_{<}(\eta) = N \int_{-\infty}^{\eta} dz\, p_{\text{rand}}(z) \ . \tag{10.16}$$

In other words, if the random generator creates random series which are distributed with $p_{\text{rand}}(\eta)$, the corresponding series $\{P_{<}(\eta_1), P_{<}(\eta_2), \ldots, P_{<}(\eta_N)\}$ is uniformly distributed over the interval $[0, 1]$. Unfortunately, this concept exhibits a slow rate of convergence. An alternative way is the application of quasirandom sequences instead pseudorandom numbers [12–20]. The quasirandom sequences, sometimes also called low-discrepancy sequences, usually permit us to improve the performance of the random algorithms, offering shorter computational times and higher accuracy.

We remark that the low-discrepancy sequences are deterministic series, so the popular notation quasirandom can be misleading. The discrepancy property is a measure of uniformity for the distribution of the points. Let us assume that the quasirandom process has generated Q points distributed over the whole hyperspace. Then, the discrepancy is defined by

$$D_Q = \sup_{\mathbb{R} \in \mathbb{C}} \left| \frac{n(\mathbb{R})}{Q} - \frac{v(\mathbb{R})}{l^n} \right| \tag{10.17}$$

where $\mathbb{R}$ is a spherical region of the hypercube, $v(\mathbb{R})$ is the volume of this region and $n(\mathbb{R})$ is the number of points in this region. Obviously, the discrepancy vanishes for $Q \to \infty$ in case the of a homogeneous distribution of points over the whole hypercube. Mainly for the multidimensional case, a low discrepancy corresponds to no large gaps and no clustering of points in the hypercube (Fig. 10.2). Similar to a pseudorandom generator, a quasirandom generator originates from the number theory. But in contrast to the pseudorandom series, quasirandom sequences offer a pronounced deterministic behavior. A quasirandom generator transforms an arbitrary positive integer I into a quasirandom number ξ_I via the following two steps. Firstly, the integer I will be decomposed into the integer coefficients a_k with respect to the basis b

$$I = \sum_{k=0}^{\infty} a_k b^k \tag{10.18}$$

with $0 \leq a_k \leq b - 1$. The coefficients form simply the representation of I within the basis b. The second step is the computation of the quasirandom number by the calculation of the sum

$$\xi_I = \sum_{k=0}^{\infty} a_k b^{-k-1} \ . \tag{10.19}$$

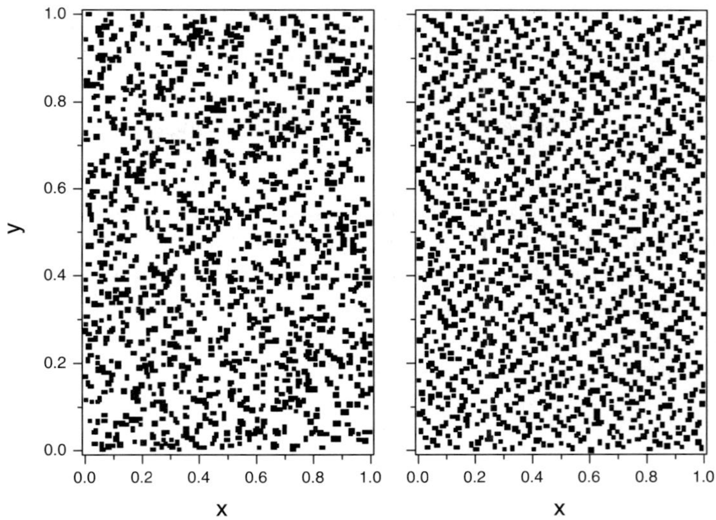

Fig. 10.2. Two-dimensional plot of pseudorandom number pairs (*left*) and quasirandom number pairs (*right*). The quasirandom number series are created with base 2 (x-axis) and with base 3 (y-axis)

For example, the first quasirandom numbers[3] corresponding to base 2 are 1/2, 1/4, 3/4, 1/8, 5/8, ... while the sequence of base 3 starts with 1/3, 2/3, 1/9, 4/9, 7/9.

The merit of a quasirandom generator method is fast convergence. The theoretical upper bound rate of convergence of the discrepancy is $\ln^n Q/Q$ where n is the dimension of the problem [21]. In contrast, the discrepancy of a pseudo-random process converges as $Q^{-1/2}$.

10.2.2 Extremal Solutions with Constraints

The standard version to determine the extremals of functions with μ constraints given by the equations

$$g_i(u) = 0 \quad \text{for} \quad i = 1, \ldots, \mu \tag{10.20}$$

is the Lagrange method. The Lagrange function

$$H'(u) = H(u) + \sum_{i=1}^{\mu} \lambda_i g_i(u) \tag{10.21}$$

can now be treated in the same way as the function $H(u)$. The n extremal equations together with the μ constraints form a system of usually nonlinear equations for the μ Lagrange multipliers and the n components of u. In principle, the above discussed deterministic and stochastic numerical methods are

[3] Corresponding to the integers $I = 1, 2, 3, \ldots$.

also applicable for the determination of extremals under constraints. A special feature is the penalty function method. Here, we construct a utility function

$$H(u,\sigma) = H(u) + \sigma \left\| g(u) \right\|^2 \tag{10.22}$$

where $\|g(u)\|$ is a suitable chosen norm with respect to the constraints. A possible, but not necessary form is the Euclidian norm $\|g\|^2 = g_1^2 + ... + g_\mu^2$. The parameter σ is the penalty parameter, where $\sigma > 0$ corresponds to a search for a minimum while $\sigma < 0$ is used for the determination of a maximum. In principle, one can determine the minimum point of the function $H(u,\sigma)$ similar to the previous section. Let us assume that a minimum point was found to be $u^*(\sigma)$. It can be demonstrated [22] that for $0 < \sigma_1 < \sigma_2$ the three relations hold

$$H(u^*(\sigma_1), \sigma_1) \leq H(u^*(\sigma_2), \sigma_2) \tag{10.23}$$

$$\left\| g(u(\sigma_1)) \right\|^2 \geq \left\| g(u(\sigma_2)) \right\|^2 \tag{10.24}$$

and

$$H(u^*(\sigma_1)) \geq H(u^*(\sigma_2)) \tag{10.25}$$

Hence, it may be expected that for a series $\sigma_i \to \infty$ the value of $u^*(\sigma_i)$ converges to the new minimum point considering the constraints.

10.2.3 Linear Programming

Linear programming deals with the optimization of linear problems. Such problems are typical for the application of Pontryagin's maximum principle to physical systems controlled by external forces and a performance independent of the control[4]. In fact, the equations of motion of such systems are given by

$$\dot{X} = \widetilde{F}(X,t) + u \tag{10.26}$$

Thus, the Hamiltonian (2.94) reads

$$H(t,X,P,u) = P\widetilde{F}(X,t) + uP - \phi(t,X) \tag{10.27}$$

and the optimization problem involves the determination of the maximum of

$$H(u) = H_0 + uP \tag{10.28}$$

In this case the maximum is at the boundary $\partial\mathbb{G}$ of the admissible region $\mathbb{G}$ of control states u. Furthermore, if this region is defined by a set of inequalities

$$\mathbb{G} = \{u \in \mathbb{U} \mid Gu \leq g \quad \text{and} \quad u \geq 0\} \subset \mathbb{U} \tag{10.29}$$

the global maximum is reached for one of the corners of the polyhedron defined by (10.29). If the region is convex, the maximum can be found by the simplex

[4] This is typical for an optimal time problem or an endpoint performance (Meier problem).

algorithm [22–26]. In principle, this algorithm starts from a randomly chosen corner $u^{(k)}$ with the initial label $k = 0$. Then, a second corner u' is chosen, so that (i) u' is the topological neighbor of $u^{(k)}$ and (ii) $H(u') > H(u^{(k)})$. If such a corner was found, we set $u^{(k+1)} = u'$ and repeat the algorithm. If no further u' can be detected so that both conditions are fulfilled, the currently reached corner corresponds to the global maximum solution. If the set $\mathbb{G}$ is nonconvex, it may be possible to separate the region to exhaustive and mutually exclusive convex subsets $\mathbb{G}_i$ and solve the linear programming problem for each $\mathbb{G}_i$ separately. The global maximum is then the maximum of all local maxima related to the subsets.

10.2.4 Combinatorial Optimization Problems

If the control state has only discrete values, we speak about a combinatorial optimization problem. Such problems play an important role for several technological and physical applications. A standard example is the so-called Ising model. Usually, the Ising model is described by a physical is described by a physical Hamiltonian given by

$$H = H_0(X) + \sum_{i,j=1}^{n} J_{ij}(X) S_i S_j + \sum_{i=1}^{n} B_i S_j \tag{10.30}$$

where S_i are the discrete spin variables, and B_i is denoted as the local field and J_{ij} as the coupling constants. The physical standard problem is the determination of the ground state of H or alternatively, the thermodynamical weight of a spin configuration $\{S_1, \ldots, S_i, \ldots\}$. This is of course a repeatedly investigated problem in the context of spin glasses [27–32], protein folding [33–35], or phase transitions in spin systems [36–38] and it is strongly connected with the concept of optimization. However, this is no real control problem.

But a control problem occurs if we are interested in the inverse situation. Usually, the physical degrees of freedom represented by the spin variables S_i are coupled with another set of internal degrees of freedom X. These quantities are assumed to be passive for the above mentioned spin dynamics, i.e., X determines the coupling constants $J_{ij}(X)$ and the spin-independent contribution $H_0(X)$, but for a given physical problem X is assumed to be fixed. But we may also ask for the dynamics of X under a certain spin configuration. Then, the Hamiltonian (10.30), leads to evolution equations of the form

$$\dot{X} = F^{(0)}(X) + \sum_{i,j=1}^{n} F_{ij}^{(1)}(X) S_i S_j \tag{10.31}$$

where we may identify the discrete spin variables as components of the n-component control u. From here, we obtain, for example via the deterministic Hamiltonian (2.94) or the corresponding stochastic version (7.65) a classical combinatorial optimization problem.

As a naive approach to such discrete optimization problems, we may solve the corresponding continuous control problem. Then the obtained result is approximated by that allowed discrete value of u which has the shortest distance to the optimal continuous result. However, this procedure fails often, especially in the case of the so-called 0–1 optimization with only two values per component u_α.

An alternative way is the direct computation of the value of the optimization function for all admissible states u. But this procedure needs an enormous amount of computation time. For example, the 0–1 optimization requires 2^n steps in order to compute the optimum.

A well-established theory exists for linear combinatorial optimization problems [23, 39–42], for example branch and bound methods [43] or modified simplex methods[5] [44]. Most of these techniques require a computation time which increases exponentially with increasing n. In the case that the control space $\mathbb{U}$ consists of only a finite number of discrete values, the optimization problem may be transformed into a 0–1 optimization problem. We remark that these special problems are also denoted as binary optimization or Bool's optimization. The transformation can be realized via

$$u_\alpha = \sum_{k=1}^{L_\alpha} u_\alpha^{(k)} s_{k,\alpha} \tag{10.32}$$

where $u_\alpha^{(k)}$ are the L_α discrete values of the component u_α while the $s_{k,\alpha}$ takes the values 0 or 1 under the constraint that

$$1 = \sum_{k=1}^{L_\alpha} s_{k,\alpha} \tag{10.33}$$

The combinatorial optimization is strongly connected with the complexity theory of algorithms. Here, the problems are classified with respect to the expected computation times so that one speaks also of time-complexity. As remarked above, the majority of the combinatorial optimization problems belong to the class of nonpolynomial (NP) problems. That means the computational time increases with increasing n faster as a polynomial of finite order. Some problems, for example the above mentioned modified simplex methods, can be reduced to polynomial problems with a computation time $T(n) \sim n^a$ with a finite exponent a. All other problems have computation times, e.g., given by 2^n or $n!$. These problems are elements of the NP class. Formally, the polynomial class is a subset of the NP class. This implies that all polynomial problems can be always expanded to a NP problem by some algebraic manipulations. But not all NP problems can be reduced to a polynomial problem. In fact, there exists a special subset of the nonlinear problems which is defined by the NP completeness. All problems of this NP complete set can be neither

[5] It should be remarked that in principle the simplex method itself can be also interpreted as a combinatorial optimization method.

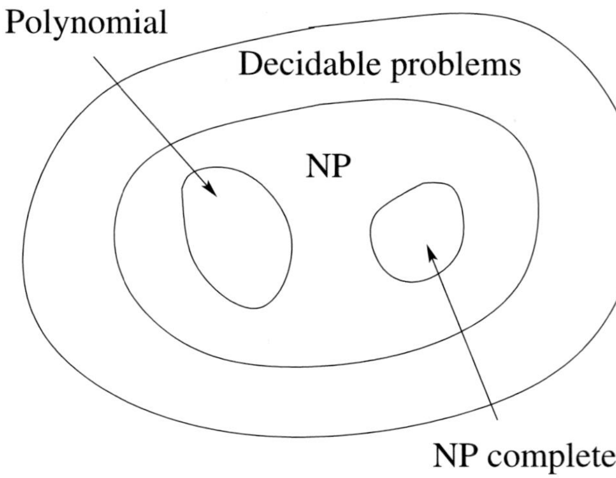

Fig. 10.3. The relations between the different classes of time complexity

reduced to a polynomial problem nor transformed into another element of the NP complete set with a computation time of polynomial lngth. Hence, both, polynomial problems and NP complete procedures are embedded in the set of NP problems (Fig. 10.3), but both classes are exclusive. We remark that the set of nonpolynomial problems is a subset embedded in the set of decidable problems.

10.2.5 Evolution Strategies

Evolution strategies [2, 4] are methods which are suggested by Darwinian paradigm of evolution. Especially the principle of variation and selection can be considered as the fundamental principle of the Darwinian theory. This principle, combined with a reproduction procedure, builds up the fundamental components of an evolutionary strategy.

The basic principle of evolution methods is quite simple. Let us assume we have a set of M different admissible vectors $u_i^{(\mu)} \in \mathbb{G}$, $i = 1, \ldots, M$. The set of the M quantities $u_i^{(\mu)}$ is called the parent set. Then the corresponding values $H_i^{(\mu)} = H(u_i^{(\mu)})$ are denoted as fitness values. The lowest (largest) value of the set of $H_i^{(\mu)}$ corresponds to the current estimation of the optimum. The index μ indicates the generation of the evolution. Initially, we have $\mu = 0$ and the $u_i^{(\mu)}$ are M (randomly) chosen quantities of the region $\mathbb{G}$.

The first step of an evolution loop is the so-called recombination procedure. There exist various techniques, e.g.

- Discrete recombination: two parent vectors, say $u_1^{(\mu)}$ and $u_2^{(\mu)}$, of the μth generation are chosen randomly. Then we choose a diagonal $n \times n$ random matrix R_1 with only 0 and 1 components, e.g.

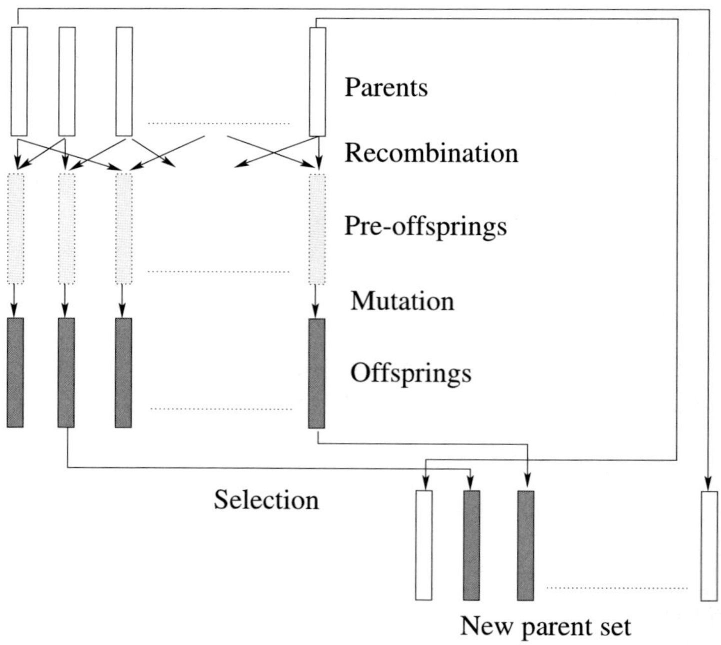

Fig. 10.4. The basic elements of evolutionary optimization strategies

$$R_1 = \begin{pmatrix} 1 & 0 & \ddots & & \\ 0 & 0 & \ddots & 0 & \\ \ddots & \ddots & 1 & & \ddots \\ & 0 & \ddots & \ddots & 0 \\ & & \ddots & 0 & 0 \end{pmatrix} \tag{10.34}$$

while the dual matrix R_2 is given by

$$R_1 + R_2 = 1 \tag{10.35}$$

Then a so-called pre-offspring corresponding to the parents is given by

$$u' = R_1 u_1^{(\mu)} + R_2 u_2^{(\mu)} \tag{10.36}$$

If u' is also admissible, $u' \in \mathbb{G}$, the pre-offspring is collected in a set $\mathbb{K}$; otherwise it will be deleted.

- Intermediate recombination: two parents are chosen randomly, and the weighted average

$$u' = \lambda u_1^{(\mu)} + (1-\lambda) u_2^{(\mu)} \tag{10.37}$$

with a randomly chosen parameter $0 \leq \lambda \leq 1$ is the offspring. This recombination procedure is always successful for convex sets $\mathbb{G}$.

- Discrete multiple recombination: L "parents", $u_{\alpha_1}^{(\mu)}, \ldots, u_{\alpha_L}^{(\mu)}$, are chosen randomly from the parent set. Furthermore, we choose L diagonal matrices of type (10.34) satisfying
$$\sum_{j=1}^{L} R_j = 1 \tag{10.38}$$
Then, the pre-offspring is given by
$$u' = \sum_{j=1}^{L} R_j u_{\alpha_j}^{(\mu)} \tag{10.39}$$
i.e., each component of the pre-offspring vector u' is equal to the corresponding component of one of its parents.
- Intermediate multiple recombination: L parents, $u_{\alpha_1}^{(\mu)}, \ldots, u_{\alpha_L}^{(\mu)}$, and L real numbers $0 \leq \lambda_j \leq 1$ with the constraint
$$\sum_{j=1}^{L} \lambda_j = 1 \tag{10.40}$$
are chosen randomly. The pre-offspring is then
$$u' = \sum_{j=1}^{L} \lambda_j u_{\alpha_j}^{(\mu)} \tag{10.41}$$
which is especially admissible if $\mathbb{G}$ is a convex set.

After determination of a set of M' pre-offsprings, these quantities are (usually slightly) changed by a mutation step, i.e., a random vector corresponding to a certain probability distribution (e.g., a Gaussian distribution or a uniform distribution) is added to each pre-offspring. The result is offsprings $u_1^{\prime(\mu)}, \ldots, u_{M'}^{\prime(\mu)}$. If some offsprings are no longer admissible, $u_j^{\prime(\mu)} \notin \mathbb{G}$, another offspring is formed by the repetition of the recombination and mutation step.

The common $(M + M')$ set of parents and offsprings,
$$\left\{u_1^{(\mu)}, \ldots, u_M^{(\mu)}, u_1^{\prime(\mu)}, \ldots, u_{M'}^{\prime(\mu)}\right\} \tag{10.42}$$
is now the input for the subsequent selection step. That means, we determine the fitness, $H(u_i^{(\mu)})$ and $H(u_i^{\prime(\mu)})$, respectively, of these components and select the best M elements. These quantities are the M parents $u_i^{(\mu+1)}$ of the next generation.

The repeated application of this procedure should drive the set
$$\left\{u_1^{(\mu)}, \ldots, u_M^{(\mu)}\right\} \tag{10.43}$$
to the optimum for $\mu \to \infty$, i.e., the lowest (largest) value of the corresponding fitness indicates the optimal solution. We remark that this expected convergence to the optimum solution is not guaranteed at all [2]. The simple $(1 + 1)$ evolution strategy (one parent, one offspring), i.e.

$$u'^{(\mu+1)} = u^{(\mu)} + \xi \tag{10.44}$$

where ξ is an admissible random vector, corresponds to the stochastic procedure discussed in Sect. 10.2.1. For more details and several applications we refer to the literature [2–4].

References

1. K.H. Elster: *Modern Mathematical Methods of Optimization* (Akademie Verlag, Berlin, 1993)
2. H.G. Beyer: *The Theory of Evolution Strategies* (Springer, Berlin Heidelberg New York, 1998)
3. M. Delgado, J. Kacprzyk, J.-L. Verdegay, M.A. Vila: *Fuzzy Optimization* (Physica-Verlag, Heidelberg, 1994)
4. B. Kost: *Optimierung mit Evolutionsstrategien* (Verlag Harri Deutsch, Frankfurt A.M., 2003)
5. C. Geiger, C. Kanzow: *Theorie und Numerik Restrigierter Optimierungsaufgaben* (Springer, Berlin Heidelberg New York, 2002)
6. C. Geiger, C. Kanzow: *Numerische Verfahren zur Lösung unrestrigierter Optimierungsaufgaben* (Springer, Berlin Heidelberg New York, 1999)
7. I. Bomze, W. Grossmann: *Optimierung-Theorie und Algorithmen* (Wissenschaftsverlag, Mannhein, 1993)
8. C. Richter: *Optimierungsaufgaben und BASIC Programme* (Akademie-Verlag, Berlin, 1988)
9. P. Spelucci: *Numerische Verfahren der nichtlinearen Optimierung* (Birkhäuser, Basel, 1993)
10. E.J. Gumbel: *Statistics of Extremes* (Columbia University Press, New York, 1958)
11. G.K. Zipf: *Human Behavior and the Principle of Least Effort* (Addison-Wesley, Cambridge, 1949)
12. J.W. Barret, G. Moore, P. Wilmott: Risk **5**, 82 (1992)
13. R. Brotherton-Ratcliffe: Risk **7**, 53 (1994)
14. K.-T. Fang: *Number-Theoretic Methods in Statistics* (Chapman and Hall, London, 1994)
15. P. Hellekalek, G. Larcher: *Random and Quasi-Random Point Sets* (Springer, Berlin Heidelberg New York, 1998)
16. C. Joy, P.P. Boyle: Manage. Sci. **42**, 926 (1996)
17. J.X. Li: Revista de Análisis Económico **15**, 111 (2000)
18. W.J. Morokoff: SIAM Rev. **40**, 765 (1998)
19. H. Niederreiter, P. Hellekalek, G. Larcher, P. Zinterhof: *Monte Carlo and Quasi-Monte Carlo Methods* (Springer, Berlin Heidelberg New York, 1996)
20. W.C. Snyder: Math. Comput. Simul. **54**, 131 (2000).
21. H. Niederreiter: SIAM, CBMS **63**, 241 (1992)
22. W. Krabs: Einführung in die lineare und nichtlineare Optimierung für Ingenieure (Teubner-Verlag, Leipzig, 1983)
23. K.H. Borgwardt: *Optimierung, Oparations Research, Spieltheorie* (Birkhäuser, Basel, 2001)

24. K. Glashoff, S. Gustafson: *Linear Optimzation and Approximation* (Springer, Berlin Heidelberg New York, 1978)
25. K. Marti, D. Gröger: *Einführung in die lineare und nichtlineare Optimierung* (Physica-Verlag, Heidelberg, 2000)
26. E. Seiffart, K. Manteufel: *Lineare Optimierung* (Teubner-Verlag, Leipzig, 1974)
27. A.K. Hartmann, F. Ricci-Tersenghi, Phys. Rev. B **66**, 224419 (2002)
28. J. Houdayer, O.C. Martin: Europhys. Lett. **49**, 794 (2000)
29. P. Palassini, F. Liers, M. Jünger, A.P. Young: Phys. Rev. B **68**, 064413 (2003)
30. A.K. Hartmann, H. Rieger: *Optimization Problems in Physics* (Wiley-VCH, Berlin, 2002)
31. J. Houdayer, O.C. Martin: Phys. Rev. E **64**, 056704 (2001)
32. M. Jünger, G. Rinaldi: Relaxation of the max cut problem and computation of spin-glass ground states. In: *Operations Research Proceedings*, ed by P. Kischka (Springer, Berlin Heidelberg New York, 1998), p. 74
33. U.H.E. Hansmann, Y. Okamoto: J. Chem. Phys. **110**, 1267 (1999)
34. S. Kirkpatrick, C.D. Gelatt, M.P. Vecchi: Science **220**, 671 (1983)
35. Y. Duan, P.A. Kollman: Science **282**, 740 (1998)
36. A. Buhot, W. Krauth: Phys. Rev. Lett. **80**, 3787 (1998)
37. J.A. Cuesta: Phys. Rev. Lett. **76**, 3742 (1996)
38. P.W. Kasteleyn, C.M. Fortuin: J. Phys. Soc. Jpn. **26**, 11 (1969)
39. A. Brink, H. Damhorst, D. Kramer, W. Zwehl: *Lineare und ganzzahlige Optimierung mit impac* (Vahlen, München, 1991)
40. R.E. Burkhard: *Methoden der ganzzahligen Optimierung* (Springer, Berlin Heidelberg New York, 1972)
41. J. Piehler: *Ganzzahlige lineare Optimierung* (Teubner-Verlag, Leipzig, 1982)
42. J. Piehler: *Algebraische Methoden der ganzzahligen Optimierung* (Teubner-Verlag, Leipzig, 1970)
43. E. Fischer, A. Stepan: *Betriebswirtschaftliche Optimierung* (Oldenbourg Verlag, München, 2001)
44. K. Neumann, M. Morlock: *Operations Research* (Carl Hanser Verlag, München, 1993)
45. K. Marti: *Stochastic Optimization Methods* (Springer, Berlin Heidelberg New York, 2005)
46. L.C.W. Dixon, E. Spedicato, G. Szegö: *Nonlinear Optimization* (Birkhäuser, Boston, 1980)

Index

Springer Tracts in Modern Physics

175 **Ion-Induced Electron Emission from Crystalline Solids**
By Hiroshi Kudo 2002. 85 figs. IX, 161 pages

176 **Infrared Spectroscopy of Molecular Clusters**
An Introduction to Intermolecular Forces
By Martina Havenith 2002. 33 figs. VIII, 120 pages

177 **Applied Asymptotic Expansions in Momenta and Masses**
By Vladimir A. Smirnov 2002. 52 figs. IX, 263 pages

178 **Capillary Surfaces**
Shape – Stability – Dynamics, in Particular Under Weightlessness
By Dieter Langbein 2002. 182 figs. XVIII, 364 pages

179 **Anomalous X-ray Scattering for Materials Characterization**
Atomic-Scale Structure Determination
By Yoshio Waseda 2002. 132 figs. XIV, 214 pages

180 **Coverings of Discrete Quasiperiodic Sets**
Theory and Applications to Quasicrystals
Edited by P. Kramer and Z. Papadopolos 2002. 128 figs., XIV, 274 pages

181 **Emulsion Science**
Basic Principles. An Overview
By J. Bibette, F. Leal-Calderon, V. Schmitt, and P. Poulin 2002. 50 figs., IX, 140 pages

182 **Transmission Electron Microscopy of Semiconductor Nanostructures**
An Analysis of Composition and Strain State
By A. Rosenauer 2003. 136 figs., XII, 238 pages

183 **Transverse Patterns in Nonlinear Optical Resonators**
By K. Staliūnas, V. J. Sánchez-Morcillo 2003. 132 figs., XII, 226 pages

184 **Statistical Physics and Economics**
Concepts, Tools and Applications
By M. Schulz 2003. 54 figs., XII, 244 pages

185 **Electronic Defect States in Alkali Halides**
Effects of Interaction with Molecular Ions
By V. Dierolf 2003. 80 figs., XII, 196 pages

186 **Electron-Beam Interactions with Solids**
Application of the Monte Carlo Method to Electron Scattering Problems
By M. Dapor 2003. 27 figs., X, 110 pages

187 **High-Field Transport in Semiconductor Superlattices**
By K. Leo 2003. 164 figs.,XIV, 240 pages

188 **Transverse Pattern Formation in Photorefractive Optics**
By C. Denz, M. Schwab, and C. Weilnau 2003. 143 figs., XVIII, 331 pages

189 **Spatio-Temporal Dynamics and Quantum Fluctuations in Semiconductor Lasers**
By O. Hess, E. Gehrig 2003. 91 figs., XIV, 232 pages

190 **Neutrino Mass**
Edited by G. Altarelli, K. Winter 2003. 118 figs., XII, 248 pages

191 **Spin-orbit Coupling Effects in Two-dimensional Electron and Hole Systems**
By R. Winkler 2003. 66 figs., XII, 224 pages

192 **Electronic Quantum Transport in Mesoscopic Semiconductor Structures**
By T. Ihn 2003. 90 figs., XII, 280 pages

193 **Spinning Particles – Semiclassics and Spectral Statistics**
By S. Keppeler 2003. 15 figs., X, 190 pages

194 **Light Emitting Silicon for Microphotonics**
By S. Ossicini, L. Pavesi, and F. Priolo 2003. 206 figs., XII, 284 pages

Springer Tracts in Modern Physics

195 **Uncovering *CP* Violation**
Experimental Clarification in the Neutral K Meson and B Meson Systems
By K. Kleinknecht 2003. 67 figs., XII, 144 pages

196 **Ising-type Antiferromagnets**
Model Systems in Statistical Physics and in the Magnetism of Exchange Bias
By C. Binek 2003. 52 figs., X, 120 pages

197 **Electroweak Processes in External Electromagnetic Fields**
By A. Kuznetsov and N. Mikheev 2003. 24 figs., XII, 136 pages

198 **Electroweak Symmetry Breaking**
The Bottom-Up Approach
By W. Kilian 2003. 25 figs., X, 128 pages

199 **X-Ray Diffuse Scattering from Self-Organized Mesoscopic Semiconductor Structures**
By M. Schmidbauer 2003. 102 figs., X, 204 pages

200 **Compton Scattering**
Investigating the Structure of the Nucleon with Real Photons
By F. Wissmann 2003. 68 figs., VIII, 142 pages

201 **Heavy Quark Effective Theory**
By A. Grozin 2004. 72 figs., X, 213 pages

202 **Theory of Unconventional Superconductors**
By D. Manske 2004. 84 figs., XII, 228 pages

203 **Effective Field Theories in Flavour Physics**
By T. Mannel 2004. 29 figs., VIII, 175 pages

204 **Stopping of Heavy Ions**
By P. Sigmund 2004. 43 figs., XIV, 157 pages

205 **Three-Dimensional X-Ray Diffraction Microscopy**
Mapping Polycrystals and Their Dynamics
By H. Poulsen 2004. 49 figs., XI, 154 pages

206 **Ultrathin Metal Films**
Magnetic and Structural Properties
By M. Wuttig and X. Liu 2004. 234 figs., XII, 375 pages

207 **Dynamics of Spatio-Temporal Cellular Structures**
Henri Benard Centenary Review
Edited by I. Mutabazi, J.E. Wesfreid, and E. Guyon 2005. approx. 50 figs., 150 pages

208 **Nuclear Condensed Matter Physics with Synchrotron Radiation**
Basic Principles, Methodology and Applications
By R. Röhlsberger 2004. 152 figs., XVI, 318 pages

209 **Infrared Ellipsometry on Semiconductor Layer Structures**
Phonons, Plasmons, and Polaritons
By M. Schubert 2004. 77 figs., XI, 193 pages

210 **Cosmology**
By D.-E. Liebscher 2005. Approx. 100 figs., 300 pages

211 **Evaluating Feynman Integrals**
By V.A. Smirnov 2004. 48 figs., IX, 247 pages

213 **Parametric X-ray Radiation in Crystals**
By V.G. Baryshevsky, I.D. Feranchuk, and A.P. Ulyanenkov 2006. 63 figs., IX, 172 pages

214 **Unconventional Superconductors**
Experimental Investigation of the Order-Parameter Symmetry
By G. Goll 2006. 67 figs., XII, 172 pages

215 **Control Theory in Physics and other Fields of Science**
Concepts, Tools, and Applications
By M. Schulz 2006. 46 figs., X, 294 pages